AFTER THE HUMAN

BOOKS BY MARK C. TAYLOR

A Friendship in Twilight: Lockdown Conversations on Death and Life (with Jack Miles)

Image: Three Inquiries in Technology and Imagination (with Thomas A. Carlson and Mary-Jane Rubenstein)

Intervolution: Smart Bodies Smart Things

Seeing Silence

Abiding Grace: Time, Modernity, Death

Last Works: Lessons in Leaving

Speed Limits: Where Time Went and Why We Have So Little Left

Recovering Place: Reflections on Stone Hill

Rewiring the Real: In Conversation with William Gaddis, Richard Powers, Mark Danielewski, and Don DeLillo

Refiguring the Spiritual: Beuys, Barney, Turrell, Goldsworthy

Crisis on Campus: A Bold Plan for Reforming Our Colleges and Universities

Field Notes from Elsewhere: Reflections on Dying and Living

After God

Mystic Bones

Confidence Games: Money and Markets in a World Without Redemption

Grave Matters (with Dietrich Christian Lammerts)

The Moment of Complexity: Emerging Network Culture

About Religion: Economies of Faith in Virtual Culture

The Picture in Question: Mark Tansey and the Ends of Representation

Critical Terms for Religious Studies

Hiding

Imagologies: Media Philosophy (with Esa Saarinen)

Nots

Disfiguring: Art, Architecture, Religion

Michael Heizer: Double Negative

Tears

Altarity

Deconstruction in Context: Literature and Philosophy

Erring: A Postmodern A/theology

Deconstructing Theology

Journeys to Selfhood: Hegel and Kierkegaard

Unfinished: Essays in Honor of Ray L. Hart

Religion and the Human Image (with Carl A. Raschke and James A. Kirk)

Kierkegaard's Pseudonymous Authorship: A Study of Time and Self

AFTER THE HUMAN

A Philosophy for the Future

MARK C. TAYLOR

Columbia University Press
New York

Columbia University Press
Publishers Since 1893
New York   Chichester, West Sussex
cup.columbia.edu
Copyright © 2025 Mark C. Taylor
All rights reserved

Library of Congress Cataloging-in-Publication Data
Names: Taylor, Mark C., 1945- author.
Title: After the human : a philosophy for the future / Mark C. Taylor.
Description: New York : Columbia University Press, 2024. | Includes bibliographical references and index.
Identifiers: LCCN 2024024923 (print) | LCCN 2024024924 (ebook) | ISBN 9780231218603 (hardback) | ISBN 9780231218610 (trade paperback) | ISBN 9780231562362 (ebook)
Subjects: LCSH: Inquiry (Theory of knowledge) | Knowledge, Theory of. | Similarity (Psychology) | Polarity (Philosophy)
Classification: LCC BD183 .T44 2024 (print) | LCC BD183 (ebook) | DDC 121/.6--dc23/eng/20240924
LC record available at https://lccn.loc.gov/2024024923
LC ebook record available at https://lccn.loc.gov/2024024924

Printed in the United States of America

Cover design: Lisa Hamm

For

Aaron   Kirsten

Selma   Elsa

Jackson   Taylor

Alethea

One ghost succeeds the other like waves on the illusory sea of birth and death. In the course of a life, there is nothing but the rise and fall of material and mental forms, while the unfathomable reality remains. In every creature sleeps an infinite intelligence, hidden and unknown but destined to awaken.

—Benjamin Labatut

I believe that in order to understand reality, we have to keep in mind that reality is this network of relations, of reciprocal information, that weaves the world. We slice up the reality surrounding us into "objects." But reality is not made up of discrete objects. It is a variable flux. Think of an ocean wave. Where does a wave finish? Where does it begin? Think of mountains. Where does the mountain start? Where does it end? . . . As humans we are that which others know of us, that which we know of ourselves, and that which others know about our knowledge. We are complex nodes in a rich web of reciprocal information.

—Carlo Rovelli

Man and all other living things must learn to live together.

—Noel A. Taylor

CONTENTS

Preface xi

Acknowledgments xv

Hors d'oeuvre xvii

1. Elemental 1
2. Lost World 26
3. Relationalism 50
4. Relativity 76
5. Entanglement 110
6. Information in Formation 145
7. Quantum Ecology 180
8. Minding the Body 208
9. Infinite Conversations 236
10. Strange Loops 279
11. After Life 327

Notes 343

Index 369

PREFACE

Philosophy is a prolonged meditation on death. Not all philosophy, but the philosophy that truly matters has always been a *memento mori*. Though death marks the end of an individual life, it is not a solitary event; when I die, the lives of others change irrevocably, and when others die, my life profoundly changes. Death is neither the opposite of life nor a singular event waiting for us sometime in the future. Constantly passing unnoticed, death is the condition of life.

Not only people die, plants and animals die, societies die, communities die, states die, nations die, civilizations die, philosophy dies, art dies, even God dies—sometimes without leaving a trace. And, of course, the ultimate fate of the universe will be heat death. Is it really possible to imagine planetary or even cosmic death? The end is the inverse image of the beginning; *creatio ex nihilo* becomes *destruction ad nihil*. Before the beginning and after the end—silence. Absolute silence. It is all too easy to acknowledge entropy without facing its profound existential implications. People who no longer believe in immortality or a personal afterlife often reassure themselves that their lives live on in those who come after them. How would your understanding of the meaning of your life change if you knew that next Friday at precisely 12:00 midnight the human race would become extinct or even the entire universe would vanish? Is it possible to imagine what might come after the human? There is no doubt the end will arrive sometime in the future. In the meantime—the time between absolute beginning and absolute ending—"our" planet is dying. *Our* planet—this is precisely the problem. Earth has never been ours alone, and the belief that it is ours leads to its death and, in turn, the eventual death of humanity. The word "human" derives from the word "humus," which is the organic component of soil, formed by the decomposition of leaves and other plant material by microorganisms. Death and life are thoroughly entangled. Bound in a parasite-host relationship, we are part of earth, which is a part of us. Though all life ends in death, not all death leads to life. Things are now so far out of balance that we may be past the point of no return. We have become an invasive species whose voracious appetites are destroying the host, without which we cannot survive.

What is to be done to delay this eventuality, and is there enough time to do it? While the answer is not clear, to throw up one's hands in despair will make the catastrophic outcome inevitable. It is obvious that corrective policies must be enacted immediately and that practical measures need to be taken. But this is not enough. I have devoted my life to reading, teaching, and writing because I believe *ideas matter*. Theories without actions are empty, actions without theories are blind. The dangers of the Anthropocene are inseparable from the anthropocentrism of modern philosophy, which begins with Descartes's collapse of truth into self-certainty and ends with Kierkegaard's declaration that "truth is subjectivity."

One of the most vexing challenges we face is the predominance of oppositional ideologies in a relational world. Thinking and Becoming or Be-ing are not opposites but are inextricably interrelated; thus, to think differently is to become different. Our age requires what early Christian theologians called *metanoia*—a spiritual and philosophical conversion to a nonanthropocentric vision that promotes hope in spite of despair. The position I develop in the following pages is *radical relationalism* in which divisive oppositions and conflictual activity give way to codependence and coemergent symbiosis where everything becomes itself in and through the probabilistic entangled play of differences.

In developing my argument, I have drawn on leading figures in modern Western philosophy whose arguments are grounded in the Jewish and Christian traditions. It is important to stress that the notion of radical relationism resonates in intriguing ways with other religious and spiritual traditions. In the following pages, I have noted the similarities Italian theoretical physicist Carlo Rovelli discerns between quantum mechanics and Mahayana Buddhism, and Chilean biologists Francois Varela and Humberto Maturana identify between autopoietic organisms and what they describe as Buddhist codependence and coemergence.

Drawing on the ancient insights from Confucius, Chinese political philosopher Zhao Tingyang develops a "methodological relationalism," which he uses to present a metaphysics of "relationology." This new philosophical vision, he argues, could form the basis of an effective political system uniquely adapted to the modern globalized world. In a passage that deeply resonates with the argument I elaborate, he writes:

> The most important idea developed by Confucianism is its philosophy of relations, better termed *relationology*, which implies *methodological relationalism*, a universal approach to understanding, analyzing, and explaining human actions and values in terms of *relations* rather than *individuals* (independent agents, subjects, or monads). I take a renewed methodological relationalism as a better horizon to discover solutions to the problems of conflicts, as well as a more reasonable and feasible approach to deal with the problematic situations of the multiversal world and/or the multicultural society with peoples of different hearts. Yet this does not mean replacing methodological individualism but erasing the mistakes or reducing the dangers caused by modern methodological individualism.[1]

Concentrating on Hobbes, Locke, and Kant, Zhao argues that Confucian relationology offers an effective alternative to the "Western" metaphysics of individualism. As the following pages make clear, the work of Hegel, Nietzsche, Heidegger, and Derrida point to a relationalism that is consistent with Zhao's position.

Multiple indigenous spiritual traditions also offer rich variations of something like radical relationalism. In their important works, Vine Deloria (*The World We Used to Live in: Remembering the Powers of Medicine Men*), Dhyani Ywahoo (*Voices of Our Ancestors, Cherokee Teachings from the Wisdom of Fire*), and John Bierhorst (*The Way of the Earth: Native America and the Environment*) explore contrasting cosmologies that bear important similarities with radical relationalism. Two of the key aspects of these traditions are their appreciation of the vitality of earth and an expanded notion of cognition that extends to plant and animal life. Robin Wall Kimmerer underscores this critical insight with the rich account of "reciprocity" in her widely influential book, *Braiding Sweetgrass: Indigenous Wisdom, Scientific Knowledge, and the Teachings of Plants*: "The relationship of gratitude and reciprocity thus developed can increase the evolutionary fitness of both plant and animal. A species and a culture that treat the natural world with respect and reciprocity will surely pass on genes to ensuing generations with a higher frequency than the people who destroy it. The stories we choose to shape our behaviors have adaptive consequences."[2] Kimmerer is a member of the Citizen Potawatomi Nation and a trained botanist who teaches at the State University of New York School of Forestry. Her effective interweaving of science, philosophy, art, and spirituality offers an exemplary model for others to follow. Finally, Jonathan Lear's remarkable *Radical Hope: Ethics in the Face of Cultural Devastation* draws profound lessons from the wisdom of Plenty Coups, who was the last chief of the Crow Nation. Facing the end of their way of life, this visionary found hope in spite of despair.

At this critical historical moment, we face a choice about how to approach the future: with either despair or hope. Despair is a self-fulfilling prophecy that closes the future by making it the necessary outworking of the past and present. Hope, by contrast, discerns in the past and present latent possibilities that leave the future open. If hope is honest, it must acknowledge how dire the situation has become, but this does not inevitably lead to despair. Radical relationalism requires radical hope, and radical hope requires radical relationalism. I firmly believe that we are not isolated individuals separated from each other and set apart from a complex world we are destined to dominate and control. No apocalypse, not here, not now—delay, defer, slow down long enough to linger and listen to what elemental voices are telling us. Reality *is* a "network of relations, of reciprocal information, that weaves the world," and if the end is to be deferred at least for a while, human beings "and all other living things *must* learn to live together" (emphasis added). *Neither* oppositions that defy unification *nor* unity that represses differences, but something in between is what must be imagined if we are to negotiate the exponential change that already is occurring.

Throughout my career, the reason I have studied, taught, and written about these complex issues is that I believe they can help us understand and negotiate existential dilemmas everyone faces. While writing this book, this lesson was driven home to me by the unexpected death of my brother, Beryl C. Taylor. In an effort to make the abstract concrete, I have interrupted my philosophical argument with personal reflections.

ACKNOWLEDGMENTS

This book has been a lifetime in the making. Along the way I have incurred countless debts to family, friends, colleagues, and students. Mentioning a few who have directly contributed to completing this book in no way minimizes the contributions of others. During my years at Williams College, I was fortunate to have remarkable colleagues who generously helped me understand complex issues in fields far from my professional training: Chip Lovet in chemistry, William Wooters in physics, Bill Lenhart in computer science, Hank Art in environmental science. Alis O'Grady always finds the books and articles I need. Elizabeth Kolbert's relentless realism keeps me honest. For many years Jack Miles has been a true friend, telling me what I sometimes do not want to hear. Herbert Allen's interest in and support of my work has enabled me to do things I otherwise would not have been able to do. Tom Carlson taught me that rocks are for climbing. Since he was young, my son Aaron has been explaining baffling geological processes to me and revealing mysteries only a thoughtful scientist appreciates; he has passed his scientific curiosity on to his children, Selma and Elsa. From the time she was a child, my daughter Kirsten's inquiries have forced me to ask myself difficult questions about her future and now the future of her children Jackson and Taylor. Dominy Gallo caught errors I missed. Both in and beyond the classroom, conversations with Alethea Harnish have clarified my thinking about the intriguing interplay between quantum mechanics and spirituality. I am especially grateful to her for her artful rendering of the diagrams in this book. Wendy Lochner not only is a gifted editor but is also a very special person of enormous insight and deep sensitivity. Finally, my wife Dinny helped me through trigonometry and physics in high school and has been tutoring me in science, mathematics, and technology ever since. This book and so much more would have been impossible without her.

HORS D'OEUVRE

I f space-time is relative, then it might be possible not just to travel to a probabilistic future, but also to travel to an improbable past. Perhaps an invisible observer could travel back in time to discover the strange entanglements without which he or she would not be the person he or she has become.

Jena, New Year's Eve, 1793. Goethe arrived from Weimar early in the afternoon to host a gala dinner party for the leading cultural luminaries of the day: Auguste Schlegel, Dorothy Schiller, Friedrich Schlegel, Caroline von Schlegel, Friedrich Schiller, Friedrich Schleiermacher, Friedrich Schelling, Friedrich Holderlin (What's with all the Friedrichs?), Novalis, Alexander von Humboldt, and Georg Wilhelm Friedrich Hegel. Has there ever been or will there ever again be a gathering of so many illustrious writers, artists, theologians, philosophers, and scientists in one place at one time? Was it chance that brought them together, or were they destined from the beginning of time to meet on precisely this evening?

This group was not given to idle chatter. Ideas that seemed to others frustratingly abstract and dauntingly complex were for the assembled luminaries deeply personal and profoundly political. Many of the guests had been arguing with each other since their university days. Goethe knew that seemingly inconsequential differences that make a difference could lead to passionate disagreements that could flare up at any moment and disrupt his carefully contrived plans for the evening. He had organized the seating arrangements to minimize the possibility of conflict. The irascible Schelling posed the biggest challenge. Goethe made sure Schelling was seated as far as possible from his erstwhile Tübingen roommate Hegel, from August Schlegel, and from Dorothy, Schelling's former wife. At the far end of the table sat the honored guest for the evening—Immanuel Kant. Only Goethe could have persuaded Kant to leave his home in Konigsberg for the first time. Goethe sat at the head of the table, where he could keep an eye on everyone and redirect the conversation if it threatened to go off the rails.

When dinner was finished, Goethe invited the group to retire to a spacious parlor where more wine and brandy were waiting. Sensitive to simmering tensions, he asked Schiller to introduce Kant. Goethe had recently read a draft of Schiller's *Letters on the Aesthetic Education of Man*, which would be published in a few months, and knew that Schiller's appreciation for Kant's *Critique of Judgment* was shared by all the guests. As Kant rose to speak, he expressed his appreciation for being included in such a prestigious gathering, though he acknowledged that the trip from Konigsberg had been a considerable strain. The title of his remarks was "Why I Could Not Bridge the Gap." Forty-three days later, Kant was dead. Did Goethe suspect or even know that he would soon die?

Kant knew his audience all too well. The warm welcome his concluding *Critique* received could not disguise the criticism that he had not gone far enough. Like Moses, he had glimpsed the Promised Land but had not been able to cross over to the other side. Each person quietly sipping wine while listening to the explanation for his hesitation had been trying in his or her own way to complete what Kant had left undone by closing the gap between idea and world.

Hegel, not-so-fondly dubbed "the old man," had remained silent and withdrawn throughout the evening. As the night wore on and the wine continued to flow freely, the scene became more a frenzied Bacchanalian revel than a sober Symposium. Hegel finally grew restive, rose to his feet, and clinked his glass repeatedly. Pausing long enough for silence to settle, he offered a toast not only to the New Year, but to a new era that he believed was dawning: "Our epoch is a birth-time, and a period of transition. The spirit of man has broken with the old order of things hitherto prevailing, and with the old ways of thinking, and is in the mind to let them all sink into the depths of the past and to set about its own transformation." Sensing that the seemingly warm reception of Hegel's remarks could not disguise tensions that continued to brew, Goethe seized the moment to thank the guests for attending, and everyone scattered into the night.

I had been listening intently all evening and furiously taking notes for hours. At the stroke of midnight, I rode the space-time curve back to the present with snippets and fragments of the heated conversations simmering in my mind. The following pages represent a modest effort to explain what I heard that night. It now seems clear to me that they knew so much more than they ever could have known they knew. But, alas, a word of caution. Among the many things I learned that night is that observation is never innocent—there is no world out there apart from the interpretation of it.

Like the Owl of Minerva looking back, I realize that from the cacophony and noise of colliding ideas, a new pattern, an unexpected order suddenly emerged. With the space-time that is never merely my own running out, I must confess that Hegel was, after all, right. What I observed that night was nothing less than the birth of a new world.[1]

AFTER THE HUMAN

CHAPTER 1

ELEMENTAL

And if you dig your fists into the earth and crumble geography, you strike geology. Climate is the wind of the mineral earth's rondure, tilt, and orbit modified by local geological conditions. The Pacific Ocean, the Negev Desert, and the rain forest in Brazil are local geological conditions. So are the slop carp pools and splashing trout riffles of any backyard creek. It is all, God help us, a matter of rocks.

—Annie Dillard

EARTH

MARK: I've been clearing out my office at Columbia and I have a problem. All the bookshelves in the house and the barn are full and I need to clear some space for the books I'm bringing home. We still have boxes filled with dirt you collected in Idaho for your doctoral research. You're never going to use that dirt so can I get rid of it?

AARON: No, you absolutely cannot get rid of my dirt. What about that dirt sculpture you created for your Mass MOCA *Grave Matters* exhibition that's hanging in the barn?[1] Dirt for books seriously? Dirt *is* my book.

MARK: Dirt is your book? Say more.

AARON: All rocks tell stories. The modest outcrop in our yard is over 500 million years old. This was the time when complex life forms were first exploding on earth. It used to sit off the coast of a large continent in the Southern Hemisphere and was most likely formed when a collection of shells and carbonite organisms accumulated. Now it sits humbly as a backdrop for your sculpture garden. That rock has seen more history than you can begin to get your mind around.

1.1 Berkshire marble, Stone Hill

Intrigued, I asked Aaron to explain the importance of his dirt.

AARON: Just as geologists can read stories in individual rocks and how they relate to one another, they can also piece together a broader understanding of how the earth has evolved over millions of years. Geologic time puts human time in a totally different perspective. To understand the questions you are asking, it is important to consider not only what comes *after* the human, but also what came *before* the human. Rocks are a time machine that lets you go back to the past so you can understand the present and predict the future.[2]

When you study rocks, you come to realize is that the earth has cycles... geologic cycles that occur over multiple millions of years. For example, the rocks will tell you that there are times when the earth has been warmer and wetter and other times when it was colder and drier. One of the factors that influences these cycles is the level of carbon dioxide in the atmosphere. All four fundamental elements are interrelated. The earth naturally regulates the amount of carbon dioxide in the atmosphere in a way that impacts the other three elements. The primary geologic source of carbon is volcanoes—the *fire* of volcanoes releases gases into the atmosphere. Once in the *air*, the carbon dioxide dissolves in *water*, where it reacts with the rocks of the *earth* and becomes bound into sediment and marine organisms that eventually form other types of rocks through a process called chemical weathering. Chemical weathering causes a rock to transform into soil by slowly dissolving. All rainwater, even absent any human pollution, is weakly acidic, and this acidity comes from the dissolution of carbon dioxide into rainwater. When the rain falls on the rock, a chemical reaction occurs and dissolves the rock and ends up binding the carbon dioxide and removing it from the atmosphere. You can think of this process as a pump that removes carbon dioxide from the atmosphere. The faster this pump runs, the more carbon is removed—under warmer and wetter conditions, the pump runs faster, creating a feedback loop. When the climate warms, the pump speeds up and removes more carbon dioxide; conversely, when the climate cools, the pump slows down, leaving more carbon in the atmosphere. As we read the records contained in certain types of rock, we can see how the speed of this chemical weathering pump has changed over time. This all occurs over hundreds of thousands of years though, so not fast enough to help with carbon sequestration on a human timeline. The challenge of the moment, of course, is that carbon from fossil fuel is an exogenous event for this system and introduces carbon dioxide at a rate that the geologic pump can't keep up with.

This is where soil becomes important. The oldest soils are only half a million years old. On the stratigraphic column of the earth, soil occupies a mere fraction on top, much like human time relative to geologic time.[3] Furthermore, human time is inextricably linked to the soil. If you dig a hole in the ground, like we did in Idaho and Wyoming, all the way down to the bedrock, you'll see the progression of a rock becoming soil. The dirt in the barn is a collection of samples taken from digging holes approximately five feet deep all the way down to the bedrock and sampling soil every six inches between the surface and the bedrock. The samples taken closest to the bedrock still largely resemble the ancient underlying rock type; as you move closer to the surface, the more easily weathered minerals disappear and you are left with more clays and silt. Gradually you approach the surface, and the nature of the dirt starts to shift—it doesn't just come from the rock below but begins to include more and more organic matter from above. By the time you reach the upper layers of the soil column, there is little sign of the original rock even if plants and other organisms continue to draw nutrients from it. Instead, you have a vibrant, interdependent community that inhabits the soil.

The minerals that are most resistant to weathering are silts and clays, which are basically the residual of the rock left when everything else has weathered away. Clays and silts create the architecture that provides the shelter for organic activity to flourish. The clays are filled with channels through which air, water, and organisms pass. These channels create a network for seeds, microbes, insects, and other creatures to circulate, each one enriching the soil in its own way. Over time, the soil becomes filled with organic matter both living and dead. As a soil matures, the quantity of organic matter it contains grows and becomes its own micro ecosystem with small animals, fungi, and plant roots all coexisting in positive feedback loops. When plants die or leaves drop, they are broken down by the various organisms living in the soil. Within a day, nearly half of the tissue of a leaf that falls to the ground has been digested by microbes and invertebrates dwelling on and in the soil through a process like mulching. As this occurs, fungi grow through the remaining plant material, releasing enzymes that break down the leaves, releasing the nutrients stored in them back into the soil. Other bacteria in the soil convert nitrogen from gas in the air to ammonia, which is a form that is useful to plants. Plants use nitrogen to grow, absorb other nutrients, and then when they die the plants are broken down to feed the bacteria again. In some areas there can be as many as two to three hundred fungal species in the top four millimeters of soil. Mycorrhizal fungi collaborate with plants, and almost every plant family has an intimate symbiotic relationship with one of the thousands of mycorrhizal species in the soil. Recent research suggests that underground fungal networks help endangered trees recover from the effects of weather extremes caused by climate change. In addition, these tangled webs play a vital role in the ability of trees to sequester CO_2. As the ecosystem recycles, it gets richer and more complex.

From the perspective of the global carbon cycle, this process is a critical sink for carbon that occurs in a human timeframe rather than the geologic carbon sink of chemical weathering. More carbon exists in soil than in the atmosphere and all plant life combined worldwide. Furthermore, soil holds as much carbon as the total amount that has been released by human activities since the Industrial Revolution. There is actually more biomass under the ground than above it. Several billion bacteria live in each square yard of soil, and, as I said, they play a critical role in the maintenance and health of the soil. On a slightly larger scale, a single acre of soil can have six million worm channels. Worms are like standalone intestines that digest, process, and enrich the soil. Problems are created by modern agricultural practices. Cultivated soil loses 50–70 percent of its carbon-storing capacity, and a lot of the good soil has been highly cultivated. For example, today just 3 percent of the tall grass prairie in North America remains. The soil underlying the prairies was among the richest in the world in terms of organic matter, and most of that carbon has been released into the atmosphere as the prairies have been transformed for agriculture. Similarly, clear-cutting forests breaks the cycle that moves nutrients among rock, soil, and trees and typically results in rapid and complete destruction of the soil. As deforestation continues around the world, the underlying soil breaks down and with it the carbon storage capacity.

But the importance of soil is about more than just carbon. The organic rich topsoil, often called humus, is a vibrant environment that is difficult to define. It is one of the most complex habitats on Earth. Because it is constantly being transformed, it is highly unstructured, with no two particles being alike. In reality, it is more like a network of relationships than a singular construct. According to some scientists, this environment is a terrestrial primordial soup in which the clays serve as a template for biosynthetic reactions that begin life. The chaotic, unstructured environment that exists in the humus is ideal for enabling the unexpected combinations and dynamic transfers of chemical and genetic information that create and modify life. Dirt, in fact, is the source of many drugs used against infectious diseases, including organisms that formed the basis for penicillin, cured tuberculosis, and show promise in treating cancer.

The soil is also a communication network. The fungi in soils that live in a symbiotic relationship with trees have been shown to transmit chemical signals that allow trees to communicate. These signals can give warnings of pests or fire nearby and cause trees to protect themselves. There are examples of larger trees with deeper roots transmitting nutrients to smaller, nearby saplings. There is even evidence for sharing resources across species. When one species is blocking light from reaching another, it can transfer nutrients to the shaded tree to help it survive. It may sound fantastical, but this soil communication network also transfers knowledge between older trees and younger trees as to what helps and what harms them and how to adapt and survive.

Unfortunately, soil is neither eternal nor immortal—like all ecosystems, it is fragile and is under growing stress. Increasing precipitation resulting from climate change accelerates erosion and results in the loss of soil. This erosion releases carbon from the soil into the atmosphere, thereby making global warming even worse. Furthermore, cultivation for agricultural purposes destroys soil's vibrant structure. Stripping crops after the harvest, removing trees, and even raking leaves and grass clippings stop providing the raw materials necessary for the soil to thrive. Industrialized farming compounds the problem by turning to alternative sources like fertilizer to provide the soil with nitrogen, phosphorus, and other nutrients. This creates further problems, ranging from destructive mining practices for chemicals to creating nutrient runoff in rainwater that results in destructive algae blooms in streams, rivers, ponds, and the ocean.

Returning to the dirt stored in the barn, it is important to understand that not all rocks are the same. Some were formed in the bowels of the earth, others on the sea floor or volcanoes, and still others, like the big outcropping in the yard, were sediment that slowly piled up and was squeezed slightly to form a rock. It turns out that the type of rock affects both how quickly it turns into soil and the composition and nature of the soils that form on top of it. The types of clay vary and, therefore, so do the structures and the networks I have been discussing. This, in turn, influences the plants and animals that succeed in any given place. The dirt in the barn comes from two very different rock types that happen to be sitting next to each other in Idaho. From these samples, we were able to learn about both how rates of weathering depend on the rock type on a geologic timeframe and how soils vary based on rock types when all other climatic conditions are held constant. It turns out that volcanic rocks consume about three times more carbon dioxide than other rock types when they weather and leave behind more complex clay structures that are conducive to richer organic layers in the soil. One partial remedy to current carbon levels in the atmosphere that has been proposed is to grind up rocks and spread them across fields to accelerate weathering above the natural process of breaking down a rock. At this point we need to try all the solutions we can, and understanding how different minerals react and absorb carbon dioxide will be an important part of making this successful.

The thing about samples like these is that if you get rid of them, you never get them back. So yes, it is like your books, but it is as if there is only one copy of each book in the world. Once it is gone, you can't read it again. So no, you can't get rid of my dirt.

MARK: Whoa! I never realized that soil, like the atmosphere, is a complex organism that must maintain such a delicate balance for human life to survive. Nor did I appreciate the intricacy of the interrelations of the four elements. Time, information, networks, circulation, communication. It's all going on right underneath our feet, and we never even notice it. I'll try to work out the implications of what you have explained for the other three elements. Mega thanks.

BTW I still have no idea where I'm going to put all my books.

WATER

Jared Diamond begins his monumental book *Collapse: How Societies Choose to Fail or Succeed* by reflecting on how fly-fishing had drawn his friend Stan Falkow, who was a professor of microbiology at Stanford University, to Montana's Bitterroot Valley. During a difficult period in Falkow's life, a trusted advisor just back from the Korean War said to him, "Stan, you look nervous, you need to reduce your stress level. Try fly-fishing." Facing increasing stress in his own life, Diamond decided to try what had worked for his friend and was soon, well, hooked on fly-fishing. Diamond recounts his first visit to Bitterroot years later:

> A dozen miles south of Missoula is a long straight stretch of road where the valley floor is flat and covered with farmland, and where the snowcapped Bitterroot Mountains on the west and the Sapphire Mountains on the east rise abruptly from the valley. I was overwhelmed by the beauty and the scale of it; I had never seen anything like it before. It filled me with a sense of peace and with an extraordinary perspective of my place in the world.... Every time I return to the Bitterroot, when I enter it on that stretch of road south of Missoula, that first sight of the valley fills me again with the same feeling of tranquility and grandeur, and that same sense of perspective on my relation to the universe.

It is precisely with this pristine environment that Diamond chooses to begin his tale of societal collapse. "Nevertheless, the Bitterroot Valley presents a microcosm of the environmental problems plaguing the rest of the United States: growing population, immigration, increasing scarcity and decreasing quality of water, locally and seasonally poor air quality, toxic wastes, heightened risks from wildfires, forest deterioration, losses of soil or of its nutrients, losses of biodiversity, damage from introduced pest species, and effects of climate change."[4]

From mountains to oceans. From dirt to water. During a conversation about how I planned to begin this book, my good friend Jack Miles told me that he had once asked Diamond what it would take to finally convince people that climate change is real. Diamond responded, "The collapse of the ocean fisheries." Worldwide more than three billion people depend on wild or farm-raised ocean fish for animal protein. In addition to signaling a disastrous environmental crisis, the collapse of ocean fisheries would create an immediate global food crisis.[5] Most of the planet we inhabit remains a dark mystery. While oceans cover 70 percent of the planet and contain 97 percent of the water on earth, only 5 percent of the waters in the ocean have been explored. This underworld is a complex ecosystem that is inseparably entangled with ecological systems above the surface. Though we know little about this dark underworld and are only beginning to understand its relation to terrestrial and celestial dynamics, we know enough to know that oceans are dying as the result of human activity. According to the National Oceanic and Atmospheric Administration, less oxygen dissolved in the water creates "dead zones" because

> most marine life either dies, or, if they are mobile such as fish, leave the area. Habitats that would normally be teeming with life become, essentially, biological deserts. [These] zones can occur naturally, but scientists are concerned about the areas created or enhanced by human activity. There are many physical, chemical, and biological factors that combine to create dead zones, but nutrient pollution is the primary cause of those zones created by humans. Excess nutrients that run off land or are piped as wastewater into rivers and coasts can stimulate an overgrowth of algae, which then sinks and decomposes in the water. The decomposition process consumes oxygen and depletes the supply available to healthy marine life.[6]

From air to water to air. Climate change is killing oceans, and dying oceans exacerbate climate change. As a deadly heat wave settled across the United States and did not lift for weeks during the summer of 2023, ocean temperatures rose to the highest level ever. About 40 percent of oceans worldwide suffered a heat wave that was equally serious. The oceans have played a major role in moderating the effects of increased greenhouse gas emissions since they absorb 90 percent of the planet's greenhouse gases. As the oceans warm, however, their ability to absorb gases decreases. Marine biologist David Taylor explains the far-reaching implications:

> The amount of atmospheric CO_2 that dissolves in the ocean is directly related to water temperature. Warm water holds fewer dissolved gases, including carbon dioxide and oxygen. Thus, the ability of the oceans to mitigate the impact of greenhouse gases is reduced as water temperatures warm (i.e., atmospheric CO_2 concentrations will accelerate because less is moving into the oceans).
>
> It is also important to understand the fate of dissolved CO_2 in the ocean. CO_2 is utilized by phytoplankton during photosynthesis—autotrophs ("self-feeding") that convert CO_2 to organic carbon. This carbon enters the food web and over time can sink to the bottom of the ocean (to be stored for 100s–1000s of years in marine sediments). This is referred to as the biological pump.[7]

Further note that ocean CO_2 concentrations have significantly increased over time (obviously in concert with atmospheric CO_2). This has altered ocean water chemistry, most notably in the form of ocean acidification. Reduced absorption by the oceans leads to increased amounts of greenhouse gas in the atmosphere, which further warms the air and water. This triggers another feedback loop by causing more rapid sea ice and glacial melting. Since sea ice mitigates planetary warming by reflecting sunlight through what is known as the albedo effect, melting sea ice means that warming trends are likely to accelerate.

>Sea ice melts.
>Water warms and absorbs less CO_2.
>Greenhouse gases increase and atmospheric temperature increases.
>Water warms and absorbs less CO_2.
>Sea ice melts faster . . .

As temperatures increase, oceanic carbon sequestration decreases. In addition to altering the chemical balance of oceans, increasing air and water temperatures change ocean currents, which, in turn, alter air currents and weather patterns. Since temperature fluctuations are greatest at the poles, higher atmospheric temperatures accelerate glacial melting. Three quarters of the earth's fresh water is stored in glaciers. Fresh water differs in density from salt water, and when there is massive glacial melting, the mix of salt water and fresh water changes. This leads to an alteration of both ocean and air currents. As the heavier salt water sinks and drifts toward the Artic, warmer water from the tropics flows northward. This creates what the late Columbia geochemist Wally Broecker labeled the "conveyor belt," which moves heat around the world and thereby plays a decisive role in weather patterns across the globe. There is growing evidence that global warming is severely disrupting this system and that the conveyor belt will shut down. This would result in a temperature drop of 9 degrees Fahrenheit throughout Europe and would create more violent storms, further damaging marine ecology.

Earth to water. While ocean water temperatures off the coast of Florida were rising to unprecedented levels, another development, which is equally important, received less attention. From the Everglades to the Mississippi to the Great Lakes, massive blooms of algae were strangling saltwater and freshwater environments. These outbreaks were caused by runoff from fertilizers with high levels of nitrogen and phosphorus. Between 1950 and 1984, a so-called Green Revolution transformed agriculture by industrializing the process of making fertilizer and changing farming methods to emphasize scale and crop uniformity. Over this period world grain production increased 160 percent. These industrialized farming techniques, however, destroy the architecture of the soil. Scientists estimate that between 24 and 46 percent of the topsoil in the Corn Belt has already eroded and blown away. Fertilizers laden with nitrogen and phosphorous seep into groundwater and run off into streams and rivers and eventually reach the ocean. Some 95 percent of the grain raised in the United States is corn, which has been cultivated with contaminated fertilizers. Most of this corn is fed to livestock like cattle, pigs, and chickens, which absorb the chemicals from their feed. When humans eat these animals, they consume the chemicals and pass them back into the soil and water in their waste. In addition to becoming contaminated, massive aquifers with ancient water are being drained at an unsustainable rate for agriculture purposes. In a *New York Times* article dated August 30, 2023, titled "The Dire Consequences of Depleting California's Groundwater," Soumya Karlamangla writes, "The dwindling of the groundwater supply could threaten America's status as a food superpower, as sustaining industrial-scale agriculture becomes more difficult. It could also slow home-building: This recently occurred around Phoenix, where officials said there wasn't enough groundwater for new homes that rely on aquifers."

In the summer of 2019, there was a massive algae bloom in the Gulf of Mexico, which was the result of fertilizer-laden water being emptied from the Bonnet Carre spillway in Louisiana to ease the rising Mississippi River. The bloom caused severe damage to coral reefs in the Gulf of Mexico. In another example, in the summer of 2023 these chemicals were ten times above the safe level in Florida's Lake Okeechobee. Water from this lake seeped into rivers, which flowed into the Everglades and eventually emptied into the Gulf of Mexico and the Atlantic Ocean. In her timely article, "Phosphorous Saved Our Way of Life—and Now Threatens to End It," Elizabeth Kolbert notes that the importance of phosphorous for all of life was discovered in 1802 during a trip to Peru by one of the guests at my Jena dinner party—Alexander von Humboldt. With characteristic wit, she writes, "He set up his instruments atop a fort on the waterfront, and then, with a few days to kill . . . wandered the docks. A powerful stench emanating from boats loaded with what looked like yellowish clay piqued his curiosity. From the locals, Humboldt learned the material was bird shit from the nearby Chincha Islands, and that it was highly prized by farmers in the area. He decided to take some with him."[8] Though it was not known at the time, what made the excrement so valuable for farmers was the phosphorous it contained, which makes plants grow faster and bigger.

Phosphorous is a rare mineral whose greatest reserves are in Morocco. With phosphorus, like everything else, it is necessary to maintain a delicate balance between too little and too much. Life is impossible both without phosphorus and with excess phosphorus. It is a component of DNA and RNA and is essential to the formation of cell membranes and to the chemical adenosine triphosphate (ATP), which enables cells to store energy. Life and death are inseparably entangled in phosphorus. Nowhere is the paradoxical effect of phosphorus more evident than in algae. Excess phosphorous makes algae grow uncontrollably, and excess algae kills coral reefs, which are essential to marine life and by extension our own. A coral reef is a living organism that forms and is formed by a delicate ecosystem that includes bacteria, plants, urchins, clams, crabs, and fish. Though still photographs can capture the colorful beauty of coral reefs, they cannot convey the sense of reefs as complex and diverse living organisms. When ocean currents shift, individual polyps respond together in gentle ripples that look like fans at a baseball game doing the wave, or a mobile flock of birds soaring through the sky in modulating patterns. Rather than acting independently as isolated individuals, polyps act as integral parts of a well-functioning whole that responds sensitively to its surroundings. Coral reefs are made up of small polyps, which form colonies that can grow very large. Millions of polyps secrete shells of calcium carbonate to create the skeletal structure of the colony. Reefs are very fragile and extremely responsive to changing environmental conditions. As in every complex system, the alteration of any part transforms the whole. Nothing is more toxic for coral than large algae blooms created by excess nitrogen and phosphorus. As macroalgae grow over corals, they smother them and prevent young corals from settling in the reef. Other factors resulting in coral death include overfishing, ocean acidification, and coral mining. Since 1950, 50 percent of coral reefs have died. The global effects of this spreading extinction event are devastating. Since the diverse environment of reefs provides the habitat for 25 percent of all marine species (that is well over one million species), they play a disproportionate role in the overall ocean ecology. The disappearance of coral sets off a domino effect that quickly spreads from sea to land. This is why Diamond cautioned that the collapse of ocean fisheries could signal societal collapse.

AIR

You can't see it, hear it, smell it, touch it, or taste it. Its effects, yes; air as such, no. Or so it seems. It is always there, surrounding us, entering us, leaving us from our first to our last breath. Ten million times a year, five million liters of air in and out. Inhaling and exhaling, talking and sneezing, we change the air that nourishes us. In his suggestive book *Heaven's Breath: A Natural History of Wind*, Lyall Watson writes, "The truth is that air supports an unseen ecology that is every bit as rich and varied as a rainforest or a coral reef." From the breath of bodies to the wind of heavens, air is defined by motion. There is no air without the atmosphere and no atmosphere without air.

> An atmosphere is the most vital prerequisite for life and mind. Without it, worlds are hard-pounded deserts, dry as bones, where alien feet fall soundlessly and even the dust is dead. With it, everything becomes possible. Orbiting rock piles organize themselves, taking on the self-contained look of living creatures, giving, and receiving information, learning the essential skills involved in managing the sun.
>
> Earth is such a creature. An organism, growing, wrapping itself in a moist and luminous membrane of air. And within this semi-transparent envelope, fed by a network of capillary breezes, energy and information from one extremity to the other, are the great arteries of the prevailing winds.[9]

Like soil, water, and the human body, air is an organism of organisms teeming with life. These interrelated organisms have a similar structure. Just as the human organism develops from a spherical blastomere with three layers known as the endoderm, mesoderm, and ectoderm, so the planetary organism is bounded by a laminated semipermeable membrane composed of five concentric spheres: troposphere, stratosphere, mesosphere, thermosphere, and exosphere.

The chemical composition of the atmosphere includes 78 percent nitrogen, 21 percent oxygen, 0.93 percent carbon dioxide, 0.04 percent argon, and a few additional trace chemicals, such as the greenhouse gases methane, nitrous oxide, and ozone. The atmosphere also has between 0.4 and 1 percent water vapor, which comes from terrestrial bodies of water. There are complex chemical interactions within each stratum and between different strata. Proper functioning requires maintaining an extraordinarily delicate balance of elements within a very narrow bandwidth. This system is a self-organizing and self-maintaining network of networks that is not static but is constantly renewing itself and adapting to fluctuating circumstances. The Goldilocks principle is operative everywhere—too much or too little results in system failure. To understand the dynamics of global climate, it is necessary to have a basic understanding of the different atmospheric layers and their interrelation.

The troposphere (Greek *tropus*, rotating + *spharia*, sphere) is closest to the earth and determines most of its weather. Its altitude varies from the equator to the poles, and its temperature, which is determined by radiation from the earth's surface, decreases with increasing altitude. Atmospheric flow is from west to east, and air currents above and below the equator do not mix. Gravity holds the atmosphere on the surface of the earth, and air pressure is determined by the total weight of molecules. At higher elevations, where gravitational forces are weaker, there are fewer molecules, causing air pressure to decrease with increased altitude. Temperature and air pressure are also linearly related, so temperature differentials also create changes in air pressure, and these vary in relation to location relative to the earth's surface. Temperature differences create wind, which causes the movement of weather systems. As air rises, it cools and moisture condenses to form clouds and, when sufficient, rain. The greater the moisture content and temperature of the air, the more the rain and the more violent the storms. High-pressure air is constantly moving toward lower-pressure areas, much like air leaving a balloon. As the heavy, high-pressure air moves in the atmosphere, the dry air above it sinks and warms, resulting in fair weather but little precipitation. If this condition persists for extended periods of time, it can lead to drought.

The next layer is the stratosphere. With the transition from the troposphere to the stratosphere, there is a temperature inversion. Rather than the temperature radiating from the earth's surface decreasing, the absorption of the sun's ultraviolet light by the oxygen molecules in the ozone layer leads to increasing atmospheric temperatures. Chlorofluorocarbons used in refrigerators, air conditioners, and aerosol sprays break down due to ultraviolet light and release chlorine, which chemically reacts with ozone and splits apart, damaging the ozone layer. One of the overlooked factors contributing to global warming is the growing use of air conditioners as a result of higher temperatures for longer periods. Increasing ultraviolet radiation damages DNA in bacteria as well as mammals, including humans. Without the protective shield of the ozone layer, life on earth would be impossible. This is why the hole in the ozone layer detected in 1985 was potentially so catastrophic. While progress has been made at reducing chlorofluorocarbon emissions, the size of the ozone hole today remains as large as it was thirty years ago.

The third atmospheric layer is the mesosphere (Greek *mesos*, middle + *spharia*, sphere), which is between the two lower and the two higher spheres. Less is known about the mesosphere than about any other part of the atmosphere. Just as the stratosphere protects earth from ultraviolet sunlight, the mesosphere protects earth from incoming meteors and meteorites. The vaporization of these extraterrestrial objects leaves a residue of atoms from iron and other minerals. At the outer edge of the mesosphere is the mesopause, below which gases are mixed by turbulence, and above which there are so few particles that little mixing occurs. The most important effect of mesospheric dynamics is the strong winds known as atmospheric tides, which dissipate waves originating in the troposphere and stratosphere. These changes influence the global circulation of air, which, in turn, alters weather patterns.

The final two layers of the atmosphere are the thermosphere and the exosphere, which include the ionosphere. The high concentration of ions and electrons in the ionosphere is essential for atmospheric electricity. Increasing heat from unfiltered sunlight accelerates the rate of chemical reactions causing the splitting of molecules, releasing positive ions and electrons. These electrons form electrical fields that create radio waves, which make global communication possible. The ionosphere is highly volatile, changing from day to night and influenced by activity both below on earth and above in distant space. Satellites orbiting in the ionosphere collect data used in long-term weather predictions. Chemical activities in the ionosphere can interrupt communication signals between satellites and disrupt electrical infrastructure on the ground.

Finally, the exosphere is the outermost boundary where the earth's atmosphere merges with outer space. The density of the exosphere is so low that molecules rarely collide and, thus, there is little chemical activity. Slower-moving particles fall to lower levels of the atmosphere, and faster-moving particles spin off into outer space.

The extraordinarily complex atmospheric system maintains a delicate balance of interrelated elements and processes. Life on earth hovers on the boundary of *neither* too much *nor* too little. Nowhere is this neither/nor more evident than in atmospheric oxygen. Neither too much (above 23.5 percent) nor too little (below 19.5 percent) oxygen is necessary for life. For humans and other mammals, too little oxygen results in convulsions, suffocation, and cardiac arrest; too much oxygen can damage cell membranes and lungs, causing seizures. While oxygen is not flammable, it oxidizes other materials. Higher oxygen levels increase flammability; above 23 percent, fires become more frequent, spread more quickly, and become much more destructive. For every 1 percent rise in the oxygen level, the chance for forest fires increases by 70 percent. Levels above 25 percent would result in fires that would destroy most plant life on earth.[10]

For the past 600 million years, the level of atmospheric oxygen has remained a relatively stable 21 percent. Maintaining this level requires a complicated chemical relationship between oxygen and the greenhouse gas methane. From the outermost atmosphere to the depths of earth and water, there are multiple sources of oxygen. In upper atmospheric layers, a process known as photolysis splits water molecules, leaving hydrogen atoms to escape into space and oxygen atoms to form bonds that make ozone, which is three oxygen molecules bonded together. A much more important source of oxygen is photosynthesis. Plants inhale carbon dioxide and exhale oxygen. This is the reason the Amazon Rainforest often is called the lungs of the planet. It absorbs about 25 percent of atmospheric carbon dioxide and produces approximately 6 percent of the oxygen in the atmosphere.[11] Some of the carbon in carbon dioxide is fixed in plants and buried in sedimentary rocks and permafrost. The remaining oxygen is released into the air.

While photosynthesis in trees and worldwide forests produces considerable oxygen, about half of atmospheric oxygen comes from an unlikely source—microscopic algae known as phytoplankton, which float on the surface of oceans. They mix sunlight with carbon dioxide from the atmosphere to produce oxygen and carbohydrates in the form of glucose, which is the primary source of energy for plants. However, the source of life is the source of death, and the source of death is the source of life. While some algae smother and kill coral reefs, disrupting ocean ecology, other algae have a positive effect by providing food for marine life and by reducing the carbon dioxide in the atmosphere and increasing the oxygen necessary for life.

Methane (CH_4), which is a greenhouse gas twenty-five times more potent at absorbing heat than carbon dioxide (CO_2), plays a crucial role in regulating atmospheric oxygen. It is "produced by bacterial fermentation in the anaerobic muds and sediments of seabeds, marshes, wetlands, and river estuaries where carbon burial takes place. The quantity of methane made in this way by microorganisms is an astonishing 500 million tons a year." Some methane rises to the stratosphere, where it oxidizes carbon dioxide and water, providing moisture to the upper atmosphere. This water eventually dissolves with hydrogen again escaping to outer space and oxygen descending to the troposphere. By contrast, in the lower atmosphere, the oxidation of methane uses oxygen. "This process," James Lovelock explains, "goes on slowly and continuously into the air we live and move in, through a series of complex and subtle reactions. . . . [I]n the absence of methane production, the oxygen concentration would rise by as much as 1 percent in as little as 24,000 years: a very dangerous change and, on the geological timescale, a far too rapid one."[12]

In complex positive feedback systems, minor changes or fluctuations can have disproportionate effects that are unpredictable. To understand more adequately the nonlinear dynamics created by radically relational atmospheric networks and systems, it is helpful to consider two codependent factors more closely—wind and heat. The composition and behavior of air are inseparable from earth and water just as the composition and behavior of earth and water are inseparable from air. Loops within loops within loops. Changing ocean currents cause airflow to shift, altering weather patterns, which lead to temperature fluctuations that bring rain that falls into lakes, rivers, and oceans. Air waves, streams, currents, eddies, and tides are always in motion. Though invisible, air is teeming with life. According to Watson:

> The census of microscopic species in the air has only just begun, but already includes a host of viruses; close to 1,000 varieties of bacteria; 40,000 species of fungi; hundreds of different kinds of algae, mosses, liverworts, ferns and protozoans in at least some stage of their life cycle; and more than 10,000 species of plants that are wind-pollinated. . . .
>
> Clouds are probably the greatest concentrators of airborne micro-life. Rising bubbles of air in convection cells bring collections of organisms to heights where some of them are frozen and thrown to earth as hail, but others simply get washed down to lower levels where rain evaporates, allowing them to be carried up once more and eventually be heaped up in greater and greater numbers at certain fixed heights.

The organisms circulating around the globe in air currents are not isolated entities but are joined in reciprocal relations that form a microclimate with a distinctive quality. "There is," Watson concludes, "good reason to accept that a distinct 'biological zone' with an indigenous fauna and flora, exists and persists in parts of the upper air."[13]

In the past few years, wearing masks has been a constant reminder of pathogens that ride worldwide currents of air. But not all microbes are toxic. There are approximately fourteen thousand species of mushroom-producing fungi, some of which release over one billion spores a day. A summer-long survey of air in rural England produced astonishing results. "The numbers of fungi varied enormously, but it was not unusual to find 12,000 in each cubic meter of air."[14] As we will see in chapter 9, fungi are essential to what Suzanne Simard describes as the wood wide web that connects trees and plants across the globe.

As Aaron explained, one of the most overlooked aspects of air is how it affects and is affected by dirt. Wind contributes to the production of soil by eroding rocks, but wind also erodes soil. According to the United Nations Environmental Program, "Each year, an estimated 24 billion tons of fertile soil are lost to erosion. That is 3.4 tons lost every year for every person on the planet." Since soils store more than 4,000 billion tons of carbon, erosion also contributes to global warming.[15] However, this is only half the picture. Dust storms and dust devils create clouds that scatter dust, which enriches soil and nourishes plants. The most remarkable and important example of this phenomenon is the flow of dust from the phosphorous-rich sands of the Sahara Desert to the Amazon Rainforest. In 2014 NASA satellites revealed that winds pick up an average of 182 million tons of dust, carry it 1,600 miles across the Atlantic Ocean, and dump it in South America and the Caribbean. "This trans-continental journey of dust is important because of what is *in* the dust.... Specifically, the dust picked up from the Bodele Depression in Chad is from an ancient lakebed where rock minerals are composed of dead microorganisms, which are loaded with phosphorus. Phosphorous is an essential nutrient for plant proteins and growth, which the Amazon rainforest depends on to flourish."[16] Without the 22,000 tons of phosphorous carried by the wind to replenish what is lost from rain and floods, the capacity of the soil in the rainforest to promote growth would decline, and the lungs of the planet would suffer.

Wind and heat are inseparably interrelated. In his timely book, *The Heat Will Kill You First: Life and Death on a Scorched Planet*, Jeff Goodell writes, "One way to think about atmospheric circulation is as a giant heat transport system, one that is constantly circulating warm air from the tropics up to the poles, bringing the cooler air from the poles down to the tropics. The main engine of this heat transport system is the jet stream, which blows from west to east in the upper atmosphere." Molecules vibrate at different rates, and temperature is the average vibrational speed of a group of molecules. The faster the vibration, the greater the heat; different rates of vibration influence the behavior of molecules. For example, the molecular structure of carbon dioxide makes it more prone to absorb sunlight. This makes CO_2 vibrate faster, thereby increasing atmospheric heat. Temperature differences determine air pressure, which causes changes in the jet stream, where these fluctuations contribute to heat waves. Increasingly frequent and severe heat waves trigger more positive feedback loops. Warmer air melts permafrost and releases methane, which further increases air temperature. Sea ice and glaciers also melt faster, creating warmer water, which further accelerates melting. Goodell explains the effect of these developments. "As the Artic warms, it's changing the temperature gradient between the poles and the tropics. That in turn weakens [atmospheric] waves, allowing the jet stream to meander and get twisty. Sometimes those twists trap hot air over a region, not allowing it to escape. The trapped air gets hotter and hotter, as a result of both the warm land below and the increasing high pressure that keeps out clouds and amplifies sunlight."[17]

As record-breaking heat waves proliferate, their negative impact on plants, animals, and people spreads rapidly. Higher temperatures increase metabolism, which affects health. While heatstroke, heart attacks, respiratory problems, and exhaustion are the most obvious effects of higher temperatures, invisible changes are even more devastating. In a *New Yorker* article titled "What a Heat Wave Does to Your Body," physician Dhruv Khullar writes, "Heat affects us on a molecular level. Excess heat interferes with the chemical bonds that help proteins to twist and fold into shape; just as a hot frying pan can denature the proteins in an egg, high body temperatures can denature the proteins in our cells, preventing them from functioning properly and even killing them off, especially in the liver, blood vessels, and brain."[18]

In addition to altering bodily processes, increasing heat has a further impact on water and fire. Higher temperatures lead to greater evaporation, which increases the moisture content of air. Hotter air holds more moisture and creates the likelihood of more frequent and more violent storms. At the same time, higher temperatures suck moisture out of the earth, creating drought conditions and making trees and plants much more susceptible to fire. According to the National Oceanic and Atmospheric Administration, an "average annual 1-degree Celsius temperature increase would increase the median burned area per year by as much as 600 percent in some types of forests."[19] Raging forest fires create terrible pollution: 260,000–600,000 people die each year from inhaling smoke from wildfires.[20] However, atmospheric dynamics once again play weird tricks. The same pollution that clouds the sky facilitates the creation of rain. "Condensation takes place only with great difficulty in clean air. Moisture needs a suitable surface on which it can form. This can be earth itself, as in the case of dew or frost, but free air condensation almost always takes place around nuclei such as salt or smoke or dust. Without some form of pollution, fair weather clouds become unlikely and rain clouds totally impossible."[21] When everything is interconnected, apparent opposites are joined in dialectical relations, and neither can be itself without the other, and all act together to create a whole that is more than the sum of its parts.

FIRE

Norman Maclean is known for water—the Blackfoot River in Montana. The first sentence of *A River Runs Through It* is one of the finest lines in the English language. "In our family, there was no clear line between religion and fly fishing." For Maclean, the Blackfoot is the river of time in which all things arise and pass away. Reflecting on his brother Paul casting on their final day of fishing together, Maclean concludes the book with words as memorable as the first line.

> Eventually, all things merge into one, and a river runs through it. The river was cut by the world's great flood and runs over rocks from the basement of time. On some of the rocks are timeless raindrops. Under the rocks are the words, and some of the words are theirs.
>
> I am haunted by waters.[22]

When he was a teenager, Maclean worked in the Bitterroot Mountains, which had drawn Diamond to Montana and changed his life forever. The following summer, he left those mountains to work for the Forest Service and never returned.

A River Runs Through It was the only book Maclean ever published. He wrestled with another manuscript for years but was never able to finish it. This was a book about a group of young smoke jumpers working for the United States Forest Service, who parachuted into Mann Gulch on August 5, 1949, and never returned. Maclean saw the fire and became obsessed with it. When he died in 1990 at the age of ninety, the book remained incomplete. Like the river, the fire was more than a fire for Maclean—it was an image of an elemental force shaping life and death. It is as if the fire were his whale, and to give up the struggle with the fire would be to give up life itself. What became *Young Men and Fire* was "a story in search of itself as a story, following where Maclean's compassion led it. As long as the manuscript sustained itself and its author in this process, it had to remain in some sense unfinished."[23] With Maclean's death, the book could finally be finished.

The year it appeared (1992), there were plans to make the book into a film. I received an invitation to join the director and several of Maclean's family members on a trip up the Missouri River to Mann Gulch, and I eagerly accepted. I had my ticket and was ready to go, but the day before my departure, my father, Noel A. Taylor, unexpectedly died and I had to cancel the trip. Since that day, the memory of the death of my father has always been associated with fire. Last summer, I returned to the North Fork of the Flathead River, which forms the western border of Glacier National Park, to visit a longtime friend. With fires raging at the edge of the park, in nearby Canada, and across the globe, Maclean's profound meditation seemed timelier than ever.

Maclean saw in the unmasterable Mann Gulch fire the limit, if not the end, of modernity. "In this story of the outside world and the inside world and the fire between, the outside world of little screwups recedes for now for a few hours to be taken over by the inside world of blowups, this time by a colossal blowup but shaped by little screwups that fitted together tighter and tighter until all became one and the same thing—the fateful blowup. Such is much of the tragedy in modern times and probably always has been except that past tragedy refrained from speaking of its association with screwups and blowups." By telling the story of thirteen young men who could not outrun death, Maclean offers a cautionary tale for a world ablaze. Through his meticulous reconstruction of the unexpected fast spreading of fire, he exposes the shortcoming of the firefighting methods guided by the principles of so-called scientific forest management and explains how the Mann Gulch fire shaped future firefighting practices. Crosses now mark the spot where fire overcame each young man. Pondering this incalculable tragedy, an old man reflects about the point where nature, mathematics, and physics meet:

> We are beyond where arithmetic can explain what was happening in the piece of nature that had been the head of Mann Gulch. Converging geometries had created something invisible like suction to carry off a natural explanation of the attraction of geometries to each other. In between these geometries for something like four minutes was a painfully moving line with pieces of it dropping out until there came an end to biology. Then it was pure geometry, and later still the solid geometry of concrete crosses.[24]

While wind is often associated with spirit (Arabic, *ruh*; Hebrew, *ruach*; Greek, *pneuma*), many believers regard fire as a theophany. The deity so revealed is simultaneously destructive and creative. Problems for the human race began when Prometheus rebelled against the Olympian gods by stealing fire and giving it to humanity. This gift has long fueled man's [sic] promethean will to mastery. This desire to master is quite literally fueling the increased frequency and intensity of fires. For decades, the conventional fire management wisdom was to suppress and put out all fires before they could spread. Over time, this caused the volume of dead timber to increase, which, when the fire eventually comes again, fuels bigger and hotter fires. Furthermore, as the planet continues to heat up from man's relentless self-assertion, both the frequency and the behavior of wildfires are changing. Baked earth, hot air, low moisture, and wind create a tinderbox waiting to explode with the slightest spark. Fires raging out of control appear to be living creatures stalking the parasites endangering the earth. Stephen Pyne, who is the most prominent historian of fire, writes that fire's "critical parts began to act in ways that rendered fire less random, that made its appearance and absence less like the roll of a roulette wheel and more like the give and take of prey and predator." Though each wildfire is unique, fires have a distinct ecology. Pyne continues: "Fire has become a selective force and an ecological factor that guides evolution, organizes biotas, and bonds the physical world to the biological.... Like storms and earthquakes, it disturbs sites; like fungi and termites, it recycles dead biomass; like sun and rain, birds and beetles, it is simply there as cause, consequence and catalyst."[25] Furthermore, fires, like people, have seasonal and diurnal rhythms. The Canadian Department of Natural Resources and Renewables has issued a warning about the effects of weather, topography, and fuels on wildfires in which it describes the "crossover phenomenon":

> There is a very distinct relationship between Temperature and Relative Humidity.... As air temperature readings increase and relative humidity readings decrease during the day, burning conditions will become more intense. Certain atmospheric conditions can produce severe fire behavior, reaching a point when the air temperature reading and the relative humidity reading will read the same (i.e., Temp. 30° C, RH 30%). At this point a condition has been achieved called *Crossover*. A fire burning under this condition will exhibit extreme behavior and should be treated with extreme caution. This condition should not be treated lightly and may continue for some time. It can become even worse if the relative humidity reading continues to drop.[26]

The increasing frequency of crossover fires is the effect of higher temperatures on earth and in water and air.

John Vaillant explores how fires are changing in his book *Fire Weather: A True Story from a Hotter World*. He focuses his scrupulous investigation on a single apocalyptic fire that occurred in the center of Canada's oil industry—Fort McMurray in May 2016. Located at the heart of Canada's oil sands and surrounded by boreal forest, Fort McMurray is the point where the fossil fuel industry collided with nature. Rapidly rising temperatures provided the spark that made this collision explosive. More than 300,000 pipelines carrying oil thread through Alberta. On the day of the fire, weather systems created the conditions for a perfect storm.[27] The temperature was 30 degrees hotter than normal, which created a 30 percent drop in atmospheric humidity to a dangerous 15 percent. At the same time, the drought index, which measures the dryness of soil, grasses, and trees, was 100 percent. During the afternoon, winds increased from 12 miles per hour to 25 miles per hour. Vaillant reports that the only other time that Alberta had experienced similar conditions was in 2001, when the infamous Chisholm fire broke out.

> In the Chisolm Fire's most explosive phase, a funnel cloud was observed, and later investigation revealed mass blowouts of mature trees within the burn area. For fires of this magnitude, we need a different scale of measurement, and, in the end, the six authors of a peer-reviewed paper entitled "The Chisholm Firestorm" resorted to megatons, the units of energy used to measure the explosive power of hydrogen bombs. The energy released during the fire's peak seven-hour run was calculated to be that of seventeen one-megaton hydrogen bombs, or about four Hiroshima bombs per minute.

Once the crossover point is reached, the fire rages out of control and cannot be slowed down or stopped. At this point, fire suddenly spreads from the groundcover and the lower limbs of trees to the treetops where atmospheric gases and higher winds cause an explosion spreading hot embers far and wide. These fires can spread as fast as 80 miles per hour. Fires of this magnitude also alter atmospheric currents, creating positive feedback loops that further feed the flames. Vaillant concludes, "This, now, is what fire is capable of on earth. The chemistry and physics of fire remain unchanged, the trees themselves are no different than they were fifty years ago, but the air is warmer and the soil is drier—enough to make the latent energy living and dying in these forests that much easier to release."[28] The Fort McMurray fire gives us a glimpse not only of the future of fire, but also of the future of a planet ablaze.

In the summer of 2023, smoke from Canadian wildfires set the city where I work aglow with an uncanny golden light and shrouded the mountains across the valley from the barn where I write in a smoky haze. On August 8, fire broke out on the Hawaiian island of Maui and quickly consumed everything in its path, killing over a hundred people. While the exact causes remain under investigation, part of the explanation appears to come down to our promethean impulse. Nonnative grasses that were used to feed livestock or for ornamental purposes were brought to the island decades ago and now pose hazardous risks because they are highly flammable. Standing in the midst of smoldering ruins, survivors who had lost everything asked why they had not been warned sooner. Local officials made lame excuses, but the real reason was that they did not understand how climate change is transforming fire. Here, as elsewhere, more is different, faster is different, and more + faster is really different.

And yet. And yet. And yet. While fire may be destructive for human beings, it is creative for forests. Wildfires regenerate forests and renew their life. "Without fire and its seemingly random but ultimately regular patterns of return, the boreal forest would collapse. There is in this cycle a kind of codependency that, when viewed from the point of view of fire, upends the notion of what a forest is, and whom it serves."[29] Higher temperatures and wind scatter seeds on ground cleared by fire, making room for new life to take root. The cones containing the seeds of some conifer species open and release the seeds only in the presence of fire. To regard fire as primarily destructive and in need of management and control is a symptom of the anthropocentrism that has created our apocalyptic condition. When the desire to control is out of control, disaster follows. As Hegel taught us long ago, if anything is pushed to the limit, it negates itself by turning into its opposite, and something new arises from the ashes. The urgent question we now face is what, if anything, will rise from the ashes.

Everything *is* connected. Four-in-one and one-in-four. Earth, air, fire, and water are not separate elements but are radically interrelated to form the elemental, which sustains life and harbors death.

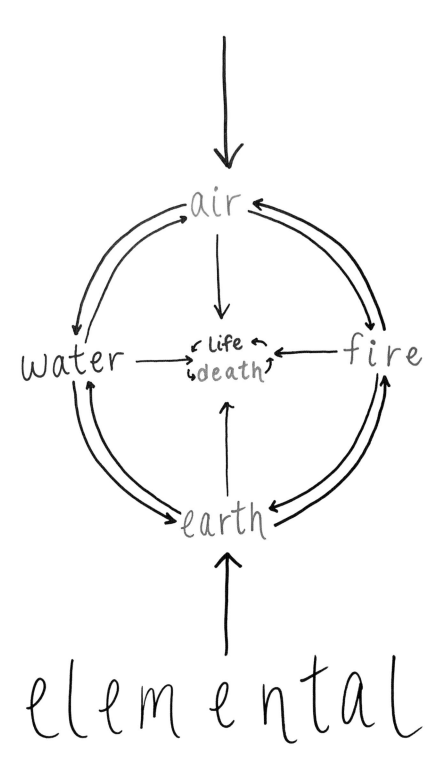

1.2 Elemental

CHAPTER 2

LOST WORLD

The world is shrinking down about a raw core of parsible entities. The names of things slowly following those things into oblivion. Colors. The names of birds. Things to eat. Finally the names of things one believed to be true. More fragile than he would have thought. How much was gone already? The sacred idiom shorn of its referents and so of its reality.

—Cormac McCarthy

SPEED LIMITS

Time is running out, running out faster and faster. Speed is the addiction that threatens what once was called the human.

The summer of 1984. Of course, 1984. Jacques Derrida published an article entitled "No Apocalypse, Not Now (full speed ahead, seven missiles, seven missives)" in a special issue of the influential journal *Diacritics* on Nuclear Criticism.

> Let me say a word first about speed.
> We are speaking of stakes that are apparently limitless for what is still now called "humanity." People find it easy to say that in nuclear war "humanity" runs the risk of its self-destruction, with nothing left over, no remainder.... There is a lot that could be said about that rumor. But whatever credence we give it, we have to recognize that these stakes appear in the experience of a

race, or more precisely of a *competition*, a rivalry between two rates of speed.... Whether it is the arms race or orders given to start a war that is itself dominated by that economy of speed throughout all the zones of its technology, a gap of a few seconds may decide, irreversibly, the fate of what is still now and then called humanity—plus the fate of a few other species.[1]

Derrida was not the first to discern the intricate interrelation of speed, modernity, science, technology, and the end of the human. No writer has understood the theological and philosophical presuppositions of modern science and technology better than Heidegger. In "The Age of the World Picture," published as World War II was erupting, he writes,

> The fundamental event of the modern age is the conquest of the world as a picture. The word "picture" [*Bild*] now means the structured image [*Gebild*] that is the creature of man's producing, which represents and sets before. In such producing, man contends for his position in which he can be that particular being who gives the measure and draws the guidelines for everything that is.... For the sake of this struggle of world views and in keeping with its meaning, man brings into play his unlimited power for the calculating, planning, and molding of all things. Science as research is an absolutely necessary form of this establishing of self in the world; it is one of the pathways upon which the modern age rages toward fulfilment of its essence, with a velocity unknown to participants.[2]

The detonation of the first atomic bomb, which brought the end of World War II, marked the beginning of the Cold War. For both Heidegger and Derrida, the specter of the apocalypse was a nuclear holocaust. What made this extinction event possible if not likely was that technology was accelerating at a rate that made timely human intervention virtually impossible. Merely five years after Derrida's apocalyptic musings, the Cold War ended with the fall of the Berlin Wall and the collapse of the Soviet Union. When Francis Fukuyama, following Hegel and Alexandre Kojève, prematurely declared the end of history and the triumph of global capitalism, he, like many others, overlooked an even more disruptive technological revolution that grew out research and development conducted to support the military industrial complex's war machine.

Neither Heidegger nor Derrida could have anticipated that in the closing decades of the twentieth century, speed itself would accelerate at an unprecedented rate. A few months after the publication of Derrida's "nuclear criticism," Apple's iconic advertisement introducing the Macintosh personal computer appeared during Super Bowl XVIII. The futuristic ad, which was directed by Ridley Scott of *Blade Runner* fame, is set in a dimly lit dystopian industrial site with a row of workers marching mechanically through a tunnel toward a large television screen displaying a huge face of a man reciting lines from George Orwell's novel *1984*, which anticipates Nikita Khrushchev's hyperbolic threat to the West, "We will bury you!"

> Today, we celebrate the first glorious anniversary of the Information Purification Directives. We have created, for the first time in history, a garden of pure ideology—where each worker may bloom, secure from the pests purveying contradictory thoughts. Our Unification of Thoughts is more powerful a weapon than any fleet or army on earth. We are one people, one will, one resolve, one cause. Our enemies shall talk themselves to death, and we will bury them with their own confusion. We shall prevail![13]

The rhythmic patter of feet is interrupted by a young woman wearing bright red track shorts and a white tank top with the image of a cubist rendering of a Macintosh computer. Outrunning the Thought Police who were pursuing her, she hurls a hammer used in track meets and shatters the screen as Big Brother confidently declares, "We will prevail!" With the screen exploding in what appears to be an atomic blast, the credits roll a voice reads,

> On January 24th,
> Apple Computer will introduce
> Macintosh.
> And you'll see why 1984
> Won't be like "1984."

This advertisement marked a shift in the battlefield from meat space to cyberspace; with this development, nuclear war morphed into cyberwar. While the foundation for the information and computer revolutions had been laid by governmental and corporate scientists and engineers during World War II and the

Cold War, the personal computer revolution would not have been possible without contributions from alienated adolescents tripping on acid while reading reports about quantum mechanics issued by unemployed PhDs hanging around Berkeley. With the war in Vietnam raging, political assassinations, and cities burning with racial conflict, many disenchanted youth saw little difference between the repression created by the centralized bureaucracy of the Soviet Union and the inequities resulting from the hierarchical regime of corporate capitalism represented by IBM's centralized mainframe computers. Gathering a few blocks from Haight-Ashbury at the Homebrew Computer Club, rebels looking for a cause saw the revolutionary potential of personal computers. By putting information in the hands of the people, personal computers were believed to hold the promise of creating a more equitable and just society in which individuals are no longer controlled by either totalitarian political systems or all-consuming institutions and corporations. This understanding of the emerging digital technologies was not incorrect, but it was incomplete. While the shift from mainframe to personal computers marked a seismic change, the connection of different machines in distributed networks was a truly radical development.

Revolutions would not be revolutionary if they were predictable. For many of us who came of age in the sixties, the technological developments during the final decades of the millennium seemed to be the realization of Marshall McLuhan's vision of the global village as well as the countercultural dream inspired by the image of the earth from space that graced the cover of the first edition of Stuart Brand's *Whole Earth Catalog* in 1968.[4] As I began to realize the far-reaching implications of personal computers networked to create worldwide webs, I foresaw the possibility of transforming higher education on a global scale. In 1998 New York investment banker Herbert Allen and I cofounded Global Education Network (GEN) to provide high-quality online education in the liberal arts, humanities, and sciences to all people in all places on earth at the lowest possible price. After twenty-seven million dollars and countless hours of intensive work by many of the most intelligent, imaginative, and dedicated people I have ever known, we failed. One of the greatest frustrations of my professional life has been my inability to get others to see what I saw so clearly more than thirty years ago. It was not just stodgy academics and wary administrators who could not see what was coming. Some of the very people who launched the digital era could not imagine the possibilities harbored by the lightning they had captured in a box. In the summer of 2000, our GEN development team gave a private demonstration of our courses and program to Bill and Melinda Gates. Without asking any questions or offering any comments, Bill, who has become a huge promoter of the Kahn Academy, said, "People will never receive their education this way."

There were many reasons why the early dream of the global village never materialized and GEN failed. The networks that were supposed to unite us now divide us; the decentralization of computer networks has led to the centralization of data controlled by fewer and fewer corporations; information in the hands of the people has not led to greater autonomy and equality but has given rise to a resurgence of authoritarian politics and corporate domination driven by the algorithmic programming of individuals; privacy has given way to surveillance, which reinforces precisely the hegemonic and totalitarian control personal computers were supposed to overthrow.

Once again, there is no single cause for these developments, but the most important factor, in my opinion, was the commercialization of the World Wide Web by Congress in 1992. What began with the National Science Foundation funding supercomputing for university research has led to two very different but related developments—what Shoshana Zuboff aptly describes as "surveillance capitalism," and what Ray Kurzweil, following Venor Vinge, labels "the Singularity." What makes capitalism such a powerful socioeconomic system is its seemingly endless capacity to adapt to the changing technologies of production and to turn them to its own ends. In retrospect, it is not surprising that many of the hippies from the counterculture have become today's uber-capitalists. Throughout the sixties a profound suspicion of governmental regulation and institutional control created a latent libertarianism that unexpectedly reinforced the neoliberal economic policies of the Reagan and Gingrich era.[5] High-speed computers connected in decentralized networks created the conditions for the implementation of the economic theories of the Austrian economist Friedrich von Hayek and his American follower Milton Friedman. Rejecting the principles of every form of socialism as well as the post–World War II Keynesian consensus, von Hayek and Friedman relentlessly criticized all forms of government regulation and enthusiastically endorsed free market policies. In a positive feedback loop, decentralized computer networks provided the technological infrastructure for deregulated markets, and policies of deregulation accelerated the spread of distributed computer networks. These interrelated developments marked a decisive change not only in global financial and economic systems, but also in what was exchanged in these networks. The production and exchange of material goods and consumer products gave way to the algorithmic exchange of immaterial financial products whose complexity increased with network speed.

A further step in the virtualization of the economic system was taken with the introduction of new technologies around the turn of the millennium, which led to the transition from financial to surveillance capitalism. The most important innovations included:

1. The further miniaturization of computational machines in mobile phones and handheld devices.
2. The linking of these devices in increasingly distributed networks.
3. The development of miniaturized sensors that can be implanted in objects as well as people to collect and transmit data.
4. The networking of personal computers and mobile phones with countless sensors, creating a condition known as ubiquitous computing.
5. The displacement of symbolic artificial intelligence (i.e., Good Old-Fashioned Artificial Intelligence) with convolutional neural networks.
6. The collection of massive amounts of data that can be used to train Deep Learning systems that take artificial intelligence to the next level, thereby enabling the development of new possibilities of manipulation and control.

Zuboff begins her widely influential book *The Age of Surveillance Capitalism: The Fight for a More Human Future at the Frontier of Power* by listing what she regards as the defining characteristics of this new form of capitalism, several of which are important in this context:

1. A new economic order that claims human experience as free raw material for hidden practices of extraction, prediction, and sales.
2. A parasitic economic logic in which the production of goods and services is subordinated to a global architecture of behavioral modifications.
 . . .
6. The origin of a new instrumentarian power that asserts dominance over society and presents startling challenges to market democracy.
7. A movement that aims to impose a new collective order based on total certainty.[6]

By stressing "hidden commercial practices of extraction," Zuboff underscores a parallel between industrial and surveillance capitalism. Whereas industrial capitalism requires the extraction of natural resources like wood, coal, oil, and minerals, surveillance capitalism requires the extraction of human resources like information and personal data. Surveillance capitalism presupposes financial capitalism's transformation of the economy into the computerized exchange of floating signifiers grounded in nothing other than themselves. Every time you use your computer, phone, or credit card, every time you ask Siri a question, order a book on Amazon, watch a show on Netflix, send an email, access a website, or do a search on Google, you produce a "behavioral surplus" that has economic value for marketers intent on targeting you with advertisements for their products.

Financial capitalism develops complicated mathematical models to predict market behavior, and surveillance capitalism processes personal data used to modify human behavior. While espousing the virtues of the free market, surveillance capitalism seeks to control individuals by algorithmically programming consumer choices. Far from cultivating free and responsible conduct, surveillance capitalism is a further expression of the will to power or will to mastery that seeks to create corporate totalitarianism by monopolizing data and using it to turn individuals into subservient automatons. The 1984 revolution that was supposed to short-circuit "1984" has unexpectedly led to a world that resembles a capitalist version of the future Orwell predicted. The most complete realization of this nightmare to date is Mark Zuckerberg's Metaverse. Pathologically adverse to uncertainty, so-called masters of the universe seek the certainty accurate control is supposed to bring. Blinkered by virtual reality goggles and reduced to manipulating extremities that feel and touch nothing, denizens of the Metaverse are increasingly cut off from other people as well as the natural world and are wrapped in solipsistic bubbles. Everywhere Metaverse voyagers turn, they see only themselves.

The widespread fascination with the notion of the Singularity throughout Silicon Valley extends the will to power from this world to the entire cosmos. For Kurzweil and his fellow believers, we are hurtling toward the convergence of multiple technological trajectories that will completely transform life and finally realize the ancient dream of immortality. He marshals endless data, charts, and graphs to argue that this revolutionary transformation is imminent. "Live long enough to live forever," he preaches.[7] For these technophiles, death is not inevitable but is merely an engineering problem that technology can solve. Jeff Bezos, Peter Thiel, Dmitry Itskov, Sergy Brin, and Larry Ellison have all invested heavily in anti-aging research. In case the technology does not develop as rapidly as predicted, several of these moguls have hedged their bets by signing up with the Alcor Life Extension Foundation, which is the leading cryogenics organization in the world.[8] "The Singularity," Kurzweil writes,

> will allow us to transcend these limitations of our biological bodies and brains. We will gain power over our fates. Our mortality will be in our own hands. We will be able to live as long as we want.... We will fully understand human thinking and will vastly extend and expand its reach. By the end of this century, the nonbiological portion of our intelligence will be trillions of trillions of times more powerful than unaided human intelligence....
>
> The Singularity will represent the culmination of the merger of our biological thinking and the existence with our technology, resulting in a world that is still human but that transcends our biological roots. There will be no distinction, post-Singularity, between human and machine or between physical and virtual reality.[9]

Such speculation transforms science and technology into religion. The technological Singularity is nothing other than the latest version of the redemption narrative that has shaped the entire Western theological and philosophical tradition. This story rests on three inextricably interrelated principles: human exceptionalism, insidious individualism, and ontological dualism. As Lynn White famously argued in his prescient essay, "The Historical Roots of Our Ecological Crisis,"

> Christianity inherited from Judaism not only a concept of time as nonrepetitive and linear but also a striking story of creation. By gradual stages a loving and all-powerful God had created light and darkness, the heavenly bodies, the earth and all the plants, animals, birds, and fishes. Finally, God had created Adam and, as an afterthought, Eve to keep man from being lonely. Man named all the animals, thus establishing his dominance over them. God planned all this explicitly for man's benefit and rule: no item in the physical creation had any purpose save to serve man's purposes. And, although man's body is made of clay, he is not simply a part of nature; he is made in God's image.[10]

White makes two crucial points: "God planned all this explicitly for man's benefit and rule," and man "is not simply a part of nature." The history of the West is the story of how the creator God dies and is reborn as the creative human subject. Just as God, through his sovereign will, created the world in his own image, so man, through the will to power or will to mastery, creates the world in his own image. Kurzweil sounds like Nietzsche's *Übermensch* when he declares, "Our progress in reverse engineering the human brain ... demonstrates that we do indeed have the ability to understand, to model, and to extend our own intelligence. This is one aspect of the uniqueness of our species: our intelligence is just sufficiently above the critical threshold necessary for us to scale our own ability to unrestricted heights of creative power—and we have the opposable appendage (our thumbs) necessary to *manipulate the universe to our will*."[11]

Such mastery presupposes that humans stand *apart from* nature and are not *a part of* the natural world. In the latest version of this dualistic ideology, the essence of life is information, which can be separated from and exist independently of any material substrate. Kurzweil admits, I am a "'patternist,' someone who views patterns of information as the fundamental reality." As Katherine Hayles argues persuasively, "the posthuman view privileges informational pattern over material instantiation."[12] Though Kurzweil casts his argument in the latest technospeak, his vision of human existence and hope for the future is nothing other than the latest version of a myth that dates back to ancient Greek philosophy and early Christian theology. Patternism is a latter-day expression of Platonism in which information, like form, is independent of and set over against matter. The foundational binary of form/matter or information/material implies a related series of oppositions that have plagued Western philosophy and theology and have had a deleterious effect on human society and culture: soul-mind/body, human beings/natural world, man/woman, reason/emotion.... If form or pattern does not require material embodiment, it might seem to be possible to escape bodily "limitations" and continue to exist in a purely in-form-ational or virtual realm. Carnegie-Mellon roboticist Hans Moravec and Transhumanist and creator of Sirius Satellite Radio Maratine Rothblatt argue that technological immortality can be achieved by uploading mind in an archive that can be accessed forever.[13] As if this were not enough, Kurzweil's ambitions extend from the personal to the cosmic.

> The law of accelerating returns will continue until nonbiological intelligence comes close to "saturating" the matter and energy in our vicinity of the universe with our human-machine intelligence. By saturating, I mean utilizing the

matter and energy patterns for computation to an optimal degree, based on our understanding of the physics of computation. As we approach this limit, the intelligence of our civilization will continue its expansion in capability by spreading outward toward the rest of the universe. The speed of this expansion will quickly achieve the maximum speed at which information can travel.

Ultimately, the entire universe will become saturated with *our intelligence*. This is the destiny of the universe. We will determine our own fate rather than have it determined by the current "dumb," simple, machinelike forces that rule the universe.[14]

It would be a serious mistake to dismiss such futuristic speculation as the idle fantasies of a few figures. Many of the people who are responsible for the digital revolution share some version of technological salvation narratives. Though differing in important ways, these myths share a profound suspicion of and hostility toward the natural world. Kurzweil insists that his vision is neither dystopic nor utopian, but it is actually both. Convinced that human life on earth is doomed by paralyzing political conflict and inevitable environmental disaster, many of the brightest scientists and engineers as well as the savviest entrepreneurs are busy plotting their strategy to escape the impending apocalypse.

Once again, this story is not new—the Singularity can best be understood as Techno-Gnosticism in which knowledge (*gnosis*) allows those who know the code to escape the evils of a fallen world that is doomed to destruction and ascend to the cloud where they will live forever.[15] With a growing sense of urgency, Jeff Bezos, Elon Musk, and Richard Bronson admonish their minions laboring to build the rockets that will allow the privileged few to achieve escape velocity, "don't look up."[16] Gazing back through rocket windows as they speed toward Mars and beyond, "Astrotopians" do not see in the luminous blue sphere floating in the dark void the image of the whole earth that nourishes life but glimpse the faceless face of death in a world that is well lost.

MACHINATIONS

The Anthropocene is coming to an end; the question that remains is whether human beings will survive or, like more than 99 percent of all organisms that have gone before humans, will become extinct.

During the last several years of my teaching career, I offered a course entitled "After the Human" in which we considered many of the technological innovations ranging from personal computers, the internet, the World Wide Web, and iPhones to neural networks, big data, ubiquitous computing, superintelligence, CRISPIR, the Internet of Things, and the Internet of Bodies. I always began the class by asking two questions: How is the human defined? and Do you think humanity is the last stage in the evolutionary process? Not surprisingly, answering the first question proved considerably more difficult than students anticipated. Most of them responded "no" to the second question, which I expected, but a significant number replied "yes," which I had not anticipated. Those who answered "no" understandably had no more idea what might come next than a dolphin could have imagined human beings; those who answered "yes" said they did not think humanity would survive long enough to evolve into another form of life. Were these students succumbing to fashionable adolescent despair, or were they expressing a hard-nosed realism about the world their elders have left them? Is the "twilight of the idols" "the eve of destruction"?

Though some of the technologies that threaten human survival are new, the beliefs that have inspired their creation are ancient. The Greek philosopher Protagoras (485–415 BCE) famously declared, "Of all things the measure is Man, of the things that are, that they are, and of the things that are not." Heidegger appropriates Protagoras's insight to disclose what he regards as the essence of modern science and technology. "Man sets himself up as the setting in which whatever is must henceforth set itself forth, must present itself [*sich präsentieren*], i.e., be picture. Man becomes the representative [*der Repräsentant*] of that which is, in the sense of that which has the character of object.... There begins that way of being human which means the realm of human capability as a domain given over to measuring and executing, for the purpose of gaining mastery over that which is as a whole." This will to mastery extends Nietzsche's will to power by transforming the entire world into a "standing reserve" for human exploitation. When the will to power is pushed to the limit, it turns back on the person exercising it, and man "comes to the point where he himself will have to be taken as a standing reserve. Meanwhile man, precisely as the one so threatened, exalts himself to the posture of the lord of the earth. In this way the impression comes to prevail that everything man encounters exists only insofar as it is his construct. This illusion gives rise to one final delusion: It seems as though man everywhere and always encounters only himself."[17] Decades before the advent of the internet,

World Wide Web, iPhones, virtual reality, and selfies, Heidegger described the alienation of human beings from the natural world, other people, and even themselves. Since self and world are inseparably interrelated, to lose one is to lose the other.

While "the term Anthropocene first appeared in 2000 when scientists Paul Crutzen and Eugene Stoermer attempted to define the environmental effects of anthropic activities," Heidegger's analysis the atomic age anticipates the central characteristics of this era.[18] The modern era begins with Descartes's (1596–1650) search for certainty and ends with Heisenberg's formulation of the uncertainty principle. Unlike most philosophers and scientists, Heidegger realizes that every theory of physics presupposes a metaphysics, and every metaphysics implies a theoretical physics. The philosophical foundation of the Anthropocene is an anthropocentric epistemology and ontology that began with Descartes's deliberate quest for certainty in the wake of the collapse of the medieval order and the dissolution of Aristotelian philosophy. "The metaphysics of the modern age," Heidegger argues,

> begins with and has its essence in the fact that it seeks the unconditionally indubitable, the certain and assured [*das Gewisse*], certainty. It is a matter, according to the words of Descartes, of *firmum et mansurum quid stabilre*, of bringing to a stand something that is firmly fixed and that remains. This standing established as object is adequate to the essence, ruling from of old, of what is as constantly presencing, which everywhere already lies before (*hypokeimenon, subiectum*). . . . Inasmuch as Descartes seeks this *subiectum* along the path previously marked out by metaphysics, he, thinking truth as certainty, finds the *ego cogito* to be that which presences as fixed and constant. In this way, the *ego sum* is transformed into the *subiectum*, i.e., the subject becomes self-consciousness. The subjectness of the subject is determined out of the sureness, the certainty of that consciousness.[19]

Descartes translates Augustine's *Dubito ergo sum* into *Cogito ergo sum*. Proceeding methodically, he doubts everything until he discovers that which is indubitable—his own act of doubting.

The philosophical foundation of the Anthropocene is the anthropocentrism of modern philosophy, which begins with Descartes's inward turn to the subject. By reducing truth to certainty and grounding certainty first in consciousness and then in self-consciousness, Descartes effectively renders truth subjective, and, as Kierkegaard would argue decades later, implies that "subjectivity is truth." Descartes's quest for certainty ineluctably leads to Nietzsche's ontological voluntarism.

> Inasmuch as Descartes seeks this *subiectum* along the path previously marked out by metaphysics, he, thinking truth as certainty, finds the *ego cogito* to be that which presences as fixed and constant.... The subjectness of the subject is determined out of the sureness, the certainty, of that consciousness.... The making secure that constitutes this holding-to-be true is called certainty. Thus, according to Nietzsche's judgment, certainty as the principle of modern metaphysics is grounded, as regards its truth, solely in the will to power, provided of course that truth is a necessary value and certainty is the modern form of truth. This makes clear in what respect the modern metaphysics of subjectness is consummated in Nietzsche's doctrine of the will to power as the "essence" of everything real.[20]

This anthropocentrism presupposes human exceptionalism, which, we have seen, lies at the heart of the Judeo-Christian biblical tradition. Descartes's search for certainty does not end with the activity of doubting; rather, he transforms the *act* of doubting into a *thing, res cogito*, which he opposes to another thing, *res extensa*. The *cogito* is limited to human being and everything else—body, brain, animals, and plants—is reduced to the status of a machine. As biologist Lynn Margulis writes in her important book, *What Is Life?*, before Descartes, "everything had been alive; now everything was inanimate, dead."[21] With a single gesture, Descartes alienates human beings from the world by taking mind out of nature thereby creating an unbridgeable fissure between self and world as well as mind and body.

> The brain, he argues, is an extended thing, a thing that occupies space or moves in it. And it is made of parts. But consciousness is not an extended thing—it is a thinking thing. And it's not made of parts but is a unity—you cannot conceive of the mind split in two as you can of material things: you cannot see the right side of what's in front of your eyes without seeing its left side too, you cannot see the shapes of objects without seeing color. So body and soul are two different substances, and there is no way that one can generate the other.[22]

This is the origin of what David Chalmers famously labeled "the hard problem" of consciousness. As Antonio Damasio makes clear, "Descartes' Error" continues to plague philosophers and mislead biologists, computer scientists, artificial intelligence engineers, neuroscientists, and Silicon Valley visionaries. "What, then, was Descartes' error? Or better still, which error of Descartes do I mean to single out, unkindly and ungratefully? One might begin with a complaint, and reproach him for having persuaded biologists to adopt, to this day, clockwork mechanics as a model for life processes."[23]

By reducing everything except human consciousness to machinic status, Descartes clears the way for human beings to exploit the natural world.

> The essence of consciousness is self-consciousness. Everything that is, is therefore either the object of the subject or the subject of the subject. Everywhere the Being of whatever is lies in setting-itself-before itself and thus in setting itself up.... The world changes into object. In this revolutionary objectifying of everything that is, the earth, that which first of all must be put at the disposal of representing and setting forth, moves into the midst of human positing and analyzing. The earth can show itself only as the object of assault, an assault that, in human willing establishes itself as unconditional objectification. Nature appears everywhere—because willed out of the essence of Being—as the object of technology.[24]

In Nietzsche's most dramatic declaration of the death of God, the Madman rages, "How could we drink up the sea? Who gave us the sponge to wipe away the horizon?"[25] For Heidegger, Descartes gave humans the sponge that modern science and technology used to drink up the sea and wipe away the horizon.

When considering Heidegger's critique of technology, it is important to note that his argument is based on the physical rather than the biological sciences. Furthermore, he limits his analysis to the implications of classical physics and does not consider alternative perspectives opened by quantum mechanics, relativity theory, and information theory. The two terms he uses repeatedly to define modern science and its application in technology are "quantification" and "calculation." The purpose of science and technology is to gain knowledge of and "mastery over that which is as a whole." Quantification privileges analysis over synthesis and presupposes that whatever exists can be *rigorously* analyzed and *exactly* described. This approach is reductive. Since mechanical parts are extrinsic to each other and the whole is nothing more than the sum of the parts, understanding the whole requires the reduction to its constitutive parts. Alternatively, the organization of the whole is extrinsic to the parts and must be imposed from without and cannot emerge intrinsically.

While scientific knowledge is supposed to be the result of objective investigation, Heidegger insists that it is a disguised expression of the subjective will to mastery. The highly touted method of establishing hypotheses, conducting experiments, and drawing conclusions involves subjective projection of "a fixed ground plan [*Grundriss*]."

> Modern science's way of representing pursues and entraps nature as a calculable coherence of forces. Modern physics is not experimental physics because it applies an apparatus to the questioning of nature. Rather the reverse is true. Because physics, indeed already as pure theory sets up nature to exhibit itself as a coherence of forces *calculable in advance*, it therefore orders its experiments precisely for the purpose of asking whether and how nature reports itself when set up in this way.[26]

Far from disinterested investigation, scientific inquiry, Heidegger argues, is the exercise of instrumental reason whose purpose is human domination and exploitation of what was regarded as the natural world. While Heidegger's interpretation of modern science is highly selective and is at odds with its assumptions, his analysis of technology correctly identifies developments that have contributed to the current climate crisis.

The scientific revolution begins with the publication of Copernicus's *On the Revolution of the Heavenly Spheres* (1543). By overturning the Ptolemaic model of the universe in which the earth is the stable center surrounded by celestial spheres and replacing it with a heliocentric model in which the earth as well as other planets revolve around the sun, the Copernican revolution would seem to be radically nonanthropocentric. There were, however, other tenets of early modern science that remained problematically anthropocentric. The mathematical quantification of a mechanical universe in the work of Francis Bacon (1561–1626), Galileo Galilei (1564–1642), and, most importantly, Isaac Newton (1643–1727) remain susceptible to Heidegger's critique.

If Descartes is the "father" of modern philosophy, Galileo is the "father" of modern science. His most important contribution was his insistence that mathematics is the proper language of science. Echoing Pythagoras's (ca. 570–ca. 490 BCE) mathematization of reality, Galileo claims, "Philosophy [that is to say, natural science] is written in this grand book, the universe, which stands continually open to our gaze, but it cannot be understood unless one first learns to comprehend the language and read the letters in which it is composed. It is written in the language of mathematics, and its characters are triangles, circles, and other geometrical figures, without which it is humanly impossible to understand a single word of it; without these, one wanders in a dark labyrinth." In his provocative book, *Galileo's Error: Foundations for a New Science of Consciousness*, Philip Goff underscores the ontological and epistemological dualism of this vision of the world. "By stripping the world of its sensory qualities (color, smell, taste, sound), and leaving only the minimal characteristics of size, shape, location, and motion, Galileo had—for the first time in history—created a material world which could be entirely described in mathematical language." Since mathematical ideality cannot appear in material reality, sensory *qualities* cannot "be captured in the purely *quantitative* language of mathematics. How could an equation ever explain to someone what it's like to see red, or to taste paprika?" Like Descartes, Galileo reinscribes the mind/matter duality, thereby "*putting consciousness outside the domain of scientific inquiry.*"[27] From this point of view, "the hard problem" of consciousness can never be solved.

Bacon's *Novum Organum* (1620) is in many ways the inverse of Galileo's scientific method. In contrast to Galileo's mathematical idealization of nature, Bacon focuses on empirical observation, experimentation, and induction. Though he rejects ontological and epistemological dualism, Bacon's scientific method shares two important assumptions that Heidegger traces to Cartesianism: first, anticipating Michel Foucault, the claim that knowledge is power, and second, the insistence that man should dominate nature. "Knowledge and human power," Bacon asserts, "are synonymous." The purpose of knowledge is to dominate nature. In one of the most notorious passages in history of modern science, Bacon writes, "My only earthly wish is . . . to stretch the deplorably narrow limits of man's dominion over the universe to their promised bounds . . . [nature will be] bound into service, hounded in her wanderings and put on the rack and tortured for her secrets." Heidegger could not have made the critical point about science and technology more clearly than Bacon when he declared, "I am come in very truth leading you to Nature with all her children to bind her to your service and make her your slave. . . . The mechanical inventions of recent years do not merely exert a gentle guidance over Nature's courses, they have the power to conquer and subdue her, to shake her to her foundations."[28] As Hegel argues, the quest for domination by repressing or even enslaving the other—be it other human beings or the natural world teeming with nonhuman forms of life—is futile. Appearances to the contrary notwithstanding, the slave is always the master of the master. Cut down all the trees, pollute the oceans, drive other species to extinction and human beings will disappear from the face of the earth.

When developing his critique of modern science and technology, Heidegger does not mention the names of individual scientists and engineers. His argument, however, makes it clear that Newton's classical physics is the paradigm for his understanding of science. In his *Mathematical Principles of Natural Philosophy* (1687), Newton brings together Galileo's mathematization of nature and Bacon's preoccupation with observation and experimentation. It is important to note that Newton was not only a scientist but also an astronomer, alchemist, and theologian, who wrote more religious tracts than scientific treatises. The cultural importance of Newtonian physics cannot be overstated; Alexander Pope spoke for many when he wrote in his "Essay on Man" (1773):

> Nature and nature's laws lay hid in night,
> God said, Let Newton be and all was light.

In the following chapters, I will consider the differences between classical Newtonian physics, on the one hand, and, on the other hand, the theory of relativity and quantum mechanics.

For Newton as well as Galileo and Bacon, the foundational gesture of scientific inquiry is the separation of subject (observer) and object (observed). While the object of inquiry can range from the micro to the macro, Newton's final object of investigation is nothing less than the cosmos itself. Rejecting the medieval belief that the laws of the heavens are different from the laws governing the earth, he insists that natural laws are universal. The scientific investigator appears to be independent of the world and to assume a God's-eye perspective on the cosmos as a whole. The object of investigation always exists prior to and independent of the investigator. Truth, which is objective rather subjective, results from the accurate *representation* of the object to the subject.

Newtonian physics is a combination of ancient Greek atomism and modern mechanism. In his *Lectures on the History of Philosophy*, Hegel argues that Greek atomism originated with Leucippus (early fifth century BCE) and was elaborated by Democritus (460–370 BCE). Atoms, Hegel explains, "are, even in their apparent union in that which we call things, separated from one another through the vacuum, which is purely negative and foreign to them, that is, their relation is not inherent in themselves, but with something other than them, in which they remain what they are." In other words, atoms are discrete entities that exist separately in a void. This void, Newton believes, is absolute, and is sensorium of God. As such, it remains independent of the entities that fill it. Time, which is modeled on space, consists of a linear series of separate points plotting the unidirectional trajectory from the past through the present to the future.

Since atoms are "separate from each other," parts are externally related, and the whole is nothing more than the sum of the parts. Whether objects in space or events in time, order and organization must be imposed from without and cannot emerge from within. "The point," Hegel argues, "is that though these atoms as small particles may be allowed to subsist as independent, their union becomes merely a combination which is altogether external and accidental. The determinate difference is missed, the one, as that which is for itself, loses all its determinateness."[29]

The ontological separation of entities and correlative externality of relations require a different form of causality. One of the defining characteristics of modern science is the displacement of Aristotle's final causality with efficient causality. Instead of interpreting events in terms of the end toward which they are directed (the future), science understands events in terms of the past from which they have come. The natural laws governing the universe are fully accessible to human reason. Linked in unbreakable causal chains by external forces, the Newtonian universe is completely deterministic. As the eighteenth-century polymath Pierre Simon Laplace explained, "An intellect which at a given instant knew all the forces acting in nature, and the position of all things of which the world consists—supposing the said intellect were vast enough to subject these data to analysis—would embrace the same formula the motions of the greatest bodies in the universe and those of the slightest atoms; nothing would be uncertain for it, and the future, like the past would be present to its eyes."[30] In contrast to Cartesian certainty, which is grounded in human consciousness and self-consciousness, the determinism of classical physics is independent of the human observer. Furthermore, according to the principle of efficient causality, effects are always proportional to their causes. Since a specific cause always has a definite effect, the future of any part of the system can in principle be predicted with absolute certainty. Furthermore, the proportionality of cause and effect excludes the possibility of runaway positive feedback.

As I have suggested, Newton's scientific investigations reinforced rather than undercut his religious beliefs. Theoretical physicist and New Age guru Fritjof Capra explains Newton's distinctive mixture of physics and metaphysics. "Newton's equations of motions are the basis of classical mechanics. They were considered to be fixed laws according to which material points move, and were thus thought to account for all changes observed in the physical world. In the Newtonian view, God had created, in the beginning, the fundamental laws of motion. In this way, the whole universe was set in motion, and it continued to run ever since, like a machine governed by immutable laws."[31] This view of the transcendent creator God who brings the world into existence *ex nihilo*, establishes the laws by which it operates, and then withdraws to let it run on its own becomes the basis of seventeenth- and eighteenth-century deism. By separating God from world, deism further exacerbates human alienation. Furthermore, by first removing the mind and then God from the natural world, early modern science and technology left the world exposed to human exploitation.

DETERRENCE

Heidegger's entire later philosophy is a sustained effort to overturn the anthropocentrism at the heart of the Anthropocene. This requires relinquishing the will to power and mastery and overcoming "the instrumental and anthropological definition of technology."[32] Rather than being consciously or unconsciously obsessed with domination and control, it is necessary to slow down and learn to listen rather than dictate. Immediately after explaining that modern man "exalts himself to the posture of lord of the earth," who "everywhere and always encounters only himself," Heidegger overturns every such lordly claim. "*In truth, however, precisely nowhere does man today any longer encounter himself, that is, his essence.* Man stands so decisively in attendance on the challenging-forth of Enframing that he does not apprehend Enframing as a claim, that he fails to see himself as one spoken to, and hence also fails in every way to hear in what respect he ex-ists, from out of his essence in the exhortation or address, and thus *can never* encounter only himself."³³ Heidegger's argument turns on a linguistic association between to hear (*hören*) and to belong (*gehören*). To overcome the alienation endemic to the modern subject, one must learn to hear the "voice" of Being, which "speaks" through *poiesis*. In this way, *poiesis* reveals that human beings along with all other beings are not alienated from the world but are integrally related to it.

Rather than simply turning away from technology, which is impossible, Heidegger reconceives it by returning to its roots in ancient Greece. The word "technology," he notes, "stems from the Greek. *Technikon* means that which belongs to *techne*. We must observe two things with respect to the meaning of this word. One is that *techne* is the name not only for the activities and skills of the craftsman, but also for the arts of the mind and the fine arts. *Techne* belongs to bringing-forth, to *poiesis*; it is something poietic." Properly understood, technology is not an expression of the will to power but is one of the ways in which Being reveals itself *through* human thinking and acting. The power of Being is not at our disposal; rather, Being is always being given—it is a gift of an other that can never be fully known or properly named. Instead of a determinate thing that can be controlled by a manipulative subject, every determinate entity is an *event* (*Ereignis*) in the midst of other events. Given his relentless critique of modern science and technology, Heidegger proceeds to make an extraordinary claim: "Technology is a mode of revealing. Technology comes to presence [*West*] in the realm where revealing and unconcealment take place, where *Aletheia*, truth, happens."³⁴ Truth is not transcendent and unchangeable but is temporal and emerges through the relational interplay of human thought and activity with an ever-changing milieu. As the mediation of self and world, technology is one of the ways in which truth appears.

The primary way in which this revealing occurs and, thus, truth happens is language (*Sprache*). Language as *poiesis* does not represent extant entities but is creative and lets the presence of what has not been come forth into appearance. As it is the "ground" of whatever is, language is "the house of Being." When language is understood poetically, human beings are the vehicles and not the agents of saying. In other words, human beings do not produce and control language, but, in a manner reminiscent of the divine Logos, language "appropriates" human beings to reveal the truth (*Aletheia*) of Being. "*Sprache spricht*"—language speaks—and by so doing displaces the centrality of *cogito ergo sum*. I do not speak, the I does not speak, to the contrary, I am, the I is spoken. The revelation of truth happens through human beings in the work (both noun and verb) of art. Since Being is always given by radical alterity, the certainty Descartes so desperately sought remains forever elusive.

Heidegger intends his critique of Cartesian self-certainty and decentering of the modern subject to overturn the anthropocentric ontology and epistemology of modern science and technology. Nevertheless, vestiges of anthropocentrism and human exceptionalism remain throughout his writings. Nowhere is this more evident than in his interpretation of the complex interrelationship of death, language, and thinking in his early and late work. In *Being and Time*, which should have been entitled *Being as Time*, Heidegger consistently associates authentic human being with being-toward-death: "Death is a possibility-of-Being which Dasein itself has to take over in every case. With death, Dasein stands before itself in its ownmost potentiality-for-Being.... As potentiality-for-Being, Dasein cannot outstrip the possibility of death. Death is the possibility of the absolute impossibility of Dasein. Thus death reveals itself as that possibility which is one's ownmost, which is non-relational, and which is not to be outstripped."[35] Only in the confrontation with death does one become aware of his or her distinctive individuality. In the awareness of death, I realize who I have become during my lifetime by recognizing what will no longer be when I no longer am. To comprehend death as one's own death is to imagine the future that is always approaching. Apart from language and the thinking it enables, it is impossible to apprehend the future. And without the awareness of the future, there is no recognition of the past, but only the total immersion in the present of which we can never be self-consciously aware. Since only human beings can speak and think, they alone can die. Plants, even animals, perish or expire, but they do not die sensu stricto.

While suffering from pancreatic cancer and acknowledging that the end could not long be deferred, Derrida became preoccupied with "the gift of death" and, correlatively, with the relation between the animal and the human. As always, he develops his argument by returning to Heidegger.

> Heidegger never stopped modulating this affirmation according to which the mortal is whoever experiences death as such, as death. Since he links this possibility of the "as such" (as well as the possibility of death as such) to the possibility of speech, [he] thereby concludes that the animal, the living thing as such, is not properly a mortal: the animal does not relate to death as such. The animal can come to an end, that is, perish [*verenden*], it always ends

up kicking the bucket [*crever*]. But it can never properly die. Much later, in *On the Way to Language,* Heidegger wrote: "Mortals are they who can experience death as death [*den Tod als Tod eifahren konnen*]. Animals cannot do this. But animals cannot speak either."[36]

While no other philosopher has given a more trenchant analysis of modern science and technology, Heidegger's critique of the anthropocentrism that is the foundation of the Anthropocene is incomplete. If, as I have asserted, philosophy is a prolonged meditation on death, and if, as Heidegger argues, death, language, and thinking are inseparably entangled, then developing a nonanthropocentric ontology and epistemology requires a more expansive understanding of death, language, and thinking. There is ample scientific evidence that animals as different as dogs, pigs, crows, and elephants are aware of impending death and therefore are capable of apprehending the future. Furthermore, as we will see in the following chapters, language, cognition, thinking, and communication are not limited to human beings but extend from animals and even plants to forms of intelligence misleadingly labeled "artificial."

To move beyond Heidegger's residual anthropocentrism, human exceptionalism, and insidious individualism, it is necessary to reconsider his limitation of science to classical physics and his failure to consider the metaphysical implications of the theory of relativity and quantum mechanics. In all the essays collected in *The Question Concerning Technology,* Heidegger mentions the influential and enigmatic nuclear physicist Werner Heisenberg only once in a fleeting reference to a 1954 article entitled "The Picture of Nature in Today's Physics." In spite of his obsession with Descartes's reduction of truth to certainty, Heidegger nowhere considers the far-reaching implications of Heisenberg's revolutionary uncertainty principle. Heisenberg, by contrast, was fully aware of the philosophical implications of quantum mechanics' critique of classical physics for the anthropocentrism of Cartesianism. In his important book, *Physics and Philosophy: The Revolution in Modern Science,* Heisenberg makes this point clearly. Given the importance of this insight for the argument developed in the following chapters, it is worth quoting the relevant passage at length.

> The division between matter and mind or between soul and body, which had started in Plato's philosophy, is now complete. God is separated from both the I and the world. God in fact is raised so high above the world and men

that He finally appears in the philosophy of Descartes only as a common point of reference that establishes the relation between the I and the world.

While ancient Greek philosophy had tried to find order in the infinite variety of things and events by looking for some fundamental unifying principle, Descartes tries to establish the order through some fundamental division. But the three parts which result from the division lose some of their essence when any one part is considered separated from the other two parts. If one uses the fundamental concepts of Descartes at all, it is essential that God is in the world and in the I and it is also essential that the I cannot be separated from the world. Of course, Descartes knew the indisputable necessity of the connection, but philosophy and natural science in the following period developed on the basis of the polarity between the "*res cogitans*" and the "*res extensa*." The influence of the Cartesian division on human thought in the following centuries can hardly be overestimated, but it is just this division which we have to criticize later for the development of the physics of our time.[37]

The critique of the seemingly unbridgeable gap between the *res cogito* and the *res extensa* requires refiguring the either/or of oppositional differences and the both/and of every *coincidentia oppositorum* through the neither/nor of codependent origination. Though never mentioning Einstein or Bohr by name, Heidegger gestures toward the weird world within worlds they describe at two points in his later writings. The first is his interpretation of truth as *Aletheia*, which is the lighting, opening, clearing, or tear that allows determinate differences constitutive of every identity and difference to appear. Since nothing is static and everything is a function of the *interplay* of determinate events, Being is inherently temporal. Being *as* time is the spacing that is a timing and the timing that is spacing. This insight leads Heidegger to the second revision of his earlier work. His recognition of the eventuality of Being requires a shift from the nonrelationality of the subject in being-toward-death to a recognition of the inherent relationality of all existence. In an important but generally overlooked passage in the seminal essay "On the Way to Language," Heidegger argues:

> For that event [*Ereignis*], holding, self-retaining is the relation of all relations [*das Verhältnis aller Verhältnise*]. Thus, *our* saying—always an answering—remains forever relational. Relation is thought of here always in terms of the event, and not long conceived in the form of a mere reference. Our relation

to language defines itself in terms of the mode in which we, who are needed in the usage of language, belong to the event.

We might perhaps prepare a little for the change in our relation to language. Perhaps this experience might awaken: All reflective thinking is poetic, and all poetry in turn is a kind of thinking. The two belong together by virtue of that saying which has already bespoken itself to what is unspoken because it is a thought as thanks [*die Gedanke ist als der Dank*].[38]

Denken ist danken—to think is to thank. If Being is always being given, man is not the master of things and things are not things but are events that can never occur by themselves but always occur in the give-and-take place in relation to others.

If understood in terms of quantum mechanics rather than classical physics, "the two processes, that of science and that of art," Heisenberg suggests, "are not very different." "Here again," he proceeds to argue, "we must not be misled by the Cartesian partition. The style arises out of the *interplay between the world and ourselves* or more specifically between the spirit of the time and the artist. The spirit of a time is probably as objective as any fact in natural science, and this spirit brings out certain features of the world which are even independent of time, are in this sense eternal."[39]

CHAPTER 3

RELATIONALISM

> *Whatever arises dependently*
> *Is explained as empty.*
> *Thus dependent attribution*
> *Is the middle way.*
> *Since there is nothing whatever*
> *That is not dependently existent,*
> *For that reason there is nothing*
> *Whatsoever that is not empty.*
>
> —Nagarjuna

REFIGURING NATURE

What made the Jena dinner party so lively were the differences among the assembled guests. Bristling with tensions and conflicts dating back to their university days and exchanging not only ideas but, in some cases, wives as well, it would take a host as socially adept and intellectually astute as Goethe to weave together the differences, which were always on the verge of slipping into hostile oppositions, into a convivial conversation. Alexander von Humboldt, who had recently returned from his voyage to the Americas, arrived late. At the time of his departure on his global voyage, "Humboldt was a well-established scientist . . . with publications in botany, physiology, and mineralogy to his credit. However, his five years in the Canary Islands, Venezuela, Cuba, Mexico, the United States, and the Andes of Colombia, Peru, and Ecuador led to his most influential scientific works, and his popular writings and lectures on his travels...established his public renown."[1] The guests, most of whom had never ventured beyond Europe, and some who had never left Germany, gathered around Humboldt to hear his tale of adventure. Having become one of the world's first celebrity scientists, Humboldt was more than eager to oblige. Looking back from more than two centuries, Humboldt could justifiably lay claim to being one of the first serious climate scientists. Concerned to move beyond the taxonomic preoccupations of botanists working in the tradition of Linnaeus, Humboldt dedicated his research to showing how "the local details of climate, flora, fauna, soil, and culture could all be seen as parts of broader regional and global patterns." He also considered the environmental impact of different social organizations and practices on plants and animals.

The two writers who most influenced Humboldt's scientific approach were Goethe, whom he met in 1795, and Kant. In *Metamorphosis of Plants* (1790), Goethe made two points that remained with Humboldt. First, "he emphasized the features that all plants held in common (rather than those which differentiated them, which Linnaeus and his followers focused on), and viewed plant genera, species, and individuals as variations of archetypal forms." Second, "he observed that the outer forms of plants and animals reflected their environment."[2] From Kant he learned the importance of empirical observation and systematic thinking. In his lectures on geography, Kant cautioned, "We have to know the objects of our experience as a whole so that our knowledge does not form an aggregate but rather a system; in a system it is the whole that comes before the parts, whereas in an aggregate the parts come first."[3] This is an important point, which, we will see, reflects Kant's foundational principle that most influenced Hegel and became the basis of Humboldt's scientific method. Careful empirical observations systematically organized create a coherent whole. In *The Order of Things*, Michel Foucault argues that this insight is symptomatic of the transition from the classical to the modern age.

> The general area of knowledge is no longer that of identities and differences, that of non-quantitative orders, that of a universal characterization, of a general *taxinomia*, of a non-measurable mathesis, but an area made up of organic structures, that is, of internal relations between elements whose totality performs a function; it will show that these organic structures are discontinuous, that they do not, therefore, form a table of unbroken simultaneities, but

that certain of them are on the same level whereas others form series or linear sequences.... The link between one organic structure and another can no longer, in fact, be the identity of one or several elements, but must be the identity of *the relation between the elements*.[4]

Nowhere is the shift from the mechanistic Newtonian and Cartesian world of the eighteenth century to the organic world of the nineteenth century more evident than in changing interpretations of nature.

To understand the significance of this development, it is necessary to underscore two of Foucault's mistakes that have plagued critical theory for decades. First, the Classical interpretive grid of identities and differences is refigured rather than negated by the holistic structure of organisms. Foucault correctly argues that "the link between one organic structure and another" cannot "be the identity of one or several elements but must be the identity of the relation between elements." But he misunderstands his own insight when he proceeds to argue that "the general area of knowledge is no longer that of identity and differences." While Humboldt was developing a systematic scientific method in which whole and parts are coemergent and codependent, Hegel was formulating a dialectical logic in which identity and difference are mutually constitutive. The recognition of this codependence transforms the interpretation of mind and nature as well as their interrelationship. In addition to this, Foucault's preoccupation with structural synchronicity leads him to overlook the diachronicity of the whole, which theorists working in different fields established during the final decade of the eighteenth century. This relational structure is not fixed and closed but is open and emergent because its differential patterns are constantly changing through coadaptive processes. This results in a dynamic morphology of becoming rather than a static morphology of being. Sylvie Romanowski elaborates this relational vision: "Humboldt imagines at a larger level what we would call today entire ecosystems."[5] In chapter 7 I will consider how the shift from tabular taxonomy to dynamic morphology eventually leads to what Richard Lewontin describes as "dialectical biology," in which organisms and environments are involved in a radically relational process.

To demonstrate his proto-ecological analysis, Humboldt plots the flora and fauna in the different regions he studies on horizontal (latitude), vertical (longitude), and altitudinal axes. He illustrates his analysis by developing an extraordinarily intricate drawing with notations including comprehensive charts entitled

Geographie der Pflanzen in der Tropen Landern. This diagrammatic representation not only anticipates but far exceeds the effectiveness of Edward Tufte's "visual display of quantitative information."

While deeply analytic, Humboldt's research and its pictorial representation are also remarkably synthetic. Consistently pursuing his "ideal of articulating the harmony of the cosmos, unity in diversity," Humboldt "provoked people to think about the globe in fundamentally new ways—as a single entity with interlinked biological, physical, and cultural properties."[6] It was as if Humboldt foresaw the image of the earth that was on the cover of Stuart Brandt's first issue of *The Whole Earth Catalog*.

Interlude

During 1971–1972, my wife, Dinny, and I lived in Denmark, while I wrote my doctoral dissertation *Kierkegaard's Pseudonymous Authorship: A Study of Time and the Self.* In the era before the internet and the 24/7 news blitz, our only contact with the "outside" world was the *International Herald Tribune* with two-day old news, three minutes of news in English on Danmark Radio, and a half-hour program broadcast at midnight on Radio Moscow. Night after night, the Moscow broadcast consisted of five minutes of "news" followed by twenty-five minutes reporting the activities of Angela Davis. Letters, typed on Aerograms, took a week to cross the Atlantic, resulting in a two-week delay between message and response. We had no television and no telephone. When my mother suffered a cerebral hemorrhage and was not expected to live through the night, my father had to call one of our Danish friends and leave a message for us to call him. I then had to go to the post office, give a clerk the number, and wait to be notified that the connection had been made and I could talk to him.

In the late winter of 1972, we visited Moscow and what was then Leningrad with a Danish tour group. In Leningrad, I bought a three-volume collection of Lenin's works and *A Short History of the Communist Party*, which, anticipating today's cancel culture, omitted any mention of Stalin. My most lasting memory from that trip is the long lines of people waiting outside small shops to purchase the essentials of everyday life. We had a car and traveled through much of Eastern Europe—Hungry, Czechoslovakia, and Yugoslavia. The conditions and people's responses to foreigners varied from country to country. With Prague Spring only four years in the past, tensions were still palpable. The most unsettling trip was through East Germany to Berlin. We followed the route Kierkegaard had taken when he went to Germany after completing *Either-Or* to hear Schelling lecture at the University of Berlin. At the time of our trip, you had to follow designated transit routes, which were not always well marked. Traveling by ferry from Denmark, we took back roads from the north rather than the main east-west transit route. As we passed through East German customs, we were repeatedly warned of the dire consequences of straying from authorized roads.

Though a veteran of the late-sixties political upheavals ignited by ill-informed students encouraged by armchair academic Marxists, and having traveled in other Eastern European countries and listened to endless riffs on Angela Davis, I still was ill-prepared for what I saw. Leaving Denmark and driving through East Germany was like going back in time to the end of the Second World War. The roads were in terrible condition, the buildings had not been painted for at least thirty years, and there were few shops. In Denmark, crops were being harvested with large tractors using the latest agricultural technology; in East Germany, there was no mechanized farm equipment, no tractors, and even very few trucks. Women in long black dresses dug beets and cut wheat by hand and loaded the crops into horse-drawn carts. Outside every town and village there was a large military installation with Russian and East German soldiers patrolling the streets. At one point a band of uniformed soldiers ordered us to stop and surrounded the car; they peered in the windows but said nothing. We had no idea what was going on; after fifteen minutes, a long flatbed truck carrying a huge missile passed by. The soldiers dispersed and signaled us to proceed. If anything had happened or if we had problems, no one knew where we were, and with no available telephones, to say nothing of cell phones, there was no way we could communicate with family or friends. In retrospect, the trip was crazy, but we were young and knew no better.

More than a wall divided Berlin—the differences between the eastern and western sectors of the city offered a stark contrast in the economic failure of communism and the economic success of capitalism. From early in my undergraduate years, I had known Alexander von Humboldt and his brother Wilhelm only as the brothers after whom what had been the University of Berlin had been named. This was the university where Hegel served as rector and where he lectured at the height of his fame until his death in 1831. I did not know at the time that when the city was divided, the Free University opened in West Berlin, and the name of the university in the East was not changed to Humboldt

University, in honor of Alexander and Wilhelm, until 1949. My primary goal in traveling to Berlin was to visit Humboldt University.

The division of Germany brought by the Cold War represented precisely the kind of oppositional differences that Humboldt and Hegel struggled to overcome. To enter East Berlin, we had to go through the infamous Checkpoint Charlie. Before clearing passport control, there was a museum recounting the history of the wall and large graphic photographs of people who had been shot while attempting to flee to the West. Once cleared, we headed straight for Humboldt University. Entering through the large front doors, I saw a tall brown marble wall where Marx's famous eleventh thesis on Feuerbach was emblazoned in large gold letters.

> Philosophers have only interpreted the world in various ways;
> the point, however, is to change it.

Having studied this text while in college, I eagerly took out my camera and snapped pictures of the inscription. As I was putting my camera back in my bag, two large security guards seized me and physically threw me out of the building. They ripped the camera from my hands, tore out the film, and told us in German even I could understand never to return. Fortunately, the guards did not realize that I had another role of film with a picture of the wall.

I did not return to Humboldt University for several decades. By that time, the wall was gone, and upscale stores and too many McDonald's restaurants lined *Under [Unter] den Linden*. On my second trip, I did something I had neglected to do the first time I was there—I visited the Dorotheenstaadt cemetery, where Hegel is buried beside his wife Marie and Johann Gottlieb Fichte and his wife Johanna. Bertolt Brecht is buried a few rows away, and nearby there is a small memorial to Dietrich Bonhoffer. Hegel's grave is covered with ivy, and I clipped a piece to smuggle back to the United States. Thirty years later the ivy is still growing in a pot beside my desk and is a daily reminder of one of the ghosts who has haunted my life.

I was never convinced that the fall of the Berlin Wall and the collapse of the Soviet Union were the result of Ronald Reagan's defense buildup and Star Wars initiative. What brought down the wall were metastasizing media promoting consumer capitalism. During my return trip in the 1990s, the bright lights and plethora of consumer goods filling fancy stores could not dispel my gnawing sense of an undercurrent, which was a toxic mixture of patriotism, nationalism, and religious fundamentalism. In recent years, there has been a disruptive return of

3.1 Humboldt University, 1971

the repressed oppositional differences—East vs. West, Communism vs. Capitalism, Authoritarianism vs. Democracy, Rich vs. Poor, Black vs. White, Red vs. Blue—still tearing us apart. Marx was right—the point *is* to change the world, but the cure he prescribed exacerbated the disease. Material conditions matter, but Hegel was right—ideas change history.

Hegel's itinerant academic career began as a private tutor and headmaster of a Nuremberg gymnasium before he received university appointments at Jena, Heidelberg, and finally Berlin, where he taught from 1818 until he died in the cholera epidemic of 1831. The University of Berlin was founded by Fichte, Schleiermacher, and Wilhelm von Humboldt only eight years before Hegel's arrival. Though Hegel was far from a dynamic lecturer, in the intellectual capital of Europe at the time, his lectures drew large crowds.

Hegel met Alexander von Humboldt during his time in Jena. Years later, with his fame growing from reports about his world travels, Humboldt delivered a series of wildly popular lectures on "The Physical Description of the World." Though Hegel did not attend these lectures, his wife, Marie, did and reported to her husband that "in one of the lectures, Humboldt delivered a thinly veiled attack on all post-Kantian philosophy," including Hegel. Humboldt began that lecture with a protest against the kind of "'metaphysics' that proceeds without a knowledge by acquaintance and experience' and advances a 'schematism' narrower . . . than that of the scholasticism of the Middle Ages."[7] Hegel was offended, and a brief controversy ensued. Humboldt's problem with Hegel's philosophy was less substantive than methodological—he objected to speculative metaphysics ungrounded in empirical investigation. Though neither could admit it to the other, Humboldt and Hegel agreed more than they disagreed. While Humboldt's lifelong quest was for a "unitary scientific vision," Hegel's lifelong goal was for a unified philosophical vision that cultivated rather than repressed differences.

Hegel begins his *Philosophy of Nature*, which is the second volume of his three-part *Encyclopedia of the Philosophical Sciences*, by asking:

> What is nature? . . . We find nature before us as an enigma and a problem, the solution of which seems to both attract and repel us; it attracts us in that spirit of presentiment of itself in nature; it repulses us in that nature is an

> alienation in which spirit does not find itself.... "What is nature?" can always be asked and never completely answered. It remains a problem. When we see nature's processes and transmutations, we want to grasp its simple essence, and force this Proteus to relinquish his transformations, to reveal himself to us, and to speak out, not so that he merely dupes us with an ever-changing variety of new forms, but so that he renders himself to consciousness in a more simple way, through language.

As we have seen, modern philosophy from Descartes to Kant involves the progressive exclusion of mind from nature. This alienation establishes irreconcilable oppositions that create an "externality that allows differences to fall apart and appear as indifferent."[8] Far from a mechanistic aggregate of separate entities, life, Hegel argues, is an integral *self-organizing process*. "Life is the union of opposites generally, not merely of the opposition of Notion and reality. Wherever inner and outer, cause and effect, end and means, subjectivity and objectivity, etc. are one and the same, there is life. The true determination of life is that the reality in the unity of Notion and reality, no longer exists in an immediate matter, no longer exists independently as a plurality of properties existing outside one another, but that the Notion is absolute ideality of their indifferent subsistence."[9] Reversing the overarching trajectory of modern philosophy, Hegel puts mind back into nature. This world-changing shift would not have been possible without Kant's critical philosophy.

FROM EITHER-OR TO BOTH-AND

By the time he had completed the *Critique of Practical Reason* (1788), Kant knew he had problems. His critical analysis of thinking and acting had led to a series of unmediated oppositions that exacerbated the very problems he had set out to solve. Rather than overcoming Humean skepticism by demonstrating the conditions of the possibility of knowledge, Kant had rendered scientific skepticism unavoidable by unwittingly showing the conditions of the *impossibility* of knowledge. He concludes that subject and object, self and world, mind and nature are irreconcilably opposed, leaving human beings isolated and alienated from the nourishing natural, social, cultural, and technological networks without which they cannot survive. To resolve these difficulties, Kant turned his attention from epistemology and axiology to aesthetics. More than any other work, the *Critique of Judgment* (1790) launches the modern world, which, until recently, seemed to be our own.

In his demanding *Science of Logic* (1812–1816), Hegel effectively identifies the heart of the Kantian revolution. "One of Kant's greatest services to philosophy consists in the distinction he has made between relative or *external*, and *internal* purposiveness; in the latter he has opened up the Notion of life, the Idea, and by so doing has done *positively* for philosophy what the *Critique of Reason* did but imperfectly, equivocally, and only *negatively*, namely raised it above the determination of reflection and the relative world of metaphysics."[10] The distinction between external and internal purposiveness or teleology (*Zweckmässigkeit*) is the philosophical articulation of the difference between mechanisms and organisms. Borrowing the image of the watch from eighteenth-century Deists, Kant explains, "One part is the instrument by which the movement of the others is effected, but one wheel is not the efficient cause of the production of the other. One part is certainly present for the sake of the other, but it does not owe its presence to the agency of that other. For this reason, also, the producing cause of the watch and its form are not contained in the nature of this material but lie outside the watch in a being that can act according to ideas of a whole which its causality makes possible."[11] As we saw in the previous chapter, insofar as parts are not intrinsically related, order and purpose must be imposed from without. The whole, therefore, is an aggregate that is nothing more than the sum of the parts. Inner teleology, by contrast, involves reciprocal relations among the parts and between the parts and the whole. "The parts of the thing combine *of themselves* into the unity of a whole by being reciprocally cause and effect of their form. For this is the only way in which it is possible that the idea of the whole may conversely, or reciprocally, determine in its turn the form and combination of all the parts." This reciprocity creates a "*neχus*" that forms a "*systematic unity* of the form and combination of all manifolds contained in the given matter."[12] Neither part nor whole exists prior to or independent of the other. In this complex interrelationship, neither causality nor temporality is linear—the parts engender the whole, which creates the parts. Each becomes itself in and through the other, and neither can be itself without the other. Within this structural neχus, everything is simultaneously cause and effect.

Kant formulates two additional ideas that become very important for later science, philosophy, and art: self-organization and systematicity. If order is intrinsic rather than extrinsic, a structure is *self-organizing*, "autotelic" (Derrida), or autopoietic (Varela, Thompson, and Margulis). The primary example Kant offers of a self-organizing structure is the biological organism. In contrast to the beautiful work of art where order is imposed by the artist, "nature organizes itself." The "principle of intrinsic finality . . . serves to define what is meant by organisms: . . . *an organized natural product is one in which every part is reciprocally both end and means.* In such a product nothing is in vain, without an end, or to be ascribed to a blind mechanism of nature." In contemporary theoretical biology, self-organization is one of the defining characteristics of life.

The second pivotal idea in the Third Critique is systematicity. If parts and whole are codependent, then neither can be understood apart from the other, and adequate knowledge must be systematic. "Every science is a system in its own right; and it is not sufficient that in it we construct according to principles, and so proceed technically, but we must also set to work architectonically with it as a separate and independent building. We must treat it as a self-subsisting whole, and not as a wing or section of another building."[13]

With these central concepts, Kant could have resolved the oppositions both between mind and world and among different entities and beings within the world. But his epistemological caution led him to back off the far-reaching implications of his ideas of inner purposiveness, reciprocal causality, self-organization, and systematicity. He concluded that these notions are merely regulative ideas or heuristic constructs that tell us nothing about the real world. Hegel fearlessly ventured where Kant dared not tread by declaring these regulative ideas to be constitutive of the world as such.

For Hegel, there is a logic to life, which his *Science of Logic* lays bare. Alternatively expressed, life is the embodiment or the incarnation of logic or the Logos. In an effort to explain this philosophical concept with a religious image, Hegel once suggested that his *Logic* articulated the mind of God before the creation of the world. But images and metaphors can be misleading because they often distort as much as they reveal. In Hegel's dialectical system, neither God nor world, mind nor matter, exists apart from the other. There can no more be a logic *of* the world apart from the existence of the world than there can be an actual world without the logical structure that makes everything what it is.

To discern the logic of life and the life of logic, it is necessary to grasp Hegel's distinction between the logic of Either-Or—understanding (*Verstand*)—and Both-And—reason (*Vernunft*). Understanding presupposes the principle of noncontradiction according to which something cannot be itself and its opposite at the same time. Everything is what it is and is not an other. "Thought as understanding," Hegel argues, "sticks to fixed determinations [*der festen Bestimmtheit*] and their differentiation from one another; every such limited abstraction it treats as having a subsistence and being of its own."[14] *Verstand* analyzes by breaking down into discrete units that are constituted by distinctions from and oppositions to each other. While Hegel recognizes the importance of this mental activity, he stresses that understanding easily becomes unreasonable and even dogmatic. "Dogmatism," he explains, "consists in the tenacity which draws a hard and fast line between certain terms and others opposite to them. We may see this clearly in the strict 'Either-Or': for instance, the world is either finite or infinite; but one of these two it must be. The contrary of this rigidity is the characteristic of all Speculative truth. There no such inadequate formulae are allowed, nor can they possibly exhaust it. These formulae Speculative truth holds in union as a totality, whereas Dogmatism invests them in their isolation with a title to fixity and truth."[15] Understanding represents the logic of the classical mechanistic universe identified by Newton and Galileo and elaborated by Descartes. "This is what constitutes the character of *mechanism*, namely, that whatever relation obtains between things combined, this relation is one *extraneous* to them that does not concern their nature at all, and even if it is accompanied by a semblance of unity, it remains nothing more than composition, mixture, aggregation and the like."[16] Having analyzed discrete differences, understanding is unable to discern integral connections.

In his first published work, *The Difference Between Fichte's and Schelling's System of Philosophy* (1801), Hegel expresses the concern that guides his entire philosophical career:

> The sole interest of Reason is to sublate such rigid opposites. But this does not mean that Reason is altogether opposed to opposition and limitation. For the necessity of bifurcation is one factor in life. Life eternally forms itself by setting up oppositions, and totality at the highest pitch of living energy [*Lebendigkeit*] is only possible through its own restoration out of the deepest separation. What Reason opposes, rather, is just the absolute fixity which understanding gives to the dualism; and it does so all the more even if the absolute opposites themselves originated in reason. When the might of union vanishes from the life of men and opposites lose their living connection and reciprocity and gain independence, the need for philosophy arises.[17]

While understanding asserts the principle of noncontradiction, which secures the opposition of identity and difference, reason insists that everything is inherently contradictory. Rather than the cause of demise, contradiction, Hegel insists, is the pulse of life.

Hegel extrapolates Kant's recognition of the reciprocal relation of opposites to form the principle of *constitutive relationality*, which establishes the union of union and nonunion and the identity of identity and difference. This insight marks the culmination of his logical investigation, which demonstrates how mind and nature mirror each other. The last section of the *Science of Logic*, entitled "The Idea," consists of two chapters, "Life" and "The Idea of Cognition."

> Life, considered now more closely in its Idea, is in and for itself absolute *universality*; the objectivity that it possesses is permeated throughout by the Notion and has the Notion alone for substance. What is distinguished as part, or in accordance with some other external reflection, has within itself the whole Notion; the Notion is the *omnipresent* soul in it, which remains simple self-relation and remains a one in the multiplicity belonging to objective being. This multiplicity, as self-external objectivity, has an indifferent subsistence, which in space and time, if these could already be mentioned here, is a mutual externality of wholly diverse and self-subsistent elements. But in life externality is at the same time present as the *simple determinateness* of its Notion; thus, the soul is an omnipresent outpouring of itself into this multiplicity and at the same time remains absolutely the simple oneness of the concrete Notion with itself. The thinking that clings to the determinations of the relationships of reflection and of the formal Notion, when it comes to consider life, this unity of its Notion in the externality of objectivity, in the absolute multiplicity of atomistic matter, finds all its thoughts without exception are of no avail.[18]

I have quoted this difficult passage at length because it represents the culmination of Hegel's analysis of logic, which is the structural foundation of his entire system. The only way to overcome the indifference of abstract individuality is through the recognition that *relations bestow specificity*. Though Hegel often uses the metaphor of the circle to describe his *Encyclopedia*, we will see in the next chapter that a careful reading of his relationalism in terms of spacetime discloses an elusive remainder that leaves the system open. In contrast to recent structuralists like Levi-Strauss, Foucault, Noam Chomsky, and their epigons, Hegel considers forms and patterns not only synchronically, but also diachronically. Rather than atemporal and static, organizational structures are *emergent* and, therefore, involve unending dynamic and temporal *processes*.

THE ONTO-LOGIC OF LIFE

Life and Idea, matter and mind, object and subject, world and self, nature and culture—*neither* identical *nor* different, but identical in their difference and different in their identity. The question is not: How can something—any thing—be simultaneously identical and nonidentical or different, but: How could something—any thing—*not* be simultaneously identical and nonidentical or different?

To answer this question, Hegel once again invokes a theological image—the doctrine of the Trinity. While the *Logic* is metaphorically the mind of God before the creation of the world, it now becomes clear that God's mind changes with the emergence of nature and history, which are its embodiment. As I have stressed, for Hegel, logic is not merely the subjective structure of the mind but also articulates the actual structure of becoming. He underscores the diachronicity of the Logos when he writes in *The Philosophy of History*, "the Trinity is the hinge upon which *history* swings." Though monotheistic, the Christian God is tripartite—three-in-one and one-in-three. In other words, the divine is a unity that includes difference, and difference presupposes unity.

Hegel was not the first person to recognize the implications of the Christian doctrine of the Trinity. In his prescient treatise, *On the Trinity*, Augustine identifies two analogies for the triune God—self-consciousness and self-love. Summarizing his argument, he writes: "The three are one, and also equal, viz. the mind itself, and the love, and the knowledge of it. That the same three exist substantially and are predicated relatively. That the same three are inseparable. That the same three are not joined and comingled like parts, but that they are of one essence, and are relatives." Augustine proceeds to explain, "But as there are two things, the mind and the love of it, when it loves itself; so there are two things, the mind and the knowledge of it, when it knows itself. Therefore, the mind, and the love of it, and the knowledge of it are three things, and these three are one; and when they are perfect, they are equal."[19]

In Hegel's dialectical vision, God is not transcendent and independent of the world but is always already incarnate as the immanent relational process of becoming. As always, he develops his argument by simultaneously borrowing from and criticizing Kant. Kant appropriately translates Aristotelian logic into the twelve categories of understanding but does not explain either their origin or their interrelation. Hegel attempts to correct these shortcomings by starting with the most abstract category that is conceivable—Being, not beings, but Being as such. What,

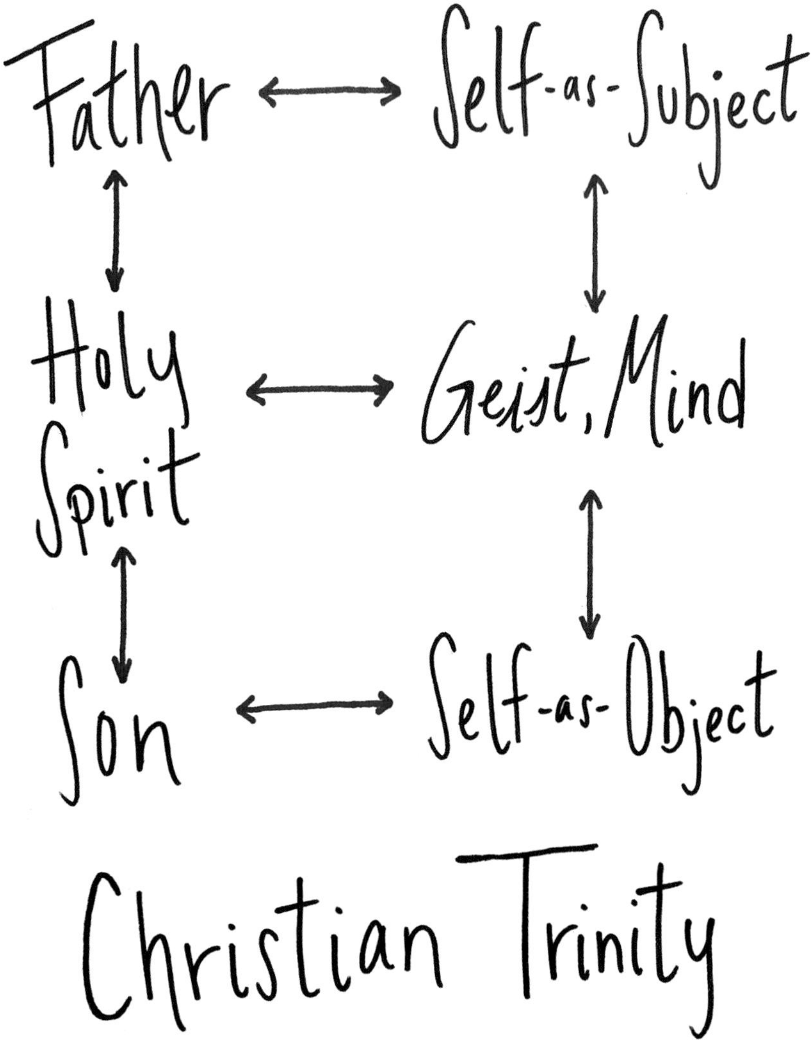

Father ⟷ Self-as-Subject

Holy Spirit ⟷ Geist, Mind

Son ⟷ Self-as-Object

Christian Trinity

3.2 Christian Trinity

he asks is "is"? From this seemingly simple point (NB) of departure, he proceeds to demonstrate the entire categorical system that structures both mind and world. When fully articulated, this emergent schema is the condition of the possibility of both human and nonhuman as well as "natural" and "artificial" cognition. Pure Being, as distinguished from determinate beings, is completely indeterminate and, therefore, is indistinguishable from nothing. Being and Nothing prove to be inverse abstractions from the sole concrete reality of Becoming.

> *Pure being* and *pure nothing* are, therefore, the same. What is the truth is *neither* being *nor* nothing, but that being—does not pass over but has passed over—into nothing, and nothing into being. But it is equally true that they are not undistinguished from each other, that, on the contrary, they are not the same, that they are absolutely distinct, and yet that they are unseparated and inseparable and that each immediately *vanishes in its opposite*. Their truth is, therefore, this movement of the immediate vanishing of the one in the other: *becoming*, a movement in which both are distinguished, but by a difference which has equally immediately resolved itself.[20]

"becoming . . . a difference." Becoming "is" *neither* Being *nor* Nothing but is the *difference between* the two, which forms their identity.

This insight is implicit in Hegel's examination of sense certainty with which he begins the *Phenomenology of Mind*. In contrast to the *Logic*, which begins with the most abstract category (that is, Being as such), the *Phenomenology* begins with the most concrete—the Here (space) and the Now (time). When the mind turns back on itself and takes itself as an object, one discovers that the here and now is never here and now—presence is *neither* present *nor* absent. This initial conceptualization of the presence seems to be thoroughly Newtonian. Time and space appear to be discrete *points* that are perceived as separate from and unrelated to each other.

> The first or immediate determination of nature is *Space*: the abstract *universality of Nature's self-externality*, self-externality's mediationless indifference. It is a wholly ideal *side-by-sideness* because it is self-externality; and it is absolutely *continuous*, because of this asunderness [*Äussereinander*] is abstract, and contains no determinate difference within itself.
>
> Time, as the negative unity of self-externality [*Äussersichseins*], is similarly an out-and-out abstract, ideal being. It is that being which, inasmuch as it *is*,

is *not*, and inasmuch as it is *not*, *is*: Becoming, directly *intuited*, means that differences, which admittedly are purely *momentary*, i.e., directly self-sublating, are determined as *external*, i.e., as external to *themselves*.[21]

What appears to be the most concrete reality turns into its opposite; it is completely in-different and in-determinant, and, therefore, is the most abstract. Just as pure Being is indeterminate, and indistinguishable from Nothing, so the here and now is never here and now. In *different* terms, the present is *neither* present *nor* absent; rather, the present is always already past and is, therefore, always yet to come. The indeterminacy of the Here and Now is a function of the self-externality of *points* "in" space and moments "in" time. In contrast to Newton, for whom absolute space and time are empty containers (objective) that are antecedent to that which "fills" them, and Kant, for whom space and time are a priori forms of intuition (subjective) that are "filled" by sense experience, Hegel argues that *space and time are relative because becoming is relational, and conversely, becoming is relational because space and time are relative*.[22]

The key to Hegel's relationalism is his consideration of the "Determinations of Reflection" at the beginning of the second book of the *Science of Logic*, where he considers the relationship between identity and difference. It is important to note that here reflection is both epistemological and ontological. Though not directly commenting on Hegel in his book *Identity and Difference*, Heidegger poses the questions Hegel probes:

> Where does the "between" come from, into which the difference is, so to speak, inserted? ... What do you make of the difference if Being as well as beings appear *by virtue of the difference*, each in its own way? To do justice to this question, we must first assume a proper position face to face with the difference. Such a confrontation becomes manifest to us once we take the step back. Only as this step gains for us greater distance does what is near give itself as such, does nearness come first to appear [*kommt Nahe zum ersten Scheinen*]. By the step back, we set the matter of thinking, Being as difference [*Sein als Differenz*], free to enter a position face to face, which may well remain wholly without object.[23]

So near, but so far. Contrary to Heidegger, the "matter" of thinking is not Being as difference but *Becoming as differential*. Furthermore, matter is not unformed, undifferentiated Ur-stuff but is a differential process that is always in formation.

If, as Gregory Bateson argues, "information is a difference that makes a difference," then the becoming of matter and mind is an endless information process.[24]

All this and much more is implicit in Hegel's interpretation of the "Determinations of Reflection," where he presents a sustained critique of the abstract Either-Or of understanding. Rejecting commonsense reflection, he maintains that "a consideration of everything that is shows that *in its own self* everything is in its selfsameness different from itself and self-contradictory, and that in its difference, in its contradiction, it is self-identical, and is in its own self the opposite of itself." To establish the codependence of identity and difference, Hegel first examines each term independently. He argues that the self-sameness of identity is usually regarded as exclusive of difference; however, on closer consideration, it becomes clear that abstract self-identity is inseparable from absolute difference.

> Identity is the reflection-into-self that is identity only as the internal repulsion [*innerliches Abstossen*], and it is this repulsion as reflection-into-self, repulsion which immediately takes itself back into itself. Thus, it is identity as difference that is identical with itself. But difference is only identical with itself insofar as it is not identity, but absolute non-identity. But non-identity is absolute insofar as it contains nothing of its other but only itself, that is, insofar as it is absolute identity with itself. Identity, therefore, is in its own self absolute non-identity.

The self-relation that forms identity is necessarily mediated by opposition to otherness. Consequently, in the act of affirming itself, identity negates itself and becomes its opposite, difference. "*Identity is difference*, for *identity* is different from difference."[25]

Conversely, difference *as* difference is indistinguishable from identity. Difference defines itself by opposition to its opposite, identity. Since Hegel has argued that identity is inherently difference, he claims that in relating to its apparent opposite, difference really relates to itself. Relation to other turns out to be self-relation. In the act of affirming itself, difference likewise negates itself and becomes its opposite, identity. Hegel contends that "difference in itself is self-related difference; as such, it is the negativity of itself, the difference not of an other, but *of itself from itself*; it is not itself but its other. But that which is different from difference is identity. Difference is, therefore, itself and identity. Both together constitute difference; it is the whole, and its moment." Identity, in itself difference, and difference, in itself identity, join in contradiction, which is the identity of identity and difference. "As this whole, each is mediated with itself *by its other* and *contains* it. But further, it is mediated with itself by the *nonbeing of its other*; hence it is a unity existing on its own account and it *excludes* the other from itself.... It is thus contradiction."[26]

It is important to recognize that for Hegel, the codependence of identity and difference results in *neither* the absorption of difference in identity *nor* the dissolution of identity in difference. He walks the fine line *between* the extremes of undifferentiated monism and abstract dualism or pluralism. By so doing, he charts the trajectory from the Either-Or of atomism and individualism to the Both-And of relationalism and communalism. Relations are not external and accidental to antecedent identity but are internal and essential to concrete differences. Mirroring each other, thinking and becoming are thoroughly relational; to be is to be related, and to fall out of relation is to fall into the indeterminateness of nonbeing. "Everything that exists," Hegel argues, "stands in relation, and this relation is the truth of each existence. In this way, the existent [*das Existierende*] is not abstractly for itself, but is only in an other. In this other, however, it is in relation [*Beziehung*] to itself, and the relation [*Verhältnis*] is the unity of relation to self and to other."[27]

The term Hegel uses to describe the interrelation of identities and differences is "dialectical." The word "dialectical" derives from the Greek *dialektikos*, which means conversation or discussion. Differences, then, become themselves in conversation with other identities. This conversation communicates information through a process of mediation (neither/nor) that translates opposition (either/or) into codependence (both/and). If information is a difference that makes a difference, the emergent logic of differential identity and self-identical differences can be understood as a distributed information process that in-forms mind (subjectivity) and matter (objectivity). Inasmuch as mind is *in* nature and history as their ontological condition, cognition is not merely anthropocentric but is distributed throughout nonhuman nature (physical systems, plants, and animals) as well as history (society, culture, and technology). Over a century before Einstein's theory of special relativity (1905), Niels Bohr and Werner Heisenberg's theory of quantum mechanics (1913), and Claude Shannon's information theory (1948), Hegel developed a theory of relationalism (Both/And) that overturned the classical world of Newton, Descartes, and Kierkegaard (Either-Or). What he saw without fully seeing is that the *Neither/Nor between* Either/Or and Both/And articulates the strange logic of the networks and webs that make "natural" and "artificial" life possible.

WEB OF LIFE

Hegel, who was dubbed "the old man" by his university classmates, refused to dance. Nonetheless, he suggestively described his *Phenomenology of Spirit* as "the Bacchanalian revel in which no member remains sober."[28] It took Nietzsche, who signed the last fragments he wrote before slipping into madness "Dionysus, the Anti-Christ," to set Hegel's austere insights to music fit for dancing. Nietzsche confessed, "I would believe only in a dancing god." "Every day that passes without dancing," he wrote, "is a day wasted." Defying what he decried as "the spirit of gravity," Nietzsche danced day after day, week after week, year after year—sometimes with partners, more often alone. He wrote frantically, often madly, in fragments and aphorisms that defy systemization. Nevertheless, the seemingly disparate parts were effectively interrelated to create poetic tissues of texts that often captured the rhythms of this Bacchanalian frenzy better than Hegel's own demanding prose. In what is, perhaps, my favorite passage in all of philosophy, Nietzsche writes:

> Did you ever say "Yes" to one joy. Oh, my friends, then you also said "Yes" to *all* pain. All things are entwined, enmeshed, enamored—
>
> Did you ever want Once to be Twice, did you every say "I love you, bliss—instant—flash—" then you wanted *everything* back.
>
> Everything anew, everything forever, everything entwined, enmeshed, enamored—oh, thus you love the world—
>
> > You everlasting ones, thus you love it forever
> > And for all time; even to pain you say: Refrain but
> > Come again! *For joy accepts everlasting flow!*[29]

Entwined, Enmeshed, and (sometimes) Enamored. "The everlasting flow" of becoming is, in Hegel's words, "the arising and passing away that does not itself arise and pass away," but is "in itself," and constitutes the actuality and the movement of the life of truth, and the truth of life.[30]

For Nietzsche, as for Hegel, oppositional dualism is thoroughly nihilistic. In the words of Georges Bataille, who heard Alexandre Kojève's lectures on Hegel and was a lifelong devotee of Nietzsche, "Hell is the opposition between heaven and hell."[31] The dualistic world that begins with Descartes's turn to the subject and culminates in Kant's critical philosophy is ontologically mistaken, epistemologically misleading, and ethically corrupt. Outwardly, the prison house of categories bars access to the world and thereby separates self from other; inwardly, the conflict between universal obligation and particular inclination divides the self from itself. In his posthumous *Will to Power*, Nietzsche argues, "Kant no longer has a right to his distinction 'appearance' and 'thing-in-itself'—he had deprived himself of the right to go on distinguishing in this old familiar way, insofar as he rejected as impermissible making inferences from phenomena to a cause of phenomena—in accordance with his conception of causality and its purely intraphenomenal validity." A thing-in-itself turns out to be no-thing. Without citing Hegel, Nietzsche offers a concise summary of relationalism in which he makes two crucial points in one brief aphorism. First, nothing stands beneath or behind the flux of appearances; and second, nothing is itself by itself. Isolated or abstract individuality, which seems to be radically concrete, is completely indeterminate. As Hegel's convoluted logic demonstrates, nothing is itself by itself but can only be itself by its relation to that which is different.

> The concept of "thing" itself is just as much as all of its qualities.—Even "the subject" is such a created entity, a "thing" like all others: a simplification with the object of defining the force that posits, invents, thinks, as distinct from all positing, inventing, thinking as such. . . .
>
> The properties of a thing are effects on other "things": if one removes other "things," then a thing has no properties, i.e., there is no thing without other things, i.e., there is no "thing-in-itself."
>
> The "thing-in-itself" is nonsensical. If I remove all the relationships, all the "properties," all the "activities" of a thing, the thing does not remain over; because thingness has only been invented by us owing to the requirements of logic, thus with the aim of defining, communication (to bind together the multiplicity of relationships, properties, activities).[32]

If "there is no thing without other things," identity *is* difference, and if becoming is relational, truth is relative, or, in Nietzsche's terms, *perspectival*. "Against positivism, which halts at phenomena—'There are only *facts*'—I would say: no, facts is precisely what there is not, only interpretations. We cannot establish any fact 'in itself': perhaps it is folly to want to do such a thing.... Insofar as the word 'knowledge' has any meaning, the world is knowable; but it is *interpretable* otherwise, it has no meaning behind it, but countless meanings.—'Perspectivism.'"

The notion of a real world beyond the dance of appearances has always been a far from innocent dream. Even if there were a real world beneath or behind, or the apparent world, it could never be known in itself. The ideal of objectivity presupposes the possibility of knowing the world as it is apart from our knowing it. As Kierkegaard insisted long ago, there is no "Archimedean point" from which to observe the world that would enable finite human beings to discern whether the world really is as it appears to be. Nietzsche, unknowingly echoing Hegel's logic, correctly concludes, "The 'in-itself' is even an absurd conception; a 'constitution-in-itself' is nonsense; we possess the concept of 'being,' 'thing,' only as a *relational* concept."[33] Mathematical physicist H. P. Stapp might well have been commenting on Hegel's relationalism and Nietzsche's perspectivism when he writes, "An elementary particle is not an independently existing unanalyzable entity. It is, in essence, a set of relationships that reach outward to other things."[34] When this insight is expanded, Heisenberg explains, "the world appears as a complicated tissue of events in which connections of different kinds alternate or overlap or combine and thereby determine the texture of the whole."[35]

The abiding problem of knowledge has been consistently misconstrued throughout the history of Western philosophy by taking as the point of departure the dualistic opposition between subject (*sub*, under + *jacere*, to throw) and object (*ob*, against + *jacere*, to throw; *Gegenstand*—*gegen*, against + *stand*, to place) and asking how they can be united. When posed in this way, there are only two alternatives: reduce subject to object (materialism), or reduce object to subject (idealism). Neither of these alternatives is satisfactory because the point of departure is mistaken. Subject and object are inextricably interrelated, and, therefore, neither can be itself without the other.

By extension, the opposition between self and world, and mind and matter is also an abstraction from the concrete reality of becoming in which everything is coemergent and codependent. We live in what theoretical physicist John Wheeler aptly describes as a "participatory universe" where nothing is immediate and everything is mediated both epistemologically and ontologically. The thing as such, like the self itself, is empty—it is no-thing, or, more precisely, it is a *nexus* of relations that makes everything what it is. Rather than being alienated from nature, the human mind is an integral part of the infinite natural world.

3.3 Subject-object

Furthermore, as we will see in later chapters, in contrast to the anthropocentricism of philosophy from Descartes to Kant, cognition is not limited to human beings but extends to physical, plant, animal, and so-called artificial realms. Through evolving forms of cognition, the world gradually becomes conscious of itself. Japanese philosopher Nishida Kitaro, who was thoroughly conversant with modern Western philosophy, effectively summarizes Hegel and Nietzsche's conclusions:

> That I am consciously active means that I determine myself by expressing the world in myself. I am an expressive monad of the world. I transform the world into my own subjectivity. The world that, in its objectivity, opposes me is transformed and grasped symbolically in the forms of my own subjectivity. But this transactional logic of contradictory identity signifies as well that the world that is expressing itself in me. The world creates its own space-time character by taking each monadic act of consciousness as a unique position in the calculus of its own existential transformation. Conversely, the historical act is, in its space-time character, a self-forming vector of the world. To bring this out I say that the temporal-spatial (that is, conscious-spatial) reflects itself within a contradictory identity. In this way each act of consciousness is a self-perspective of the dynamic, spatial-temporal world.[36]

Nishida's use of the term "monad" is carefully chosen to recall Leibniz's influential monadology. Far from an isolated atom, the monads of Leibniz and Nishida are thoroughly relational—each reflects the universe is a particular way. *Neither* simply discontinuous *nor* seamlessly continuous, they form something like quantum bits whose interrelation forms the fabric of life.

Though enormously influential, Hegel's philosophy has been subject to radically different interpretations. In recent decades, the two most important readings of Hegel for philosophy and critical theory have been Alexandre Kojève's *Introduction to the Reading of Hegel*, which focuses exclusively on the master-slave relationship in the *Phenomenology*, and Jean Hyppolite's *Genesis and Structure of Hegel's Phenomenology of Spirit*, which is more inclusive and is organized around the theme of unhappy consciousness. It is possible to chart the course of twentieth-century continental philosophy and literary criticism in Europe and the United States through the influence of these two interpreters of Hegel. While Sartre, Lacan, Merleau-Ponty, and Bataille followed Kojève's lectures, Derrida, Foucault, and

Deleuze studied with Hyppolite. Though less well-known than his commentary of the *Phenomenology*, Hyppolite's short book, *Logic and Existence*, which would have been better titled *Logic of Existence*, is more insightful because it demonstrates the centrality of Hegel's *Science of Logic* for his system as a whole. Hyppolite's nuanced reading shows how Hegel's logic both anticipates contemporary structuralism and implies poststructural critiques of structuralist thinking.[37] Reading Hegel's logic through Heidegger's interpretation of language, Hyppolite argues, "nature reveals itself as Logos in human language." Insofar as human beings are part of, rather than alienated from, nature, Hegel's logic shatters what had appeared to be the prison house of language. "Self-consciousness is not human self-consciousness, but Being's self-consciousness across human reality. Absolute knowledge is not an anthropology (one need only read Hegel's *Logic* in order to realize this); it is the knowledge which has sublated the opposition of self-consciousness and being, but this absolute knowledge is what appears in history." Writing at the height of the Cold War and facing the threat of nuclear disaster (1953), Hyppolite understood that the oppositional logic of Either-Or is not only philosophically mistaken but also self-destructive. Like so many European and American intellectuals at the time, Hyppolite placed misguided hope in the possibility that Marxism would resolve global tensions and usher in a more equitable and just society. Thus, he joined his erstwhile colleague Kojève and declared the end of history. Hyppolite continues, "Why not turn this revelation into a genuine end of history? Why not make this end of history coincide with the realization of the human essence?"[38] Here he turns a blind eye to the harsh reality I discovered driving through East Germany in 1971. Several decades later, this reality became unbearable—the Berlin Wall fell, the Soviet Union collapsed, and Eastern Europe was liberated. It was left to a student of Kojève, Francis Fukuyama, to declare once again "the end of history" and the decisive triumph of global capitalism.[39] But the specter of a certain Marxism has not disappeared, and with the violent resurgence of Russia and the growing threat of China, Fukuyama's bet on global capitalism looks as naïve as Kojève and Hyppolite's bet on global Marxism.

Hyppolite's political naivete did not, however, prevent him from discerning a remarkably prescient insight with far-reaching implications of a world he never could have imagined. In what Hegel described as the "immanent plasticity" of dynamic relationalism, Hyppolite foresaw what several decades later would become the networked world of the World Wide Web. Kant's self-organizing neχus and Nishida's mirroring monads become Hegel's nodes stitched in a universal network.

> Speculative logic therefore takes up all the nodes of determinations experienced in their isolation. But it does not turn them into rules or instruments. Speculative logic grasps them in itself and for itself, as moments of the universal, which is the base and the soil of their development. These determinations are no longer object (*Gegenstand*), as in the sensible world, the *a posteriori of experience*; they are phases of an absolute genesis.... [T]he philosopher who expounds this logic adds to it historical commentaries, reflections external to thing itself. He indeed strives to rediscover all the categorial nodes in their immanent order. But, in this regard, his work will be perfectible since the nodes are moments of an infinite (and yet closed upon itself) network.[40]

In the wired world of ecstatic information, "natural" and "artificial" become indistinguishable. Hegel's Bacchanalian revel becomes Nietzsche's "*Dionysian* world of eternally self-creating, the eternally self-destroying . . . world beyond good and evil."

> This world: a monster of energy, without end . . . set in a definite space as a definite force, and not a space that might be "empty" here or there, but rather as force throughout, as a play of forces and waves of forces, at the same time one and many, increasing here and at the same time decreasing there; a sea of forces flowing and rushing together, eternally changing, eternally flooding back, with tremendous years of recurrence, with an ebb and flow of its forms.[41]

Five years after Nietzsche's death, this entangled world of waves that are particles and particles that are waves was transformed from philosophical speculation to scientific theory. In 1905 Einstein published his theory of special relativity, and eight years later Niels Bohr and Werner Heisenberg proposed the Copenhagen theory of quantum mechanics. From a 1793 dinner party in Jena to twentieth-century classrooms and laboratories, a circuitous thread connects inner teleology, self-organization, dialectical logic, relativity theory, and quantum mechanics. Apparent circles become recursive spirals as relationalism morphs into entanglement, which is the web of life that weaves together the opposites that threaten to tear the world apart.

CHAPTER 4

RELATIVITY

> *It is impossible to get outside the system we're studying when that system is the entire universe.*
> —Lee Smolin

A MATTER OF TIME

It's always a matter of time, which unexpectedly turns out to be the time of matter. Sitting in our tiny apartment in Copenhagen in the winter of 1971, I quoted Augustine's meditation of time to begin my dissertation, which eventually became *Kierkegaard's Pseudonymous Authorship: A Study of Time and Self*. Augustine's recognition of the temporality of selfhood led him to write the first autobiography in history, *The Confessions* (381). "What then is time?" he asks, "I know what it is if no one asks me what it is; but if I want to explain it to someone who has asked me, I find that I do not know." As Augustine ponders this puzzle, he turns inward, where he discovers a memory palace in which time is inseparable from space. "And I come to the fields and spacious palaces of memory, where lie the treasures of innumerable images of all kinds of things that have been brought in by the senses." Discovering that time is something like what has become known as space-time, Augustine turns his reflection from the past to the future:

> But in what sense can we say that those two times, the past and the future, exist, when the past is no longer and the future is not yet? Yet if the present were always present and did not go by into the past, it would not be time at

all, but eternity. If, therefore, the present (if it is to be time at all) only comes into existence because it is the transition toward the past, how can we say that even the present *is*? For the cause of its being is that it shall cease to be. So that it appears that we cannot truly say that time exists except in the sense that it is tending toward nonexistence.

Centuries later, scientists would name the arrow of time tending toward nonexistence *entropy*.

Naming this nothingness does not, however, solve the enigma of becoming. Rather than concluding that the passing present is never present in sensu stricto, Augustine argues that the present is actually omnipresent. "It is now, however, perfectly clear that *neither* the past *nor* the future are [sic] in existence, and that it is incorrect to say that there are three times—past, present, and future. Though one might perhaps say: 'There are three times—a present of things past, a present of things present, and a present of things future.' . . . The present time of things past is memory; present time of things present is sight; the present time of things future is expectation."[1] This privileging of the presence of the present has shaped the entire Western theological and philosophical tradition.

I read Augustine's *Confessions* for the first time in my freshman humanities course at Wesleyan University in 1964. Three years later I encountered Hegel and Kierkegaard in two seminars, "The Dialectic of Alienation and Reconciliation in Hegel, Feuerbach, and Marx," and "Kierkegaard's Dialectic of Existence." Even at that early stage of my education, I realized that Hegel's *Phenomenology of Spirit* is something like a rewriting of Augustine's *Confessions*, and his *Philosophy of History* translates *The City of God* into a modern idiom. What I did not realize until many years later was that while I was beginning what has become a lifelong conversation with Hegel and Kierkegaard, on the other side of the Atlantic, Jacques Derrida was publishing his seminal essay "Différance." While 1968 is remembered for world-changing political upheaval, few people realize that it also marked the changing of an era in philosophy and critical theory. That is the year that Alexander Kojève and Jean Hyppolite died, and when structuralism and poststructuralism were launched on the world stage. One year earlier, Derrida published three major works that transformed continental philosophy and critical theory: *Speech and Phenomena*, *Writing and Difference*, and *Of Grammatology*, all of which engage Hegel on the question of time. Though Derrida did not directly discuss Augustine for more than two decades, when he finally rewrote the *Confessions* as his own autobiography, *Circumfession* (1991), his "presence" is everywhere indirectly "present" in "Différance." For Derrida, the pivotal issue has always been time, or more precisely, time, space, and death. Augustine's interpretation of the three tenses of time as three modalities of the present represents one of the foundational arguments for what Derrida, following Heidegger, labels the Western ontotheological tradition in which being is interpreted in terms of presence. From this point of view, to be is to be present here and now, *hic et nunc, hier und jetz, ici et maintenant*. This seemingly simple assertion raises more questions than it answers. Is the present ever truly present, or has it always already passed and, hence, is ever yet to come? Might the present "be" nothing—nothing more than a site of transition or passing that is *neither* present *nor* absent? What if there "is" no Being but only Becoming? What if Becoming were Be-ing? Becoming, binds and rebinds (*re-ligare*) past and future in a (k)not that cannot be undone.

When I first read *Of Grammatology* and "Différance" in the fall of 1978, I was intrigued by the possibility that Derrida's Neither/Nor marks and remarks the joint that simultaneously connects and separates Hegel's Both/And and Kierkegaard's Either/Or. Even then the critical issue for me, as it was for Derrida, was the matter of time. Derrida's interpretation of Hegel in these two works led me to reread his *Science of Logic* (1812) and *Jena Logic* (1804–1805). As we will see in more detail, in "Différance," Derrida focuses on a term Hegel uses only twice—*differente Beziehung*, different or differentiating relation. "In the Jena *Logic*," Derrida writes, "Hegel uses the word *different* precisely where he treats time and the present." Reading Hegel against the grain, Derrida recognizes that Hegel's account of sense certainty in the *Phenomenology* has far-reaching implications that Hegel himself did not fully grasp. Instead of omnipresent, the present *neither* is totally present here (space) and now (time), *nor* is it completely absent. As the (k)not of past and future, the present "is" a *differentiating relationship* that can only be itself by not being itself. The relationality of becoming is the differentiating relationship whose endless separating and joining gives both presence and the present. When interpreted in this way, Hegel's *differente Beziehung* anticipates Derrida's *différance*, which transforms both the present and past by rendering the future undecidable and hence probabilistic rather than deterministic. In this opening, clearing, tear, interval, the identity and difference of everything that "is" and "is" not arise and pass away.

Early twentieth-century scientists unknowingly translated Hegel's emergent relationism into the theory of the relativity of time and space and the thoroughgoing entanglement of the elements constituting the microcosmos. To understand the seemingly unlikely trajectory from Hegel through Derrida to Einstein and Bohr, it is necessary to return to Kant by way of Heidegger and the much-overlooked Alexandre Koyré.

TIMING-SPACING

In an interview conducted in 1971, Derrida confessed, "We will never be finished with the reading or rereading of Hegel, and, in a certain way, I do nothing other than attempt to explain myself on this point. In effect, I believe that Hegel's text is necessarily fissured; that it is something more and other than the circular course of its representation."[2] How is this fissure to be understood? Since the principle of negation is at the heart of Hegel's logic, Derrida understood that to oppose him would confirm his argument. In a manner similar to Kierkegaard, he had to proceed indirectly by exposing aporia (that is, non-Hegelian contradictions) "within" Hegel's argument that create an irreducible opening he either overlooked or repressed. Derrida believes he finds this aporia in Hegel's own notion, or non-notion, of the *differente Beziehung*. He approaches Hegel indirectly through his interpretation of Freud's account of the unconscious, Saussure's analysis of language, Nietzsche's play of forces, and, most important, Heidegger's ontological difference between Being and beings. Except for Heidegger, these writers were not directly commenting on Hegel, but their arguments nonetheless help to explain the puzzling differentiating relation that forms the neχus of Hegelian relationality.

In his *Course in General Linguistics*, Saussure lays the foundation for twentieth-century structuralism by distinguishing language (*la langue*) from speech (*la parole*). Language is the universally shared syntactical *structure* that is the condition of the possibility for the *event* of speech. In what could serve as an explanation of Hegel's dialectical logic, Saussure writes:

> Everything I have said up to this point boils down to this: in language there are only differences. Even more important: a difference generally implies positive terms between which the difference is set up; but in language there are only *differences without positive terms*. Whether we take the signified or the signifier, language has neither ideas nor sounds that existed before the

> linguistic system, but only conceptual and phonic differences that have issued from the system. The idea or phonic substance that a sign contains is of less importance than the other signs that surround it. Proof of this is that the value of a term may be modified ... solely because a neighboring term has been modified.[3]

In his explanation of the importance of Saussure's insight for his notion of *différance*, Derrida focuses on the interplay of space and time.

> But first let us remain within the semiological problematic in order to see the *différance* as temporization and *différance* as spacing cojoined.... Now Saussure first of all is the thinker who put the *arbitrary character of the sign* and the *differential character* of the sign as the very foundation of general semiology, particularly linguistics. As we know, these two motifs—arbitrary and differential—are inseparable in his view. There can be arbitrariness only because the system of signs is constituted solely by the opposition in terms, and not by their plenitude. The elements of signification function due not to the compact of force of their nuclei but rather to the *network of oppositions* that distinguishes them, and then relates them to one another.

As structural linguistics, anthropology and literary criticism make clear, Saussure's formulation of the relation between identity and difference suggests a static synchronic grid rather than a dynamic diachronic network.

Nietzsche's genealogical analyses of the play of forces sets identity and difference in motion in a relational web. Derrida explains that for Nietzsche, "force itself is never present; it is only a play of differences and quantities. There would be no force in general without the difference between forces; and here the difference of quantity counts more than the content of the quantity, more than absolute size itself."[4] This notion of play can be traced to Kant's account of beautiful works of art and biological organisms in terms of inner teleology or purposelessness. While work is utilitarian and, therefore, purposeful, play is nonutilitarian and, thus, has no *extrinsic* purpose. When the world and life in it are understood as a play of forces, everything that seemed to be transcendent or beyond, be it above or below (space), past or future (time), collapses into an immanent process of becoming. Describing his final vision, Nietzsche describes "a becoming that knows no satiety, no disgust, no weariness; this is my *Dionysian* world of the eternally self-creating, the eternally self-destroying, this mystery world of the two-fold voluptuous delight, my 'beyond good and evil,' without goal, unless the joy of the circle is itself a goal."[5]

Though Nietzsche does not mention Hegel, whom he neither read carefully nor understood adequately, Hegel anticipated his account of the play of forces in the explanation of force and understanding in his *Phenomenology*. Hegel's dialectical interpretation of force (*Kraft*) is one of the clearest examples of the complex interrelation of identity and difference at the heart of his relational ontology and epistemology. There can never be just one force because force requires resistance, which is a counterforce. Always already double,

> force is equally in its own self what it is *for an other*. In order, then, that force may in truth *be*, it must be completely set free from thought, it must be posited as the substance of these differences, that is, *first for itself, and then in its differences* as *possessing substantial being*, or as moments existing on their own account. Force as such or as driven back into itself, thus exists on its own account as an *exclusive One*, for which the unfolding of the [different] matters is *another* subsisting essence; and thus two distinct independent aspects are set up. But force is also the whole, that is, it remains what it is according to its notion; that is to say, these *differences* remain pure forms, superficial *vanishing* moments. . . . [T]here would be no force if it did not exist.

There is no force as such apart from the interplay of at least two forces (*das Spiel der beiden Krafte*). Force and counterforce—neither one nor two, but two-in-one and one-in-two.

To understand the relevance of Hegel, Nietzsche, and Derrida's argument for Einstein's theory of relativity, it is very important to distinguish Hegel's account of force from Newton's position. As we have seen, Newton's world consists of distinct entities distributed in absolute (that is, independent and antecedent) space without any *inherent* relation to each other. Relations, therefore, are *extrinsically* imposed. For Hegel, by contrast, nothing *is* itself by itself, and everything *becomes* itself through the "reciprocal interchange of determinatenesses [*Austauschung der Bestimmeiten gegeneinander*]." Since there is no force apart from the expression of counterforces, force is never abstract. The differentiating interplay of force is the "universal medium in which the moments subsist as 'matters' [*als Materien*]."[6] Stated concisely, *force is relative*.

Derrida sees in this play of forces a prefiguration of *différance*. "Force is never present; it is only a *play of differences* and quantities. There would be no force in general without the difference between forces."[7] Neither simple nor passive, *différance* is the complex and active intersection of spacing and timing. To understand this critical point, it is necessary to turn to Heidegger's perplexing notion of the ontological difference between Being and beings. While his preoccupation with time is nowhere more apparent than in his magnum opus, *Being and Time* (1927), Heidegger offers helpful clues to his interpretation of temporality in two little-known essays whose English translation bears the simple title *Hegel*: *Die Negativitat—Eine Äuseinandersetzung mit Hegel aus dem Ansatz in der Negativität* (1938-1941) and *Erläuterung der 'Einleitung' zu Hegel's Phänomenologie des Geistes* (1942). Here, as elsewhere, Heidegger distinguishes "originary time" from tensed time (that is past, present, and future). In a surprisingly non-Heideggerian aphorism worthy of Nietzsche, Heidegger concisely summarizes the point of his fragments.

Beyng—the "in-between" and "beings"?? (no Platonism; no inversion Of the same, no inversion of Metaphysics but *an-nihilation*)	Unique—Singular *the error?* Of that which always is, and what is effective precisely in this way; appreciation of the refusal *as strife* Of world and earth.

To differentiate [*unterscheiden, die Unterscheidung*]:

1. *to carry apart?* Or *to ascertain only after the fact*, namely, the passage and the transition of?, "*the between*"
2. *to make equal*
3. *to abstain and look away* (mindlessly).[8]

This cryptic remark on differentiating suggests the bridge between Heidegger's early and late philosophy. The central notions of his mature thought all involve the rhythms of differentiating: clearing or lighting (*Lichtung*), opening (*eroffnen*), origin (*Ursprüng*), originary (*ursprünglich*), tearing (*Zerissenheit*), no(-)thing (*Nichts*), decision (*Entscheidung*), event or happening *(Ereignis)*, and, above all, truth (*Aletheia*). Like a bridge spanning two banks of a river, differentiation simultaneously separates and connects. The "in-between" "is" the place or nonplace of the repeated oscillation or alternation of the "originary time-space of becoming."

> This clearing cannot be explained from beings; it is the "between" [*Zwischen*] and in-between [*Inzwischen*] (in the time-spatial sense of the originary time-space).... The clearing is the a-byss as ground, the nihilating counterpart to all that is [*das Nichtende zu allem Seienden*] and thus the *heaviest thing*. It is thus the "ground" that is never "present-at-hand" and that is never found, the "ground" that refuses itself in the nihilation as clearing—the *supporting-founding* one that *decides* the one that e-vents—the e-vent.... *The a-byss: the nothing* what is most a-byssal—beying itself; not because the latter is what is most empty and general, and what fades most, the last fumes—but the richest, the singular, the middle that does not mediate and thus can never be taken back.[9]

So understood, differentiating is the endless activity that establishes the relationality and, thus, the relativity of spacing-timing.

Heidegger refined his understanding of spacing-timing in courses on Kant he taught around the time *Being and Time* appeared. These seminars were published in 1929 with the title *Kant and the Problem of Metaphysics*. Heidegger focuses his reading on Kant's interpretation of the imagination. The German word for imagination—*Einbildungskraft*— suggests the multiple layers of Heidegger's argument. *Ein*—one + *bild/ung*—image; culture, education + *Kraft*—force. According to Kant, the imagination is the power to synthesize the multiple differences of sensible intuition into a comprehensible whole. Samuel Taylor Coleridge, who heard Fichte's lectures and transmitted a version of German idealism to English romantics like William Wordsworth and American transcendentalists Ralph Waldo Emerson and Henry David Thoreau, aptly dubbed Kant's imagination the "esemplastic power."[10] Heidegger's interpretation of Kant's theory of the imagination is the key to his late philosophy, which anticipates Derridean deconstruction.

As we have seen, the purpose of the Third Critique is to resolve the unmediated oppositions established both within and between the first and second critiques. In its theoretical deployment, the activity of the imagination occurs along the margin *between* the sensible manifold of intuition and the forms of intuition (space and time) as well as the categories of understanding. In the First Critique, Kant translates classical ontology into epistemology by recasting Plato's Demiurge, who creates the world by bringing together form and matter, as the imagination, which creates the world by forming, ordering, or structuring the matter of sensation. This process is not limited to human mental activity. Kant effectively de-anthropomorphizes the imagination by arguing that this transcendental activity is a universal structural function that is both a differentiating unity and a unifying differentiation through which distinct beings are articulated, and, thus, the world is created. As such, the imagination is the formative activity (*Bilden*) that first distinguishes Being and beings and then figures images (*Bilde*) by clearing (*lichten*) the space *between* subjects and objects, which is their creative milieu (*Mitte*). It is precisely at this *point* that time comes into play. This opening is what Heidegger labels "originary time (*ursprüngliche Zeit*)." This argument turns on a characteristic Heideggerian play with language. *Ursprünglich* suggests multiple meanings: *Ur* means "primitive, primordial, original," and *Sprüng* means "crack, fissure, fault, flaw, beak [break], split, leap, spring." While "origin" implies a past static state, "originary" suggests an interstitial dynamic process of e-mergence that opens onto an uncertain future. For the Western metaphysical tradition, which identifies Being with presence, this margin of difference appears to be an unfathomable abyss that sweeps the ground from beneath one's feet. And yet, it is precisely this "groundless ground" that "is" the gift of time. Heidegger argues, "If the transcendental power of the imagination, as pure, forming faculty, itself forms time—that is, allows time to spring forth [*entspringen*]—then we cannot avoid the thesis stated above: *the transcendental power of the imagination is original time* [*ursprüngliche Zeit*]."[11] Original or originary time gives the present without ever being present.

In the concluding section of this chapter, I will consider the differences among linear, cyclical, narrative, and probabilistic time. For the moment (NB), it is important to distinguish Heidegger's "originary time," which can best be understood as temporality, from tensed time, which, as we have seen Augustine insists, consists of three modalities of the present. For Kant, the forms of intuition, which schematize the categories of understanding, function as something like fixed algorithms that program data of experience. The sensible manifold of intuition is the undifferentiated sensual flux (a posteriori), which is first processed by the forms of intuition, space, and time, and then organized by the twelve categories of understanding (a priori).[12] Kant underscores the necessity of both moments of this process with his telling maxim, "concepts without percepts are empty, percepts without concepts are blind." In contemporary terms, a program cannot run without data, and unprocessed data makes no sense.

While Kant translates ontology into epistemology, Heidegger extrapolates the ontological implications of Kantian epistemology. The here and now of sensible intuition are not original but are the product of a process of differentiating and connecting that eludes both perception and conceptualization. Heidegger labels this differentiating unification and unifying differentiation the (in)formative activity of "presencing [*Darstellung*]." As the condition of the possibility of the presence of every thing (be it subject or object), presencing is no thing, or, more precisely, is the nothing that is *neither* simply present *nor* absent. In the poetic words of Wallace Stevens's "Snowman," presencing is "the nothing that is not there and the nothing that is." This no-thing resembles the emptiness of Sunyata that discloses the coemergence and codependence of every-thing. While Heidegger's early work culminated in *Being and Time*, the *point* of his later work could be described as the interpretation of Being *as* time, which "is," of course, Becoming. As the condition of the possibility of presence of all determinate beings, this presencing is *neither* present *nor* absent, but resembles an elusive shadow that ghosts every present. The spacing-timing of originary temporality leaves the trace of a radical past that never was, is, or will be present but is always *à-venir* (to come, *avenir*, future).

> Time as pure intuition is the forming intuiting of what it intuits in *one*. This gives the full concept of time for the first time. Pure intuition, however, can only form the pure succession of the sequence of nows [*das reine Nacheinander der Jezfolge*] as such if in itself it is a likeness-forming prefiguring, and reproducing power of imagination. Hence, it is in no way permissible to think of time, especially in the Kantian sense, as an arbitrary field which the power of imagination just gets into for purposes of its own activity so to speak. Accordingly, time must indeed be taken as pure sequence of nows in the horizon within which we "reckon with time." The sequence of nows, however, is in no way time in its originality [*iher Ursprünglichkeit*]. On the contrary, the transcendental power of imagination allows time as sequence of nows spring forth [*lasst die Zeit als Jetzfolge entspringen*], and this letting-spring-forth is therefore original time.[13]

This "letting-spring-forth," which lets beings be, transforms Being into Becoming or Be-ing. In the ebb and flow of becoming-be-ing, what had appeared to be discrete things turn out to be transitory happenings or e-vents that resemble particles appearing and disappearing in the oscillation of forceful waves.

Heidegger concludes that the present is not original because it is always already given and, therefore, is secondary to something else, something other. In different terms, the present is never original because always pre-sent; what had long been understood as Being is always being given, and, thus, is becoming, which is never our own and always remains a gift that arrives from an elsewhere that is never present. At this point the fundamental question that runs throughout all of Heidegger's writings becomes: What gives? Heidegger's cryptic answer, which raises more questions, is, "*Es gibt*"—"It gives, there is." What "is" this "*es*"— this "it"—that is *neither* masculine *nor* feminine, yet engenders all that is and is not? What gives, Heidegger concludes, is time—not tensed time as a sequence of points that are here-and-now, but a more primordial time that is an "originary [*ursprünglich*]" temporality that is a "letting spring forth [*lasst entspringen*]." Life and death are gifts of time. Letting-spring-forth is the leap that lets beings be and thereby transforms Being into Becoming-Be-ing. Everything and everybody e-merges from and will return to primordial temporality. In the ebb and flow of becoming, things are not static entities but are *dynamic happenings or e-vents* (*Ereignisses*).

Recalling Kierkegaard while commenting on Hegel, Heidegger adds a surprising twist by associating the letting-be of originary temporality with decision.

THE DIFFERENTIATION OF BEING AND BEINGS
Differentiation as de-cision [*Ent-scheidung*]

De-cision—here, that which takes out of the mere *separation* and differentiation of what can be pregiven.

Beyng is itself the *decision*—not *something that is differentiated* from beings for a representing, supervening, reifying differentiation that levels them.

Being de-cides as an e-vent in the *e-venting* of man and of the gods into the need for the essence of mankind and of divinity.—This e-venting lets the strife of the world and the earth arise to striving,—the strife which alone the open clears, in which beings fall back to themselves and receive a *weight*.[14]

To decide (*de*, off + *caedere*, to cut) is to separate (*Entscheidung—ent*, original + *Scheidung*, separation). Time and de-cision are united in their difference. Decisive separation is the originary leap (*Sprung*) that constitutes the becoming *of* time.[15]

As Kierkegaard insisted, every decision is an undetermined leap that constitutes a decisive change. But his understanding of decision remains anthropocentric.

While Descartes reduced truth to certainty, Kierkegaard collapses objectivity into subjectivity when he declares, "truth is subjectivity." In ways that are not immediately obvious, Kant and Heidegger reverse this trajectory. Just as Kant had de-anthropomorphized the imagination, Heidegger generalizes or de-anthropomorphizes decision by making it the genesis of all becoming. From this point of view, human decisions are folded into a larger decision-making process that extends from physical through plant, animal, and human to sociopolitical and technological systems and networks.

When Being *and* time becomes Being *as* time, the axis of temporality shifts. Probability opens the future in a way that transforms the present and past. Heidegger's emphasis on decision as separation often obscures the alternative rhythm of connection. Derrida's *différance* complicates the margin of difference by stressing what is implicit but not always explicit in Heidegger's ontological difference. Derrida marks and remarks this difference with a subtle word play that shifts the "e" (*différence*) to the "a" (*différance*), which can be written, but not spoken.

> From this point of view, that the difference marked in the "differ()nce" between the *e* and the *a* eludes both vision and hearing perhaps happily suggests that here we must be permitted to refer to an order which no longer belongs to sensibility. But *neither* can it belong to intelligibility, to the ideality which is not fortuitously affiliated with the objectivity of *theorein* or understanding. Here, therefore, we must let ourselves refer to an order that *resists the opposition*, one of the founding oppositions between the sensible and the intelligible.[16]

Throughout this essay, Derrida uses multiple images to suggest this "between": weaving, interlacing, play, kernel, middle voice, arche-writing, interval, cleavage, trace, dance, and network. Noting that *différance*, like *différer*, derives from the Latin verb *differre*, which, in turn, can be traced to the Greek *diapherein*, Derrida stresses that his neologism has two meanings—to differ (space) and to defer (time).

> I would summarize here in a word I have never used but that could be inscribed in this chain: *temporization*. *Différer* in this sense is to temporize, to take recourse consciously or unconsciously, in the temporal and temporizing mediation of a detour that suspends the accomplishment or fulfillment of "desire" or "will," and equally effects the suspension of a mode that annuls or tempers its own effect. And we will see, later, how this temporization is also the temporalization and

spacing, the becoming-time of space and the becoming-space of time, the *originary* of time and space, as metaphysics or transcendental phenomenology would say, to use the language that is here criticized and, displaced.[17]

Rather than antecedent to separate entities, *différance* is inseparable from the differences, which can become oppositions, it establishes and sustains.

Turning Derrida's deconstructive strategy back on itself, his argument establishes the reciprocal relation of differences constituted and reconstituted in the neχus of *différance*. With this twist, Derrida's *différance* appears to explain the strange logic of Hegel's relationalism.[18] A few pages after this explanation of *différance* as spacing-timing and timing-spacing, Derrida turns to Hegel by quoting a passage from the Jena *Logic* published in a seminal essay written by Alexander Koyré—"Hegel à Jéna."

> The limit or moment of the present (*der Gegen-wart*), the absolute "this" of time, or the now, is of an absolute negative simplicity, which absolutely excludes itself from all multiplicity, and, by virtue of this, is absolutely determined; it is not whole or a *quantum* which would be extended in itself (and) which, in itself, also would have an undetermined diversity which, as indifferent (*gleichgultig*) or exterior in itself, absolutely different from the simple (*sondern es ist absolut differente Beziehung*).

Derrida then adds, "Koyré most remarkably specifies in a note: 'different Relation: *differente Beziehung*.' One might say 'differentiating relation.' And on the next page, another text of Hegel's in which one can read this: '*Diese Beziehung ist Gegenwart als seine different Beziehung* (This relationship is [the] present as a different relationship).' Another note of Koyré's: 'The term *different* here is taken in an active sense.'"[19] In this *differente Beziehung*—differentiating relation—Derrida sees the fourth and most suggestive anticipation of *Différance*.

Born in Russia in 1892, Koyré moved to Paris in 1908 to study with Henri Bergson. Three years later he left Paris to study in Göttingen with Edmund Husserl and the well-known mathematician David Hilbert. After World War I, he began teaching at the École pratique des hautes études, and in the early 1930s, Koyré taught an influential course entitled "The Religious Philosophy of Hegel." Though little noticed when it was published in 1934, his essay "Hegel à Jéna" was enormously important for Derrida. When Koyré left Paris to take a teaching position in Cairo, his fellow Russian immigrant Alexandre Kojève took over the course. The overarching goal of Koyré's essay was to understand a note Hegel scribbled the margin of his 1803–4 lecture notes: "*Geist ist Zeit*"—Spirit-Mind is time, which Hegel reformulates in the *Phänomenologie des Geistes* as "Time is the Dasein of the Concept." Kojève also takes up this question in his influential seminar, later published in *Introduction to the Reading of Hegel*, which includes "A Note on Eternity, Time and the Concept."

Koyré first translates and then comments on the central section of the Jena *Logic*. Given its importance for Derrida's entire *oeuvre* and the argument to follow, it is worth quoting Koyré's argument at length.

> The present—sublating [*supprimer, Aufheben*] itself in such a way that it is rather the future [*avenir*], which is engendered in it—is itself avenir [*à-venir*]; or this yet-to-come itself is not *à-venir*, *it is what sublates the present, but insofar as it is, this simple [something], which is an action of absolute negation, is rather the present, which is yet, in its essence, as much nonbeing of itself, or yet-to-come.*
>
> In fact, there is *neither* present *nor* yet-to-come (that is, the future), but only this mutual *relation* between the two, equally negative relation to each other, and this negation of the present self-negates itself as well; the difference between the two reduces itself in the rest of the *past*. The now has its own nonbeing in itself, and immediately becomes for itself an other, but this other yet-to-come in which the present transforms itself is immediately the other of itself, for it is not present. However, it is not this first "now," this notion of the present, but rather a now that has engendered itself from the present through the future [that is, the yet-to-come] a now in which the future and the present have equally suppressed and absorbed each other, a being that is a nonbeing of both, an activity, overcome and absolutely in rest, of the one over the other. The present is only the simple limit, self-negating itself, which, in the separation of these negative moments, is the relation of its [action of] exclusion to that which excludes [itself]. *This relation is [the] present, as a different relation in which both are conserved*; but if they do not conserve each other, they just as well reduce themselves to an equality to themselves in which both are not, and are absolutely destroyed.[20]

This reading of the Jena *Logic* reverses the flow of time—rather than the past necessarily unfolding in the present to prefigure the future, the uncertain future interrupts the present and refigures the past. Commenting in a note in Koyré's argument, Derrida argues, "*Rapport différent*: *differente Biezhung*. We could say differentiating *relation*." This differentiating relation is the *intertwining* of future, present, and past, which forms the diachronic *ne χus* of becoming.

> It is not "from the past" that that time comes to us, but from the future. Duration [*la durée*] does not extend from past to the present. Time forms itself by extending itself, or better by exteriorizing itself from the "now" or better yet prolonging itself by lasting. It is instead from the yet-to-come that it [time] comes to itself in the "now." The prevalent "dimension" of time is the yet-to-come that is, in some way, anterior to the past. It is this insistence on the future, the primacy given to the future over the past, which constitutes ... Hegel's greatest originality.[21]

Pause to ponder this puzzling claim: What can the assertion that the future is "anterior to the past" possibly mean?

Koyré, like Kojève, though in a different way, reads Hegel through Heidegger's account of primordial temporality. Koyré bends Heidegger's account of the activities of differentiating and de-cision back on itself to argue that far from the culmination of the Western ontotheological tradition that identifies Being with presence, Hegel's interpretation of the interrelation of the tenses of time in the Jena *Logic* disrupts the metaphysics of Being as presence and points to the *eventuality of Becoming*. Rather than the past unfolding in the present and prefiguring the future, decision dis-closes an open future that interrupts the present and refigures the past. *Temps* becomes *contretemps*. It is undeniable that the chiasmatic relation of past and future, which infinitely defers the presence of the present, harbors irreducible uncertainty. For Derrida, the uncertainty, even the undecidability of the future, does not cause despair that leads to nihilism. To the contrary, he concludes "Différance" by invoking Nietzsche, who "puts affirmation into play, in a certain laughter and a certain step [*pas*, not] of the dance. From the vantage point of this laughter and this dance, ... what I call a certain Heideggerian *hope*, comes into question."[22]

WARPING SPACE-TIME

Recalling Augustine's meditation on time in his *Confessions*, theoretical physicist and head of the Quantum Gravity group at the Centre de Physique

Theorique of Aix-Marseille University Carlo Rovelli stresses the enormous progress physics has made, but admits, "The nature of time is perhaps the greatest remaining mystery." To support this observation, he cites a comment by his fellow theoretical physicist, John Wheeler: "Explain time? Not without explaining existence! To uncover the deep and hidden connection between time and existence . . . is the task for the future."[23] As we will see in the next chapter, Einstein remained resistant to quantum theory because the notion of quantum entanglement seemed to violate his principle that nothing can exceed the speed of light. Rovelli offers an interpretation of quantum theory that reconciles it with relativity theory and quantum mechanics through a notion of constitutive relationality. "If the strangeness of quantum theory confuses us," he explains, "it also opens new perspectives with which to understand reality. A reality that is more subtle than the simplistic materialism of particles in space. *A reality made up of relations rather than objects.*"[24] In *Reality Is Not What It Seems*, he identifies three fundamental principles of quantum theory: (1) information is finite, which I will consider in chapter 6; (2) indeterminacy, which I will consider in the next section of this chapter; and (3) reality is relational. Here I will focus on Rovelli's third point.

> The theory [that is, quantum mechanics] does not describe things as they "are:" it describes how things "occur," how they "interact with each other." It doesn't describe *where* there is a particle but how the particle *shows itself to others*. The world of existent things is reduced to a realm of possible interactions. Reality is reduced to interaction. Reality is reduced to relation.
>
> In a certain sense, this is just an extension of relativity, albeit a radical one. . . . Speed is not a property of an object on its own: it is the property of the motion of an object *with respect to another object*. . . . Quantum mechanics extends relativity in a radical way: *all* variable aspects of an object exist only in relation to other objects. It is only in interactions that nature draws the world.[25]

After studying Hegel's relationalism, Nietzsche's perspectivism, Heidegger's originary temporality, and Derrida's becoming-time-of-space and becoming-space-of-time, there is something strangely familiar about both relativity theory and quantum mechanics as well as their interrelation.

The year 1905 was when the world and our understanding of it changed in ways that are still not fully understood. Einstein published four papers in which he proposed:

1. An explanation of the random movement of gas molecules known as Brownian motion
2. A theory of the photoelectric effect, which is crucial for quantum theory
3. His famous equation for the relationship between mass and energy— $E = mc^2$
4. The theory of special relativity

The theory of special relativity, published in an article entitled "On the Electro-dynamics of Moving Bodies," overturned the interpretation of the world based on the principles of classical Newtonian physics.

As we have seen, Newton, in effect, updated Democritus's atomistic universe with variations of mathematical formulae reminiscent of Pythagorean calculations. In this scheme, reality consists of discrete entities related through universals that can be expressed in abstract laws. Apart from the external imposition of these laws, particular entities remain separate from each other. Law, in turn, is independent of the entities it relates. Newton's justification for these claims is theological rather than scientific. "It seems probable to me," he writes, "that God, in the beginning, formed matter in solid massy, hard, impenetrable particles of such sizes and figures, and with such properties, and in such proportions to space."[26] In this vision, space and time form God's *sensorium*; they are homogeneous and independent of each other and are, therefore, universal and absolute. As such, they are originally empty and "exist" prior to the determinate entities and events that fill them. When Kant transforms space and time into subjective forms of intuition, they become the a priori conditions of the possibility of both objects and objectivity.

This theoretical construction of space and time continues to have significant practical implications. The earliest measurements of time were local and, therefore, were inseparable from place and space. Time varied from village to village, and city to city. Temporal standardization grew out of spatial acceleration on sea and land. A naval disaster in 1707 led the British Parliament to establish the Board of Longitude, which offered a reward to any person who could accurately determine longitude. English clockmaker John Harrison realized that it is necessary to use time to determine spatial location. "For longitude is not merely a matter of spatial position. It is a matter of where one is at a certain time—'mean' or 'local' time—relative to the time it is then at the prime meridian."[27] In 1735 Harrison built a chronometer that accurately measured longitude by determining the precise position east or west of the prime meridian in Greenwich, England. In Great Britain, British railways standardized time in 1847 by adopting Greenwich Mean Time, known as "railway time," as the national standard. In the United States, different railroads continued to determine their own time until 1883. All time was local until Congress passed the Standard Time Act in 1918. This standardization of time facilitated the synchronization of activity required for industrialization.

Without its theological underpinning, Newton's interpretation of absolute universal space and time becomes difficult, if not impossible, to defend. According to Newton, God's transcendence makes it possible for him (sic) to view the entire universe *sub specie aeternitatis*. Vestiges of this vision persist even today. According to Rovelli, "Usually, we call 'real' the things that exist *now* in the present. Not those which existed once or may do so in the future. We say that things in the past or the future 'were' real or 'will be' real, but we do not say that they 'are' real." This seemingly commonsense position is implied in what some contemporary philosophers and physicists label "Presentism." Presentism is an ontological theory according to which only the present and present things exist. Though the language is very different, the presentism of commonsense and theoretical physics is consistent with what Heidegger and Derrida following him describe as the "Western metaphysics of presence." Rovelli underlines the problematic assumptions and implications of presentism. "Twentieth-century physics shows, in a way that seems unequivocal to me, that our world is not described well by presentism: an objective global present does not exist. The most we can speak of is a present *relative* to a moving observer. But then, what is real for me is different than that which is real for you, despite the fact that we would like to use the expression 'real'—in an objective sense—as much as possible. Therefore, the world should not be thought of as a succession of presents."[28] This point can be concisely summarized: *there is no outside of the universe*, or, as Lee Smolin puts it in the epigraph to this chapter, "It is impossible to get outside the system we're studying when that system is the entire universe."[29] Thus formulated, the point is so obvious that it is puzzling how long it took to acknowledge its implications.

These developments suggest contradictions in Newton's position that created the opening for Einstein's theory of relativity. Contrary to Newton's assumptions, space and time are not separate but are *relative* to each other. If the *now* of the present is inseparable from the *here* of place, then the present of time can be neither universal nor absolute but must be particular and relative. The relativity of space and time insinuates itself into Newton's purportedly absolute space and time at two other points. First, velocity is relative rather than absolute. "That is, there is no meaning to the velocity of an object with respect to itself: the only velocity that exists is the velocity of an object with respect to another object."[30] Second, rather than singular or absolute, gravity is, as Hegel and Nietzsche realized, a *relative play of forces*. Hegel's analysis of force and understanding in the *Phenomenology* points to unexamined convolutions of Newton's law of gravity. Since force always requires counterforce, it can never be itself by itself and only *becomes* itself in *relation* to an other. The interrelation of forces forms a field that transforms Newton's "solid, massy, hard, impenetrable particles" into *emergent events*. Since space and time are relative, they are bound in a "differentiating relation" that Einstein defines as spacetime and Derrida describes as a spacing that's a timing and a timing that is a spacing.

While Einstein remained suspicious of quantum mechanics, Rovelli persuasively argues that his interpretation of space-time relativity and the relationalism of quantum entanglement are reconcilable. The transition from understanding reality in terms of the present being of discrete entities or particles to understanding reality in terms of the becoming of emergent events grew out of Michael Faraday and James Clark Maxwell's discovery of electromagnetic *fields*. Forces do not act between two originally separated objects that are subsequently joined but are relational fields through which codependent events emerge. The notion of the field provides a way to understand action-at-a-distance, which Newton identified but could never adequately explain. Rovelli stresses the importance of this development: "Introducing the new entity, the field, Faraday departs radically from Newton's elegant and simple ontology: the world is no longer made up only of particles that move in space while time passes. A new actor, the field, appears on the scene. Faraday is aware of the importance of the step he is taking.... He is conscious that he is suggesting nothing less than a modification of the structure of the world, after two centuries of uninterrupted success for Newtonian physics."[31] The notion of the field led to Faraday's electromagnetic theory of light. In his article "Thoughts on Ray Vibrations," he theorizes on the basis of his recognition of the interrelation of electromagnetism and light that light must be a *vibration* of electric and magnetic lines of force. It was left for Maxwell to develop mathematical equations to support Faraday's insight. With the work of Faraday and Maxwell, "the world has changed: no longer made up of particles in space but of particles in fields in space."[32] Without the shift from understanding the world in terms of separate atomistic particles joined by external forces to understanding the world in terms of intersecting waves and particles, neither Einstein's relativity theory nor quantum mechanics would have been possible.

The change from extra-worldly to intra-worldly theory leads to a view of the universe that conforms to Nietzsche's perspectivism. A few years after his premature death (1900), scientists theorized the "play of forces and waves of forces" that Nietzsche had poeticized. Force, we have seen, is constitutively relational, and, therefore thoroughly relative. In his 1915 extension of the theory of special relativity to general relativity, Einstein argued that what Newton had understood as absolute space is, in fact, a *gravitational field* whose ceaseless fluctuation renders space inseparable from time. Space, therefore, can no longer be plotted on a static flat grid but must be understood as a dynamic relational field. Nor can time be plotted as a linear series of present moments. Gravity warps space-time in a way that undercuts its universality and homogeneity. Contrary to common perception, the rate of time's passage is not uniform but varies with spatial location. "Place a watch on the floor and another on a table: the one on the floor registers less passing of time than the one on the table. Why? Because time is not universal and fixed; it is something that expands and shrinks, according to the vicinity of masses. Earth, like all masses, distorts space-time, slowing down time in its vicinity. Only slightly—but two twins who have lived respectively at sea level and in the mountains, when they meet up again, one will have aged more than the other."[33] Plotting space-time would require something like Humboldt's chart of the geography of plants with at least longitudinal, latitudinal, and altitudinal coordinates.

The heterogeneity of space-time has baffling consequences. If time varies with space, *simultaneity is impossible* and the global present posited by presentism vanishes. When spacing is minimal, temporal delay is virtually imperceptible, but when it is extended, the delay becomes undeniable. The impossibility of simultaneity resulting from the timing of space and the spacing of time transforms both the present and presence. As a result of the relativity of space-time, rather than a punctual moment, the present is an "intermediate zone *between* the past and the future."

> Between the past and the future of an event . . . there exists an "intermediate zone," an "extended present"; a zone that is *neither* past *nor* future. This is the discovery made with special relativity.
>
> The duration of this "intermediate zone" (that is, the set of events at a spacelike distance from a reference event), which is neither in your past nor in your future, is very small and depends on where an event takes place relative to you . . .: the greater the distance of the event from you, the longer the duration of the extended present. At the distance of a few meters from your nose, the duration of what for you is the "intermediate zone," neither past nor future, is no more than a few nanoseconds: next to nothing. . . . But on the moon the duration of the "extended present" is a few seconds, and on Mars it is a quarter of an hour. This means we can say that events on Mars are events that in this precise moment have already happened, events that are yet to happen, but during which things occur that are neither in our past nor in our future. . . . They are *elsewhere*.[34]

How is this "elsewhere" to be understood?[35]

The present is never simply itself but is the interstitial zone that is *neither* past *nor* future. This neither/nor deepens nonsimultaneity by turning the noncoincidence of the present back on itself. Since the identity of the present is its difference from the past and the future, the present is never totally present, but is the aftereffect of a relational process that is never present as such. Neither/nor is, then, the relational neχus in which the present never "is" itself by itself; rather, it becomes itself in and through an infinite web of reciprocal synchronic (spatial) and diachronic (temporal) interactions. "We can think of the world as made up of *things*. Of *substances*. Of *entities*. Or we can think of it as made up of *events*. Of *happenings*. Of *processes*. Of something that *occurs*. Something that does not last, and that undergoes continual transformation, that is not permanent in time."[36] While Hegel's relationalism transforms substance into subject (that is, process), Einstein's relativity transforms subjects and objects into emerging networks of events.

In his brilliantly provocative book, *Time Reborn: From the Crisis in Physics to the Future of the Universe*, theoretical physicist Lee Smolin explores the implications of what he describes as "the relational revolution" for our understanding of time and, by extension, reality itself. If, as Heidegger argues, the Western ontotheological tradition has constituted itself by "forgetting" Being, the Western scientific tradition, Smolin argues, has been a prolonged forgetting of time. Ever since

Plato's forms and Newton's paradoxically atemporal interpretation of the cosmos, both philosophy and science have repressed temporality in search of abstract universal natural laws. To correct this error, time itself must be temporalized through something like Hegel's *differente Biezhung*, Nietzsche's perspectivism, Heidegger's originary temporality, and Derrida's *différance*. Smolin concisely summarizes the conclusion to which these timely meditations lead:

> In a relational world (which is what we call a world where relationships precede space), there are no spaces without things. Newton's concept of space was the opposite, for he understood space to be absolute. This means that atoms are defined by where they are in space but space is in no way affected by the motion of atoms. In a relational world, there are no such asymmetries. Things are defined by their relationships. Individuals exist, and they may be partly autonomous, but their possibilities are determined by the *network of relationships*. Individuals encounter and perceive one another through the links that connect them within the network, and the networks are dynamic and ever evolving.[37]

If Being is Becoming, and space-time is relational, then to be is to be connected—every thing and every body is codependent and coemergent.

FUTURES OF THE PAST

Questions linger. Does the past have a future—a future or more than one future in which it can become something other than what it has been? Physics can never be completely separated from metaphysics any more than theory can be separated from theology. Newton's idealized conception of absolute space is a scientific and mathematical translation of the longstanding belief in divine omniscience. Since God is eternal, "his" knowledge cannot develop and, thus, cannot change. This transcendent God comprehends everything that ever was, is, and will be from what Kierkegaard labels the Archimedean point, which constitutes in an eternal now that is the *nunc stans* of contemporary Presentism.

In ways that are not immediately obvious, different theological visions of time entail different understandings of causality. Smolin begins *Time Reborn* with a quotation from the play *Arcadia* in which Tom Stoppard probes the puzzles of

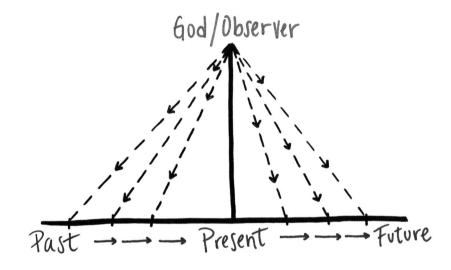

4.1 Archimedean point

time in contemporary physics: "If you could stop every atom in its position and direction, and if your mind could comprehend all the actions thus suspended, then if you were really, really good at algebra you could write the formula for all the future; and although nobody can be so clever to as to do it, the formula must exist just as if one could." In Newton's formulation of what became the classical physics that dominated scientific thought for more than two centuries, Smolin explains, "everything that happens in the universe is determined by a law, which dictates precisely how the future evolves out of the present. The law is absolute and, once present conditions are specified, there is no freedom or uncertainty in how the future will evolve."[38] Divine omniscience, then, implies predestination and vice versa.

In classical Greek philosophy, there are two basic types of causality, efficient and final, which lead to two different interpretations of time—linear and circular, which, in some cases, is cyclical. According to efficient causality, the present and future can be understood by tracing them to their origin in the past. In contrast to this archaeological hermeneutic, final causality understands the past and the present in terms of the future to which they lead. Throughout the Western theological tradition, these different interpretations of causality have been deployed in three traditional arguments for the existence of God: ontological, cosmological, and teleological.[39] The ontological argument derives from Platonic and Neoplatonic philosophy in which God is identified with Being as such (*Sein*), and all finite beings (*Seiende*) exist by virtue of their *participation* in the divine essence. While Being is one—(S)*ein* (*ein*, one), beings are many. The cosmological and the teleological arguments for God's existence presuppose the efficient notion of causality. Rather than the immanent essence of Being, God is a transcendent singular being who creates the world *ex nihilo*. Reflecting the divine Singularity, finite beings are separate individuals who have no intrinsic relations with each other. In efficient causality, the cause is antecedent to the effect; furthermore, the principle of causality is externally imposed on the cosmos as a whole as well as the separate entities in the cosmos. This is the mechanistic clockwork universe of eighteenth-century Deism, which is fully consistent with the scientific principles of Newtonian physics. Time in this scheme consists of a series of discrete points (here and now) that are externally joined to form a line.

The teleological argument, also known as the argument from design, was given its classic formulation by the British theologian William Paley in his *Natural Theology* (1805). As we will see in chapter 7, Paley's work was extremely influential

Past → Present → Future

Linear Time

4.2 Linear time

in Darwin's lifelong effort to become the "Newton of Biology." Paley gives the common deistic image of the mechanical clock a different twist. He opens his treatise by imagining a person who finds a watch while walking on a heath.

> When we come to inspect the watch, we perceive (what we could not discover in the stone) that its several parts are framed and put together for a purpose, e.g., that they are so formed and adjusted to each other as to produce motion, and that motion so regulated as to point out that the hour of the day; that, if different parts had been differently shaped from what they are, or placed after any other manner, or in any other order, than that in which they are placed, either no motion at all would have been carried out on the machine, or none which would have answered the use that is not served by it.... [B]eing once observed ... observed and understood, the inference, we think, is inevitable, that the watch must have had a maker: that there must have existed, at some time, and at some place or other, an artificer or artificers who formed it for the purpose which we find it actually to answer; who comprehended its construction and designed its use.[40]

Both the cosmological and this version of the teleological argument move from effect to cause. While the cosmological argument proceeds from the existence of the world to God as creator, Paley's version of the teleological argument proceeds from the design of the world to the necessity of a cosmic designer. There is, however, a significant difference between these two arguments. In contrast to separate entities externally related in the mechanistic model, the teleological argument interprets the world as designed in such a way that parts are adapted to each other to create an integral whole. More precisely, there are no parts without the whole and no whole without the parts. The relation among the parts as well as between the parts and the whole is reciprocal, and, therefore, everything is simultaneously means and end. In other words, the purpose of the parts is the whole, and the purpose of the whole is the parts; *neither* parts *nor* whole can exist or be understood apart from the other.

Kant's analysis of purposelessness or inner teleology represents a significant modification of the teleological argument for the existence of God. Rather than order being imposed from without, the interrelations among parts and between parts and whole is immanent or intrinsic. "An organized being," Kant argues, "is, therefore, not a mere machine. For a machine has solely *motive power*, whereas an organized being possesses *inherent formative* power, and such, moreover, as in it can impart to material devoid of it—material which it organizes. This, therefore, is a self-propagating formative power.... Nature, like art, *organizes itself*."[41]

Though Kant intends the Third Critique to mediate the First and Second Critiques, he does not explicitly consider the epistemological, axiological, and ontological implications of his interpretation of inner teleology. Nowhere is this shortcoming more evident than in his failure to rethink space and time on the basis of self-organization. The a priori forms of intuition are, in effect, an internalization of Newton's absolute space and time. In both cases, space and time are independent of particular entities or sensible intuitions upon which order is externally imposed. The inextricable interrelationship of parts as well as parts and whole entails a reciprocity that displaces a linear interpretation of time with a circular or cyclical temporal vision. While linear time is an open-ended series of infinite points, circular/cyclical time is a closed recurrent circuit.

I have argued that Hegel completes what Kant began by translating his heuristic interpretation of inner teleology into an ontological principle that informs both matter (nature, objectivity) and mind (spirit, subjectivity). The image of philosophy as a circle in which alpha and omega are one recurs repeatedly throughout Hegel's writings. One of the clearest examples of what seems to be a circular interpretation of time and history appears in the preface to the *Phenomenology*, where Hegel explains how his philosophy developed from its predecessors.

> The more conventional opinion gets fixated on the antitheses of truth and falsity, the more it tends to expect a given philosophical system to be either accepted or contradicted; and hence it finds only acceptance or rejection. It does not comprehend the diversity of philosophical systems as the progressive unfolding of truth, but rather sees in it simple disagreements. The bud disappears in the bursting-forth of the blossom, one might say that the former is refuted by the latter; similarly, when the fruit appears, the blossom is shown up in its turn as a false manifestation of the plant, and the fruit now emerges as the truth instead.... Yet at the same time their fluid nature makes them moments of an organic unity in which they not only do not conflict, but in which each is as necessary as the other; and this mutual necessity constitutes the life of the whole.[42]

It is, however, too simple to claim that Hegel's view of time and history is merely circular; rather, he attempts to combine linearity and circularity in what can best

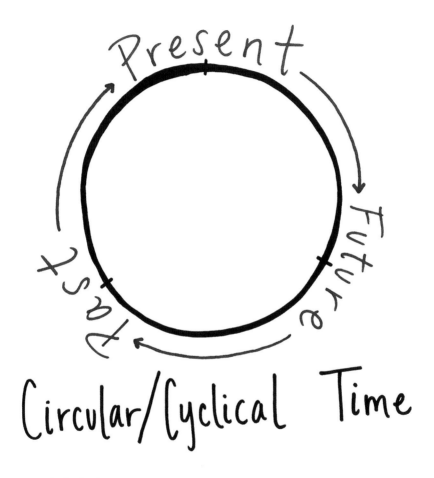

4.3 Circular/cyclical time

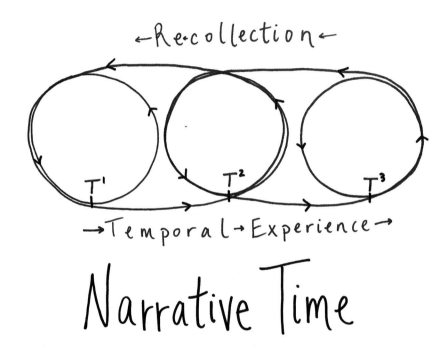

4.4 Narrative time

be described as a narrative view of time. The primary purpose of the *Phenomenology* is to transform the linearity of experience into the circularity of knowledge through a process of recollection. *In medias res*, experience is just one damn thing after another; time often seems out of joint with each moment disconnected from all others, and therefore, events remain external to each other. Looking back, however, it becomes possible to discern a thread that weaves together fragmentary moments into a coherent narrative. Whether this narrative is discovered or created (that is, constructed) remains an important question. Hegel, following Plato and Augustine, argues that knowledge and memory are inseparable—knowledge is a process of re(-)collection or re(-)membering. Hegel's term for this process is *Erinnerung*, which suggests the activity of inwardization. In re-collection, the moments of time are no longer scattered but now appear to be integrally related in such a way that one leads to another, and earlier moments are preserved, though displaced, by later moments. In this way, there is an irreducible "narrative quality of experience."[43]

Though linear and circular time as well as the effort to integrate the two in narrative time differ, they all tend to be deterministic, albeit in different ways. Whether through the unbreakable law of efficient causality or the archaeo-teleological unfolding of final causality, the present and the future appear to be the necessary outworking of the past. The understanding of the past, then, becomes the prophecy of the future. If the future is closed by its predictability, time, in the final analysis, is illusory. Kierkegaard makes this point in a section of *Philosophical Fragments* suggestively entitled "Interlude":

> If the past had become necessary it would not be possible to infer the opposite about the future, but it would rather follow that the future also was necessary. If necessity could gain a foothold at a single point, there would no longer be any distinguishing between the past and the future.... The past has come into existence; coming into existence is the change of actuality brought about by freedom. If the past had become necessary it would no longer belong to freedom, i.e., it would no longer belong to that by which it came into existence.... Freedom itself would be an illusion and coming into existence no less so; freedom would be witchcraft and coming into existence a false alarm.[44]

Decision is the activity through which possibilities are realized. The interpretive challenge is to understand how the coemergence and codependence of temporal moments do not negate unpredictability and uncertainty of becoming. The theory of probability suggests a possible solution to this vexing problem.

To understand the importance of probability in this context, it is necessary to return to Maxwell's discovery that light is electromagnetic radiation. In developing his ideas, Maxwell made use of atomic theory to argue that matter is made up of countless tiny particles that are in constant motion. In a radical departure from the principles of classical physics, he applied statistical methods and probability, which had been developed in the analysis of average behavior by sociologist Adolphe Quetelet, to physical phenomena. This innovation, according to Nobel laureate Ilya Prigogine, "was to introduce probability in physics not as a means of approximation but rather as an explanatory principle, to use it to show that a system can display a new type of behavior to which the laws of probability could be applied."[45] This was the first time that a law of nature was cast as probable rather than absolute. This insight has far-reaching implications for human existence as well as physical systems. If the future is not open to some extent, neither freedom nor novelty is possible. However, freedom, innovation, and novelty come at a price. Uncertainty is not the result of ignorance or the partiality of human knowledge but is a characteristic of the world itself. In the next chapter, I will consider the importance of uncertainty and probability in quantum mechanics. For now, I return to the question I posed at the beginning of this section: Does the past have a future or more than one future through which it can become something other than what it has been? Contrary to the common understanding of time, the past has multiple futures.

Never the necessary result of a past determined by efficient or final causality, probability reflects an open future that invades the present and transforms the past. This is not to say that the future is completely open—the future is *neither* absolutely determined *nor* wholly undetermined and, therefore, remains irreducibly unpredictable. The future is probable because the actuality of the past lends the present a trajectory that influences without completely determining the outcome of the future. The future, in turn, *pre-sents* the present with a cloud of possibilities that might or might not be realized through a process Kierkegaard, Heidegger, and, surprisingly, quantum physicists Heisenberg and Wheeler label *decision*. Whether conscious or nonconscious, de-cision marks the cutting off of many possibilities in order to actualize one possibility. Far from dead and gone, the past is redeemable because it is always in the process of becoming through the retroactive effect of decisions made in response to the in-breaking future.

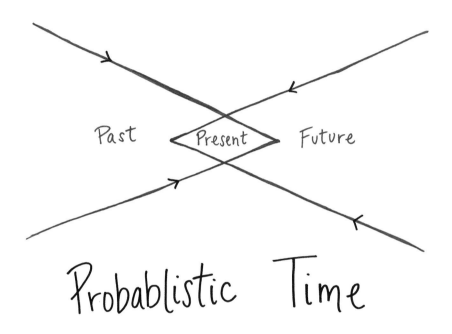

4.5 Probabilistic time

4.6 The Moment, II

Neither past *nor* future, the present "is" the liminal spacing-timing and timing-spacing that is never present but is the arising and passing away that does not arise and pass away. Being is Becoming though Becoming "is" not Being. The neχus of this neither/nor is originary temporalization, which, though never itself present, marks the emergence of tensed time. In a probabilistic world, the past is not destiny because it is never fixed or finished and is always becoming something other than it has been. De-cisions are leaps (*Ursprünge*) made in the twinkling of an eye (*Øjeblik*, *Augenblick*), which recast past moments to create unexpected novelty. In every moment there is something new under the sun. The openness of the future is the dis-closure of the past as unending becoming, which makes hope possible.

CHAPTER 5

ENTANGLEMENT

The elements of a system were themselves systems, and one could begin with the social sphere, with its groups of individuals, continue with the individual themselves, and move on to the brain and the brain cells—ever downwards as if descending a staircase, until reaching the very smallest elements in the depths of the strange world of quantum physics, where the insight that nothing exists in its own right, only in relation to something else, presumably had to be taken literally, at least according to certain physicists who believed that such particles simply did not exist at all until the moment they entered into a relation. They are nothing in themselves, however strange that may sound. They possess no properties of their own, and they only exist in relation to others.

—Karl Ove Knausgaard

COPENHAGEN

In 1978-1979 I spent a second year in Denmark writing *Journeys to Selfhood: Hegel and Kierkegaard*, for which I received a Doktorgrad in philosophy from the University of Copenhagen. During our previous visit in 1971-1972, Dinny and I lived in the heart of Copenhagen, near Nyhavn Canal and just three blocks from the Royal Theater, where Kierkegaard mingled with the theatergoing crowd in between fits of mad writing. By 1979 we had two children, Aaron and Kirsten, and could not afford a city flat, so we had to live in the working-class suburb of Albertslund, which was a planned community that represented everything Kierkegaard foresaw and rebelled against. The homogeneous modular concrete structures were so similar that it was all but impossible to tell one apartment from another. Planned streets and walkways guided residents from home to school, stores, train, and work like rats caught in a maze. Deviation from the straight and narrow posed a considerable challenge.

Though I was once again conducting research at the Kierkegaard Institute, the apartment where we lived was owned by the Niels Bohr Institute. At the time, I knew much less about Bohr and quantum theory than I should have known. I had no idea that Kierkegaard's existential analytic had played a significant role in Bohr's theoretical breakthrough. Bohr borrowed Kierkegaard's notion of decision as a leap to express his world-shattering theory of subatomic quantum leaps. Philosophy and physics—seemingly separate trajectories—turn out to be thoroughly entangled. At the beginning of my career, I did not know what I did not know; now, as my career, and, indeed, my life, are nearing their end, I am slowly coming to understand what I do not understand. When I made my own leap into quantum theory, I was reassured to read Richard Feynman's well-known quip, "If you think you understand quantum mechanics, you don't understand quantum mechanics." I was doubly reassured to learn that Bohr once said, "If quantum mechanics hasn't profoundly shocked you, you haven't understood it." Immersing myself in quantum theory, my head began to spin like two entangled particles simultaneously turning in opposite directions.

Struggling to make sense of what was for me a new field, I stumbled on Werner Heisenberg's book, *Philosophy and Physics: The Revolution in Modern Science*.[1] The son of a classics professor and himself knowledgeable in Western classics, Heisenberg's interests were broad, and he had an unusual ability to explain complicated scientific ideas to readers who cannot recognize, let alone solve, a differential equation. His humanistic sensibility led him to write in a way that is accessible to readers who lack the necessary mathematical knowledge to grasp the intricacies of theoretical physics. *Philosophy and Physics* is not precisely physics for poets, but it is poetry for physicists.

Whether for the reasons Heisenberg suggests or for other reasons, artists—particularly playwrights—have been attracted to the enigmas of quantum theory. In his head-spinning play *Hapgood* (1988), Tom Stoppard stages a performance of doubles, duplicity, and deceit that explores the uncertainties of quantum theory as an espionage drama between a woman whose name suggests chance, Hapgood, and twin Russian KGB agents. The play opens with a farcical enactment of the famous two-slit experiment to test the wave-particle theory of light.

Extrapolating from the micro to the macrolevel in an exchange between two agents who cannot believe their eyes, Stoppard summarizes the crucial insight of what is known as the Copenhagen interpretation of quantum theory.

> BLAIR: I want to know if you're ours or theirs, that's all.
> KERNER: I'm telling you but you're not listening. Now we come to the exciting part. We will watch the bullets to see how they make waves. This is not difficult, the apparatus is simple. So we look carefully and we see the bullets, one at a time. Some go through one gap and some go through the other gap. No problem. Now we come to my favorite bit. The wave pattern has disappeared. It has become particle pattern again.
> BLAIR: (*obliging*) All right—why?
> KERNER: Because we looked. Every time we don't look, we get a wave pattern. Every time we look to see how we get the wave pattern we get particle pattern. The act of observing determines what's what.

The explanation proves more baffling than the question, so Kerner tries again. "Positive and negative. I'm not going to help you, you know. Yes-no, either-or.... You have been too long in the spy business, you think everybody has no secret but one big secret, they are what they seem or they are the opposite. You look at me and think: *Which is he*? Plus or minus? If only you could figure it out like looking into me to find my root. And then you still wouldn't know. We're all doubles. Even you."[2] Either/Or ... Both/And ... Neither/Nor. What if every root/route is connected by division and is, therefore, neither simply one nor two but is the play that is an interplay of doubles?

There was always something duplicitous about Heisenberg; everything about him was uncertain. In 2000 British novelist and playwright Michael Frayn's *Copenhagen* premiered in London's National Theater before appearing a few years later on Broadway, where I saw it. While Hapgood is commonly regarded as Stoppard's least successful play, probably because few people understand it, Frayn's play won a Tony award in 2000. Like Stoppard, Frayn is fascinated by the existential implications of Bohr's quantum theory and Heisenberg's uncertainty principle. Unlike Stoppard, however, he is interested in the complicated relationship between the enigmatic Heisenberg and his mentor Bohr. Having reached an impasse in his understanding of the atom, Bohr invited Heisenberg, who at the time was an assistant to Max Born in Göttingen, to join him at the University of

Copenhagen's Institute for Theoretical Physics. Suffering severe allergies, Heisenberg retreated to the desolate North Sea Island of Helgoland to ponder the puzzle Bohr had challenged him to solve. On June 7, 1925, he had a revelation that not only overturned classical physics but literally changed the world.

> When the first terms seemed to come right [given Bohr's rules], I became excited, making one more mathematical error after another. As a consequence, it was around three o'clock in the morning when the result of my calculation lay before me. It was correct in all terms.
>
> Suddenly I no longer had any doubts about the consistency of the new "quantum" mechanics that my calculation described.
>
> At first, I was deeply alarmed. I had the feeling that I had gone beyond the surface of things and was beginning to see a strangely beautiful interior and felt dizzy at the thought that now I had to investigate this wealth of mathematical structures that Nature had so generously spread out before me.[3]

This insight led to Heisenberg receiving the 1932 Nobel Prize in Physics for "the creation of quantum mechanics."

During the war, Bohr, who was half Jewish, stayed in Copenhagen and helped refugees even through the German occupation. With the outcome of the war in doubt, Bohr agreed to flee to Sweden and allow himself to be "kidnapped" in a British commando raid and taken out of occupied Denmark. He was transported to England, where he was received personally by Churchill. From there he traveled to the United States to teach a generation of young physicists how to use the new quantum theory to make atomic bombs. Heisenberg, who was a Lutheran Protestant, took the opposite path. He returned to his homeland to lead Germany's nuclear program. The uncertainty surrounding Heisenberg's contribution to Germany's effort to build a nuclear bomb is the focus of Frayn's *Copenhagen*. The lingering uncertainty was whether Heisenberg did all he could to deliver the bomb to Hitler or whether he deliberately sabotaged Germany's research efforts. The crucial question concerned the amount of fissile material necessary to start a nuclear chain reaction. Basing his judgment on natural uranium rather than the isotope U-235, Heisenberg concluded that a ton or more of material would be necessary rather than the correct amount of only several kilograms. As a result of this mistake, he failed to do the calculation that held the key to constructing the bomb. It remains unclear whether Heisenberg's failure was a mistake or was deliberate, thereby allowing the Allies to create the bomb first and win the war. Bohr did not initially do the calculation either, but he and his wife Margrethe suspect that despite Heisenberg's claim to the contrary, their motives were different.

> HEISENBERG: Though in fact you made exactly the same assumption! You thought there was no danger for exactly the same reason that I did! Why didn't *you* calculate it?
> BOHR: Why didn't *I* calculate it?
> HEISENBERG: Tell us why *you* didn't calculate it and we'll know why *I* didn't!
> BOHR: It's obvious why *I* didn't!
> HEISENBERG: Go on.
> MARGRETHE: Because he wasn't trying to build a bomb!
> HEISENBERG: Yes. Thank you. Because he wasn't trying to build a bomb. I imagine it was the same with me because *I* wasn't trying to build a bomb. Thank you.[4]

But Bohr was not convinced and suspected that his former mentee was attempting to salvage his reputation. In his postscript, Frayn explains that Heisenberg "wanted to distance himself from the Nazis, but he didn't want to suggest that he had been a traitor. He was reluctant to claim that he had failed them simply out of incompetence."[5]

Copenhagen combines historical fact and fiction to explore the complexity of reality by applying quantum mechanics and the uncertainty principle to the lives of the scientists responsible for their formulation. The play centers on a meeting between Bohr and Heisenberg that took place in 1941, when Heisenberg returned to Copenhagen purportedly to discuss the unexpected consequences of the theories. There are only three characters—Bohr, Heisenberg, and Margrethe, who contributed to her husband's work. By the time of the play, all three have died, but they return as ghosts for a final conversation; they are as spooky as the ghostly events their theories conceive. Heisenberg complains that the more he tried to explain the reasons for his trip to Bohr as well as to government authorities, the more uncertain his motives became. Much of the dialogue consists of actual words spoken or written by the three historical actors. The genius of the play is the way it *enacts* quantum theory and the uncertainty principle through the interaction of Bohr, Heisenberg, and Margrethe. Three

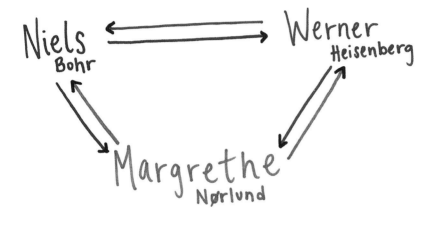

5.1 Mirror effects

116　ENTANGLEMENT

people, three complementary couples circling each other and interpreting the data of experience from three interrelated perspectives. With no Archimedean point from which to judge their judgments, everything remains uncertain.

In this play, which is an interplay, speculation (Latin, *speculum*, mirror) becomes scientific theory in which mirrors mirror mirrors.

> HEISENBERG: I look at the two of them looking at me, and for a moment I see the third person in the room as clearly as I see them. Their importunate guest, stumbling from one crass and unwelcome thoughtfulness to the next.
>
> BOHR: I look at him looking at me, anxiously, pleadingly urging me back to the old days, and I see what he sees. And yes—now it comes, now it comes—there's someone missing from the room. He sees me. He sees Margrethe. He doesn't see himself.

Before turning to science "proper," a supplementary detail. If neither Bohr nor Heisenberg did the calculation, who did?

> BOHR: An Austrian and a German.
> HEISENBERG: So they should have been making their calculations for us, at the Kaiser Wilhelm Institute in Berlin. But instead they made it at the University of Birmingham in England.
> MARGRETHE: Because they were Jews.[6]

In the wake of the quantum revolution, even scientists have been forced to admit that neither their research nor their theory is truly objective. Everything is perspectival, and, therefore, relative. Research that claims to be disinterested and objective is funded by universities, governments, and increasingly corporations with economic and political interests of their own. Knowledge, therefore, is never completely free of prejudice. The scientific breakthroughs between 1905 and 1927 were resisted throughout much of the twentieth century. The ostensible reason for this was that a quantum cosmos contradicts not only classical Newtonian physics, but also the commonsense view of the world. There is, however, another, darker reason for this resistance. In some quarters, quantum physics was known as "Jewish science." During the war, more than a hundred leading German scientists and mathematicians fled continental Europe for the United Kingdom and United States. Many of these exiles contributed to the Allied program to build the atomic bomb. The political exigencies and the practical demands they faced during World War II and the Cold War left little time for theoretical speculation. In his revealing book *How Hippies Save Physics: Science, Counterculture, and the Quantum Revolution*, David Kaiser persuasively demonstrates that quantum theory was revived in the 1970s by a band of unemployed physics PhDs tripping on acid and calling themselves the Fundamental Fysics Group who gathered—where else?—in Berkeley. These acid freaks experienced the "dizziness" Bohr described when first encountering theory. Learning the complicated formulas that were the passwords for entering the quantum world was like passing into an alternative reality where nothing is what you believed it had been.

ALTERNATE FACTS

Even though Wesleyan was a hotbed for political and social protest during the 1960s, the drug revolution did not hit campus until 1969, the year after I graduated. While I have never taken LSD, that does not mean that I did not experience something like Bohr's dizziness back in the sixties. When students occasionally ask me if I have ever dropped acid, I respond, "Hegel and Nietzsche have long been my acid." They usually look at me incredulously, but I am completely serious. Hegel's and Nietzsche's philosophies not only mess with your head as much as dope, they also are uncannily similar to Einstein's theory of relativity, Bohr's quantum theory, and Heisenberg's uncertainty principle.

My head would not have been set spinning by Hegel and Nietzsche if I had not already been unsettled by my introduction to Kant in a seminar on the *Critique of Pure Reason* in the spring of 1967. Louis Mink, who was one of the three best seminar teachers I ever had, began the first class by saying, "I became a philosopher because I have worn glasses ever since I was a child." Puzzled, the class asked, "What do you mean?" I have never forgotten his response and have quoted it countless times as I have struggled to interpret Kant's philosophy for my students. "From a young age, I knew," Louis responded, "that the world is not as it appears to be." He went on to explain that for Kant, we can never know the "thing-in-itself" (that is, the world as such) because experience is always filtered by the forms of intuition and categories of understanding. Though Louis did not make the comparison, I now understand that the two lenses of eyeglasses, like the two slits in the double-slit experiment, are the apparatus through which the world appears.

Hegel totally disagreed with Kant's limitation of knowledge and countered that rather than overcoming Humean skepticism, Kant had alienated subject and object as well as self and world, thereby making knowledge impossible. The thing-in-itself, he insisted, is nothing but an empty concept constructed to expose the supposed limitations of all concepts. The *Phenomenology of Mind* replays the emergence of *knowledge as it appears*. It was left for Nietzsche to draw the conclusion implicit in Hegel's phenomenology. In *Twilight of the Idols* he writes, "The reasons for which 'this' world has been characterized as 'apparent' are the very reasons which indicate its reality; any other kind of reality is absolutely indemonstrable."[7] Here oppositional dualism collapses into the entangled interplay of differences. The real world is apparent, and appearances are real. This is also the conclusion of the Copenhagen interpretation of quantum mechanics.

In his informative book, *Beyond Weird: Why Everything You Thought You Knew About Quantum Physics Is Different*, Philip Ball writes, "The question now hurtling towards us might sound like pedantic, navel-gazing philosophy, but really there is no escaping it: What do we mean by 'is'?"[8] Physics has always been metaphysics—quantum mechanics harbors an implicit ontology, which turns out to be inseparable from epistemology. This radically new scientific ontology/epistemology is strangely familiar to anyone who knows the Western ontotheological tradition. Three theoretical physicists who best understand the important relationship between physics and philosophy are Heisenberg, Rovelli, and Wheeler. As I have noted, Richard Feynman observed that if you think you understand quantum mechanics, you don't understand it. Responsible physicists interpret quantum mechanics in very different ways. In the following discussion, I have taken Heisenberg, Rovelli, and Wheeler as my primary guides. My goal is not merely to represent their arguments as well as I can, but also to borrow their insights to develop my own constructive argument. I realize that where I see similarities, the three theorists undoubtedly would see differences.

During the first two decades of the twentieth century, Einstein's theory of relativity (1905), Bohr's quantum theory (1913), and Heisenberg's uncertainty principle (1927) overturned Newton's classical physics, which had been accepted for more than two centuries. "The change in the concept of reality manifesting itself in quantum theory," Heisenberg claims, "is not simply a continuation of the past; it seems to be a real break in the structure of modern science."[9] Quantum theory is, in other words, a quantum leap in our understanding (epistemology) of reality (ontology).

The interpretation of the world in classical physics is a refinement of Democritus's atomism and is remarkably similar to what is for many people the common-sense understanding of reality. By the nineteenth century, science and technology joined to create a utilitarianism that subjects the natural world to "man's" desires and interests. Heisenberg explains that for classical physics, "the world consisted of things in space and time, the things consisted of matter, and matter can produce and can be acted upon by forces; every event is the result of the cause of other events. At the same time the human attitude toward nature changed from a temporal to the pragmatic one. One was not so much interested in nature as it is; one rather asked what one could do with it. Therefore, natural science turned into technical science; every advancement of knowledge was concerned with the question as to what practical use could be derived from it."[10] Given the devastating consequences of this development, the question is whether it is possible to develop a theory that *neither* alienates human beings from nature (objectivism) *nor* subjects nature to human exploitation (subjectivism). A relational interpretation of quantum mechanics suggests a middle way between these two extremes.

Quantum mechanics overturns these primary assumptions of classical physics:

1. Space is an eternal structure that is independent of the entities that "fill" it.
2. Atoms are separate solid balls.
3. Objects, which are made up of atoms, are independent of each other.
4. At the microlevel atoms and at the macrolevel entities are connected through the imposition of external forces.
5. Classical systems are closed and tend toward equilibrium.
6. The real world can be described without any reference to us; that is to say, objectivity is possible.
7. The position and the velocity of atoms and entities can be simultaneously determined.
8. With precise measurement of the initial conditions of a system, it is possible to calculate the future states of the system; that is to say, systems are deterministic.

It is important to understand that quantum mechanics does not completely dismiss classical physics. While classical principles are predictively accurate at the macrolevel, they do not apply at the microlevel. In terms I have previously used, the macroworld is Either/Or and the microworld is Both/And. The question that remains is whether quantum theory provides the Neither/Nor that simultaneously connects and distinguishes the micro and macroworlds.

While Bohr and Heisenberg are credited with launching the quantum revolution, the classical model had been gradually eroding for years. Even in the early twentieth century, the structure and operation of the atom were not well understood. As I noted in the previous chapter, the publication of Maxwell's article, "A Dynamical Theory of the Electromagnetic Field" (1865), and Faraday's mathematical formulation of that theory marked a shift in the understanding of the fundamental elements of the universe. Prior to Newton, philosophers assumed that changes in individual physical objects could be caused only by direct contact. By introducing the law of gravity, Newton provided an alternative causal explanation by proposing action at a distance. However, he never made the law of gravity itself a focus of investigation. By explaining fields of force, Maxwell transformed discrete entities into fuzzy interrelated waves. Heisenberg explains the significance of this shift:

> Now in the theory of fields of force one could come back to the older idea, that an action is transferred from one point to a neighboring point, only by describing the behavior of the fields in terms of differential equations. This proved actually to be possible, and therefore the description of the electromagnetic fields as given by Maxwell's equations seemed a satisfactory solution to the problem of force. Here one had really changed the program given by Newtonian mechanics. The axioms and definitions of Newton had referred to bodies and their motion; but with Maxwell the fields of force seemed to have acquired the same degree of reality as the bodies of Newton's theory.[11]

A second development that contributed to quantum theory grew out of Max Planck's investigation of how objects radiate heat. At the end of the nineteenth century, there was still no experimental evidence for the existence of atoms. Drawing on the work of Maxwell, Planck modified the notion of waves to theorize that what he called "oscillators" vibrate at different frequencies. His surprising discovery was that oscillators only emit and absorb radiation in discrete packets. He formulated this discovery in Planck's constant (1900), which states that electromagnetic radiation from heated bodies is emitted in discrete units or quanta instead of a continuous flow. This discovery cannot be reconciled with the principles of Newtonian physics. Rather than a continuous flow and flux, "reality" appears to be discontinuous, or, in Rovelli's terms, "granular."

Planck's constant proved to be crucial for the theory of relativity. Einstein used the notion of quanta to explain light. He "proposed that quantization was a real effect, not just some sleight of hand to make the equations work. Atomic vibrations really do have this restriction. Moreover, he said, it applies also to the energy of light waves themselves: their energy is parceled up into packets, called photons. The energy of each packet is equal to [a multiple] of the light's frequency (how many wave oscillations it makes each second)."[12] Einstein's understanding of light in terms of discrete quanta raised further questions about the interpretation of light waves.

Before turning to Bohr's appropriation of these insights, it is necessary to consider a third crack in the classical façade. The Newtonian model, as I have noted, presupposes that objective knowledge of the world is possible. It also assumes that it is possible to determine accurately both the position and the velocity of every particle and entity. This, in turn, enables scientists to calculate deterministic causal relations in systems. In 1927 Heisenberg introduced his uncertainty principle, which calls into question all these assumptions. "One could speak of the position and the velocity of an electron as in Newtonian mechanics," he writes, "and one could observe and measure these quantities. But one could not fix both quantities simultaneously with an arbitrarily high accuracy. Actually, the product of these two inaccuracies turned out to be no less than Planck's constant divided by the mass of the particle. Similar relations could be formulated for other experimental situations. They are usually called relations of uncertainty or the principle of indeterminacy."[13] In his postscript to *Copenhagen*, Frayn explains the uncertainty surrounding Heisenberg's term that is usually translated "uncertainty":

> Heisenberg and Bohr used several different German words in different contexts. Bohr . . . sometimes referred to *Unsicherheit*, which means quite simply unsureness. In Heisenberg's original paper he talks about *Ungenauigkeit*—inexactness—and the most usual term now in German seems to be *Unschafe*—blurredness or fuzziness. But the word he adopts in his general conclusion, and which he uses when he refers back to the period later in his memoirs, is *Unbestimmtheit*, for which it's harder to find a satisfactory English equivalent. Although it means uncertainty in the sense of vagueness, it's plainly derived from *bestimmen*, to determine or ascertain. . . . "Undeterminedness" would be closer still, though clumsy. Less close to the German, but still closer to the reality of the situation, would be "indeterminability."[14]

Unsicherheit: insecurity, precariousness, unsteadiness, uncertainty; *Ungenauigkeit*: inaccuracy, inexactitude; *Unschafe*: blurred, hazy, poorly defined, out of focus, not sharp; *Unbestimmtheit*: indefiniteness, uncertainty, vagueness, indecision. As always, the question of proper translation remains *undecidable*. Nonetheless, the notions of "blurred" and "fuzzy" are suggestive for describing waves that are particles and particles that are waves. Whether interpreted in terms of uncertainty, indeterminability, or undecidability, Heisenberg's principle undercuts Newtonian determinism and replaces it with quantum probability.

Before proposing his quantum theory, Bohr had developed formulas that accurately predict the properties of chemicals by measuring the frequency of light emitted when they are heated. Bohr had used Planck's constant to argue that the stability of the elements even after chemical bonding is a function of discrete energy quanta, the lowest of which is the normal state. But neither Bohr nor others understood how electrons maintained their orbits or why they *leaped* from orbit to orbit in discontinuous increments. Experiments by Bohr and others confirmed his findings but raised more questions about the wave-particle duality. "How could it be," asked Heisenberg, "that the same radiation that produces interference patterns, and therefore must consist of waves, also produces the photoelectric effect, and therefore, must consist of moving particles?"[15] The attempt to answer this and related questions in terms of classical physics led to further paradoxes and contradictions. Though the discontinuities of quantum leaps suggest Kierkegaard's Either/Or, Bohr tries to resolve the wave-particle dilemma with a Hegelian Both/And. It is not necessary to choose between a wave or a particle explanation, he argues, because light waves, for example, seem to be both waves and particles until they are observed. But further questions remained.

The principle of complementarity according to which objects have pairs of complementary properties that cannot be observed simultaneously leads to the interrelated notions of probability and uncertainty at the microlevel. Heisenberg realized that the shift from the calculability and certainty of classical physics to the incalculability and uncertainty of quantum physics requires a different method for testing the probability of a system.

Therefore, the theoretical interpretation of an experiment requires three distinct steps:

1. The translation of the initial experimental situation into a probability function.
2. The following up of this function in the course of time.

3. The statement of a new measurement made of the system, the result of which can be calculated from the probability function.

> For the first step the fulfillment of the uncertainty relations is a necessary condition. The second step cannot be described in terms of the classical concepts; there is no description for what happens to the system *between* the initial observation and the next measurement. It is only in the third step that we change over again from the "possible" to the "actual."[16]

I will consider the actualization of possibility in what follows. For the moment, it is important to understand that between one state and a different or subsequent state, there is an interstitial gap that eludes observation. This interval is something like a bridge that simultaneously separates and joins two "actual" states. How is this gap, interval, ellipsis, tear, margin, limen, hinge, cleavage, void, emptiness, spacing to be understood? This is the question Heisenberg withdrew to Helgoland to contemplate. He begins his revolutionary article, written when he was only twenty-four, by expressing his extraordinary ambition. "The objective of this work is to lay the foundations for a theory of quantum mechanics based exclusively of the relations between quantities that are in principle observable." Rather than trying to understand the quantum leap in terms of familiar or even new forces, Rovelli explains, Heisenberg argues, "Let's change, instead, our way of thinking about the electron. Let's give up describing its movement. Let's describe *only what we can observe*: the light it emits. Let's base everything on quantities that are *observable*." Rovelli leaves no doubt about the importance of this insight. This revolutionary idea "could only be had with the unfettered radicalism of the young. This idea would transform physics in its entirety—together with the whole of science and our very concept of the world. An idea, I believe, that humanity has not yet fully absorbed."[17]

In contrast to classical physics in which subject and object are separate and science is the objective description of the world as such, Heisenberg insists that we deal only with what is observable. It is not only, as Kant argued, that there is no knowledge of the world in itself, but it is impossible even to imagine the world apart from our awareness of it. The scientific goal of objectivity involves the illusion of knowing the world as it is apart from our knowing it. As we will see in the final section of this chapter, the quantum critique of objectivity does not necessarily lead to solipsistic subjectivism. What is clear is that classical physicists and those who knowingly or unknowingly subscribe to its foundational principles are mistaken when they set objects against objects, subjects against subjects, and subjects against objects. Quantum theory reveals objects and subjects to be inextricably interrelated and, therefore, relative to each other; in other words, they are coemergent and codependent. With these insights, I return to the interrelation of ontology and epistemology.

Inasmuch as quantum mechanics deals only with phenomena, the Copenhagen interpretation is, in effect, a *phenomenology of mind*. It is noteworthy that during the same years that scientists were discovering that the "objects" of their investigations are phenomenal, Edmund Husserl was launching the influential philosophical movement known as phenomenology. In 1912 he issued the *Yearbook for Philosophy and Phenomenological Research*, which was the bible of the movement until 1930. Twentieth-century phenomenology, which traces its roots to Hegel's *Phenomenology of Spirit/Mind* (1807), was very influential for both Heidegger and Derrida, as well as philosophers as different as Maurice Merleau-Ponty and Jean-Paul Sartre. While twentieth-century phenomenology is not a unified movement, David W. Smith explains that it can be generally defined as "the study of structures of experience, or consciousness. Literally, phenomenology is the study of 'phenomena': appearances of things, or things as they appear in our experience, or the ways we experience things, thus the meanings things have in our experience. Phenomenology studies conscious experience as experienced from the subjective first-person point of view."[18]

Though the interpretations of phenomena in phenomenology and quantum physics differ, these two perspectives share a common methodological preoccupation with objects *as they appear in consciousness*. Wheeler concisely summarizes the operative distinction: "No elementary phenomenon is a phenomenon until it is a registered (observed) phenomenon."[19] With the question of registration, we return to the issue of Louis's glasses. Another way to make Wheeler's point is to say that experience and, therefore, knowledge is always mediated. Elsewhere Wheeler expresses what he describes as "nature's revolutionary pistol" in a characteristic epigram, "No question, no answer."[20] Ball explains the far-reaching implications of Wheeler's quip for philosophy as well as physics:

> Today most scientists would accept that our reliance on sensory data puts us at one remove from any *Ding an sich*: all our minds can do is to use those data to construct its own image of the world, which is inevitably an approximation and idealization of what is really "out there." Stephen Hawking has written that "mental concepts are the only reality we can know." There is no model-independent test of reality.... Bohr, influenced by Kant's ideas, went further.

> He said that the world revealed by experience—which is to say, measurements—is the *only* reality worthy of the name.[21]

Like the glasses I wear, the experimental apparatus through which measures are made mediates or filters experience by posing questions that phenomena answer. Unlike Kant, but in a manner similar to Hegel, this exchange leaves neither questioner/observer nor respondent/observed unchanged. In this way the relationship between the two is reciprocal—each transforms the other.

To illustrate the way registering apparatuses mediates experience, it is instructive to consider the well-known double-slit experiment. Long before quantum theory, British polymath Thomas Young developed this experiment to demonstrate that light and matter display wave properties (1809). Over a century later, Louis de Broglie extended this analysis to theorize that electrons and all quantum particles also have wave-like properties. Between 1924 and 1927 Clinton Davisson and Lester Germer conducted a series of experiments demonstrating that quantum particles do, indeed, have wave-like properties. They detected wave-like patterns in beams of electrons emitted from a heated metal electrode.

In the double-slit experiment a beam of particles, which can be electrons or photons, is manipulated in such a way that only one particle passes through the slots at a time. It is important to note that while this experiment is designed to investigate quantum effects, the apparatus and the procedures can be described only in classical terms. In addition to this, the experiment conforms to classical principles by isolating the "objects" studied from their milieu and setting the objects investigated apart from the investigator. In spite of these restrictions, the results of the experiment cannot be explained in classical terms.

> The electrons are detected one particle at a time, but over time, the pattern that builds up where they are striking proves to be a series of parallel bands in which a high density hits alternates with low density.... We can't explain this result in terms of particles, but only in terms of electron waves.... It is hard to understand how one-by-one passage of what seem to be particles (judging from the discrete bright spots that appear on the screen) can produce wave-like interference. We are forced to conclude that "wave-like electrons" can interfere with *themselves*. But that requires us to believe that each individual electron passes through *both slits*—for there must be two sources of electron waves on the far side if there's to be

interference. What's going on? Why should the electron act like a particle before and after encountering the slits but become a spread-out wave as it passes through them?[22]

With this puzzle, we return to Stoppard's play of duplicity and deceit—an individual "intelligence agent" is observed entering one door of the locker, and someone appearing to be the same agent appears to come out the other door. What goes on in between remains invisible and, thus, is unknowable to the observer. If the registration device is turned off, something equally if not more surprising occurs—the fuzziness caused by interfering waves reappears.

The conclusion seems both unavoidable and implausible—the act of observing (measuring) determines what is observed (measured). In other words, experience is always mediated through a variety of apparatuses, and we can never take off our glasses and see the world as it is in itself. Bohr summarizes the fundamental principle of the Copenhagen interpretation of quantum mechanics: "There is no quantum world. There is only an abstract quantum description. It is wrong to think that the task of physics is to find out how nature *is*. Physics concerns only what we can *say* about nature."[23]

Once again, purported answers raise more questions. Are the claims of Bohr, Heisenberg, and their colleagues epistemological, ontological, or both? To begin to respond to this question, it is necessary to consider two additional aspects of quantum theory—superposition and entanglement. These puzzling theories cannot be understood in classical terms. As the double-split experiment demonstrates, quantum "objects" are not discrete and cannot be separated from each other, and the "object" observed is not separate from the observer.

Inasmuch as quantum phenomena appear to be wave-like, they are not sharply delineated and separated but are fuzzy. Ball offers a helpful description of the difference between classical and quantum "objects."

> We might then be inclined to point to features that classical objects like coffee cups have but which quantum objects don't necessarily have: well-defined positions and velocities, say, or characteristics that are localized on the object itself and not spread out mysteriously through space. Or we might say that the classical world is defined by certainties—either *this* or *that*—while the quantum world is (until a classical measurement impinges on it) no more than a tapestry of probabilities, with individual measurement outcomes determined

5.2 Classical and quantum "objects" and "subjects"

by chance. At the root of the distinction though lies the fact that quantum objects have a wave nature—which is to say . . . that they should be described *as if* they were waves, albeit waves of a peculiar, abstract sort that are indications only of probabilities.

It is this waviness that gives rise to the distinctly quantum phenomena, like interference, superposition and entanglement. These behaviors become possible when there is a well-defined relationship between quantum "waves"; in effect when they are in step. This co-ordination is called *coherence*.[24]

Contrary to classical physics, for quantum mechanics, "objects" are not separate from their environment but are integrally related to it; rather than independent, "objects" are *interdependent*. To understand this shift in perspective, it is necessary to stop thinking of "objects" as things or quasi-things and to conceive of them as *relational happenings or events*. What appears at the macrolevel to be separate objects and separate subjects *emerge* from *reciprocal* events at the microlevel.

Three different concepts—superposition, entanglement, and coherence—are, well, completely entangled. "If an object's wave-like nature is split in two, then the two waves may coherently interfere with each other in such a way as to form a single state that is a superposition of the two states. This concept of superposition is famously represented by Schrödinger's cat, which is both dead and alive at the same time when it is in its coherent state inside a closed box." When coherence is lost (decoherence), quantum properties disappear, and objects conform to classical principles. In entanglement, by contrast, the states in a superposition are the shared states of two entangled particles rather than those of two split waves of a single particle. "The intrigue of entanglement lies in the fact that the two entangled particles are so intimately correlated that a measurement of one particle instantly affects the other particle, even when separated by a large distance."[25]

To understand this argument, let us return to Heisenberg's three steps required to interpret an experiment:

1. The translation of the initial experimental situation into a probability function.
2. The following up of this function in the course of time.
3. The statement of a new measure to be made of the system, the result of which can be calculated from the probability function.

The first and third steps involve actualities and their measurement; what occurs between steps 1 and 2 cannot be observed. In Heisenberg's words, "there is no description of what happens to the system between the initial observation and the next measurement." What lies between Actuality 1 and Actuality 2 is something like a "cloud of possibilities." "It is only with the third step," he continues, "that we change over again from the 'possible' to the 'actual.'" The spacing and timing between is the interstitial Neither/Nor of becoming. The transition between Actuality 1 and Actuality 2 is a quantum leap in which possibility is actualized. This reduction of possibility to actuality involves something like a two-way communication process that enables a reciprocal "decision," which leads to the event-ual emergence of codependent particles that form an identity-in-difference and difference-in-identity. This actualization of possibility is *neither* completely determined *nor* completely indeterminate; rather, it is probabilistic. This Neither/Nor is the spacing-timing and timing-spacing of the aleatory. Probability gives chance a chance and, thus, keeps the future open. While reality is not totally chaotic, God does, to a certain extent, let the universe play dice with itself. While not all things are possible, not everything is determined. Interrelated actualities and possibilities close some doors (or slits) and open others.

With quantum entanglement, the weird gets even weirder. "Quantum superpositions," Ball explains, aren't fragile. "On the contrary, they are highly *contagious* and apt to spread out rapidly.... If a quantum system in a superposed state interacts with another particle, the two become linked into a composite superposition. That... is exactly what entanglement is: a superposed state of two particles, whose interaction has turned them into a single entity."[26] Two-in-one, one-in-two. This relativity does not presuppose spatial and temporal proximity—entanglement delocalizes interrelationships to create a global, indeed, a *cosmic* communications network. Two quantum particles on opposite sides of the world, or even on earth and in a distant galaxy, are entangled in such a way that the measurement (observation) of one (for example, its spin) *instantly* yields information about the other.[27] Nietzsche's prescient aphorism helps to untangle this complex interrelationship. "The properties of a thing are effects on other 'things': if one removes other 'things,' then a thing has no properties, that is, there is no thing without other things, that is, there is no 'thing-in-itself.'"[28]

Einstein was troubled by the implications of quantum mechanics and never completely accepted its conclusions. He expressed his suspicions by dubbing entanglement "spooky action at a distance." Einstein formalized his misgiving in an article he coauthored with Nathan Rosen and Boris Podolsky, "Can Quantum Mechanical Description of Physical Reality Be Complete?" (1935). Their answer is a resounding "No!" In a thought experiment, they imagined two particles separated by a significant distance and argued that an action taken on one particle could not *instantly* provide information on the other particle, as quantum mechanics theorized, because this would require information to be transmitted faster than the speed of light, which is impossible according to the theory of relativity. To complete quantum theory, Einstein and his colleagues posited a "hidden variable." If such a variable could be identified, it would resolve Einstein's other reservation about quantum mechanics—probability. Rather than the completion of quantum theory, such an explanation would be its negation.

The article by Einstein and his colleagues was uncharacteristically obtuse, and the issue remained unresolved until 1951, when David Bohm carried out their thought experiment. By this time, much more was known about photons, and it was even possible to manipulate both their spin directions and their polarizations. Bohm imagined two particles moving in opposite directions, each with unknown properties. "Once they have traveled for some time, we measure the respective property (polarization, spin) of one of them. We don't know what we'll measure until we measure it—but once we know the outcome, we can be sure also that the other particle has the opposite value." To understand the significance of this conclusion, it is necessary to recall Bohr's and Heisenberg's discovery that quantum phenomena do not exist apart from their interaction with the measuring device and the observer. This means that the spin direction and polarization *do not exist prior to their measurement*. If knowledge of particle A is knowledge of particle B, no matter how great the distance or time separating them, then information must be transmitted instantaneously, that is, greater than the speed of light.

This updated version of the thought experiment seemed to prove the truth of entanglement, but Bohm resisted his own conclusion. In the era of the Cold War, his credibility was not enhanced by his Marxist sympathies, which led to an itinerant academic career across several continents. By the time that he wrote *Wholeness and the Implicate Order* (1980), the hidden variable that was supposed to complete quantum theory had become the undifferentiated and, therefore, unknowable origin of both mind and matter.

> Intelligence and material process have thus a single origin, which is ultimately the unknown totality of universal flux. In a certain sense, this implies that

what have been commonly called mind and matter are abstractions from the universal flux, and that both are to be regarded as different and relatively autonomous orders within the one whole movement.... It is thought responding to intelligent perception which can bring about an overall harmony.... If the thing and the thought about it have their ground in one undefinable and unknown totality of flux, then the attempt to explain their relationship by supposing that thought is in reflective correspondence with the thing has no meaning, for both thought and thing are forms abstracted from the total process. The reason why these forms are related could only be in the ground from which they arise, but there can be no way of discussing reflective correspondence in this ground, because reflective correspondence implies knowledge, while the ground is beyond what can be assimilated in the content of knowledge.[29]

Bohm's implicate order is nothing other than what the ontotheological tradition has long described as the ground of being, which lies beneath or behind all becoming. Rather than completing quantum theory, this hidden variable negates it.

The tension between quantum mechanics' explanation of entanglement and nonlocality and the notion of an underlying hidden variable was unresolved for more than a decade. Einstein remained committed to an understanding of causality that quantum mechanics regarded as limited to the macrolevel until his death in 1955. In 1964 Irish physicist John Bell devised an experiment to test the difference between the predictions of the two theories. Every time he ran the experiment, he found that the only way a hidden variable could confirm the predictions of quantum mechanics is if they were nonlocal and, thus, the particles interact instantaneously. With this confirmation of quantum entanglement and nonlocality, the hidden variable theory became untenable.

RELATIONAL QUANTUM MECHANICS

While his genius as a mathematician and physicist is beyond doubt, Einstein's knowledge of theology was minimal. Nevertheless, his resistance to quantum theory rests on a religious conviction he expresses scientifically, or a scientific conviction he expresses religiously in his well-known claim that "God does not play dice with the universe." The point of this assertion is his rejection of the probability and, thus, the uncertainty involved in quantum relations. Though obviously intended metaphorically, Einstein's declaration presupposes the same type of theistic God that is widespread among believers and governs Newton's deterministic universe. There are, of course, other gods as well as other interpretations of the Judeo-Christian God that are not necessarily at odds with the metaphysic implicit in quantum physics. On July 16, 1945, Robert Oppenheimer observed the detonation of an atom bomb at the Trinity Test Site in New Mexico. Responding to the unprecedented destructive force he and his team of scientists had unleashed upon the earth, Oppenheimer, who had studied Sanskrit, turned to the Bhagavad Gita rather than the Hebrew bible. "Now I have become Death, the destroyer of worlds." Kali is only one of the alternative masks of what many call God. Carlo Rovelli comes closest to the religious or spiritual implications of quantum theory when he suggests its similarities to Mahayana Buddhism. "The quantum world is more tenuous than the one imagined by the old physics; it is made up of happenings, discontinuous events, without permanence. It is a fine texture, intricate and fragile as Venetian lace.... Events are punctiform, discontinuous, probabilistic, relative." Rovelli draws on Anthony Aguirre's *Cosmological Koans* to explain the implications of his insight. "We break things down not into smaller and smaller pieces, but then the pieces, when examined, are not there. Just the arrangements are. What then, are *things*, like the boat, or its sails, or your fingernails? What *are* they? If things are forms of forms of forms of forms, and if forms are order, and order is defined by us...they exist, it would appear, only as created by, and in relation to, us and the Universe. They are, the Buddha might say, emptiness."[30] In a relational world, every thing is no thing—no thing in itself or by itself—but is only something in and through its relations with other things.

While Einstein remains drawn to the certainty of Newton's deterministic universe and resists the uncertainty of quantum probability, Rovelli is convinced that Einstein's criticism of Newtonian space and time points toward the possibility of reconciling relativity and entanglement. As we have seen, in contrast to Newton, for whom space and time are absolute and independent of the things that "fill" them, Einstein's spacetime is inseparable from the interrelations of determinate spatiotemporal entities. The key to Rovelli's argument is his thoroughgoing *relationalism*. Entanglement, he argues, is "the web of relations that weaves reality." The Neither/Nor of differentiating relationality mediates the Either/Or of classical physics and the Both/And of quantum physics. "The world that we observe," Rovelli argues, is continuously interacting. It is a dense web of *interactions*. Individual objects *are* the way in which they interact. If there was an object that had no interactions, no effect upon anything, emitted no light, attracted nothing and repelled nothing, was not touched and had no smell . . . it would be as good as nonexistent. To speak of objects that never interact is to speak of something—even if it existed—that could not concern us. It is not even clear what it would mean to say that such objects "exist." The world that we know, that relates to us, that interests us, what we call "reality," is the vast web of interacting entities, of which we are a part, that manifest themselves by interacting with each other. It is with this web that we are dealing.

Understanding the radical implications of quantum mechanics requires a fundamental shift in perspective from interpreting entities (both objects and subjects) as separate and stable to interpreting them as coemergent and codependent. "If we think in terms of processes, events, in terms of *relative* properties, of a world of relations, the hiatus between physical phenomena and mental phenomena is much less dramatic. It becomes possible to see both as natural phenomena generated by complex structures of interaction."[31]

This interpretation of quantum mechanics involves a significant expansion of the position of Bohr and the Copenhagen school beyond the atomic and subatomic level to all interacting systems. In a seminal paper entitled "Relational Quantum Mechanics," published in 2004 and revised in 2019, Rovelli explains that Relational Quantum Mechanics (RQM) "is essentially a refinement of the textbook 'Copenhagen' interpretation, where the role of the Copenhagen observer is not limited to the classical world but can instead be assumed by any physical system."

This position de-anthropomorphizes the role of the observer by arguing that any system—even a physical system—can play the role of observer. This absolutely essential point deserves to be quoted in full:

> In textbook presentations, quantum mechanics is about measurement outcomes performed when an "observer" makes a "measurement" of a quantum system. What is an observer, if all physical systems are quantum? What counts as a measurement? Common answers invoke the observer being macroscopic, onset of decoherence, irreversibility, registration of information, or similar. RQM does not utilize anything of the sort. Any system, irrespective of its size and complexity, can play the role of the textbook's quantum mechanical

5.3 Relational systems

observer. The "measured outcomes" of a given observer, however, refer only to values of variables of the quantum system relative to that system. In particular, they do not affect events relative to other systems. In the same spirit, textbook presentations of quantum refer to "mechanics outcomes." In the relational interpretation, any interaction counts as a measurement, to the extent one system affects the other and this influence depends on a variable of the first system. Every physical object can be taken as defining as *perspective*, to which all values of physical quantities can be referred.[32]

A system—any system—never is itself by itself but *becomes* itself in and through its interplay with other systems. This means that every system from microlevel to macrolevel is *contextual*. Furthermore, systems are never merely passive in relation to other systems. Interactions, therefore, are always multidimensional.

The relational reading of quantum mechanics has both ontological and epistemological implications. Ontologically, Rovelli's quantum theory confirms Nietzsche's claim that "the properties of a thing are effects on other 'things.'"

> Revised in this way, Bohr's observation captures the discovery that forms the basis of the theory: the impossibility of separating the properties of an object from the interactions in which these properties manifest themselves and the objects to which they are manifested. The properties of an object *are* the way in which it acts upon other objects; reality is the web of interactions. Instead of seeing the physical world as a collection of objects with definite properties, quantum theory invites us to see the physical world as a net of relations. Objects are nodes.... [T]*here are no properties outside of interactions*.[33]

Rovelli does not limit constitutive relationality to two or three systems but extends it to all reality. In "the web of relations that weaves reality," every thing and every body is related.

In a relational world, subjects cannot be separated from objects, and, therefore, ontology and epistemology are inextricably interrelated. As I have argued, a relational ontology entails a relativistic epistemology. While Nietzsche insists that "there are no facts, only interpretations," Rovelli claims, "Facts are relative.... Facts relative to one observer are not facts relative to another. It is a shining example of the relativity of reality." Nietzsche and Rovelli agree that since becoming is relational, knowledge is necessarily perspectival. Perspective is not merely a human

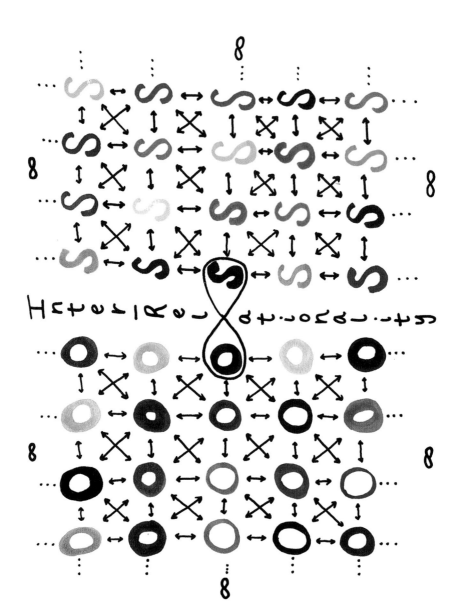

5.4 Interrelationality

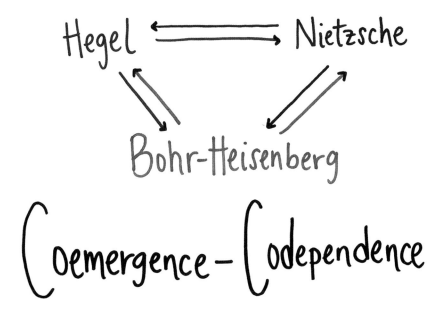

5.5 Coemergence-codependence

phenomenon—every particle, every thing *is* a perspectival response to another particle or thing that is responding to it. Once again, there is no Archimedean point from which to view the universe, so our knowledge remains incomplete. Though entanglement is global, our vision is not. "The world," Rovelli explains, "fractures into a play of points of view that do not admit of a univocal, global vision. It is a world of perspectives, of manifestations, not of entities with definite properties or unique facts. Properties do not reside in objects, they are bridges between objects. Objects are such only with respect to other objects, they are nodes where bridges meet. The world is a perspectival play of mirrors that exist only as reflections of and in each other."[34]

This *play* of mirrors reflects both Hegel's speculative (*speculum*, mirror) system and Nietzsche's network of *seemingly* fragmentary aphorisms. If becoming as well as knowledge is perspectival or relative, then everything changes from context to context. Hegel is different because of his relation to Nietzsche, and vice versa; and Hegel-Nietzsche are different because of their relation to Bohr-Heisenberg, and vice versa.

If meaning, like becoming, is relative and hence contextual, then the meaning of a passage from Nietzsche's *Will to Power* cited in chapter 2 will have a different meaning in this context.

> The perspective therefore decides the character of the "appearance"! As if a world would still remain after one deducted the perspective! By doing that one would deduct relativity!
>
> Every center of force adopts a perspective toward the entire remainder, that is, its own particular valuation, mode of action, and mode of resistance. The "apparent world," therefore, is reduced to a specific mode of action on the world, emanating from a center.
>
> Now there is no other mode of action whatever; and "world" is only a word for the totality of these actions. Reality consists precisely in this particular action and reaction of every individual part toward the whole.[35]

In this context, Nietzsche effectively bridges his perspectivism, Hegel's relationalism, and Rovelli's relational quantum mechanics in a way that mediates Einstein's theory of relativity and the Copenhagen interpretation of quantum mechanics.

In this expanded version of general relativity theory and quantum mechanics, the opposition between mind and matter, self and world, and subject and object becomes fuzzy without dissolving into an undifferentiated flux where differences are lost in identity. Rovelli unexpectedly cites Bertrand Russell to support his point. "The raw material out of which the world is built up is not of two sorts, one matter and the other mind; it is simply arranged in different patterns by its inter-relations: some arrangements may be called mental, while others may be called physical."[36] When interpreted relationally, quantum theory is *neither* idealism, which reduces matter to mind, *nor* materialism, which reduces mind to matter. As such, it is *neither* monistic *nor* dualistic. Descartes and all those who have followed him with devastating effects still tearing the world apart were wrong—mind and matter are not separate and opposed but are inextricably interrelated in webs of intersecting nodes.

The certainty for which Descartes longed is impossible in this relational world. Far from provoking distress, this insight is liberating—the only thing more dreadful than the uncertainty of the future is its certainty. The probabilistic universe keeps the future open without falling into a chaotic abyss. Rovelli's relational quantum mechanics puts mind back into nature and nature back into mind to reveal a world in which the alienation of the human from the natural is not inevitable. In his most concise formulation of his deepest conviction, Rovelli writes, "I believe, and the keystone of the ideas in this book [*Helgoland*] is the simple observation that scientists, and their measuring instruments as well, are part of nature. What quantum theory describes, then, is *the way in which one part of nature manifests itself to any other single part of nature.*"[37] Alternatively expressed in Hegelian terms, our knowledge of the world is the world's knowledge of itself. Rovelli turns east rather than west to express his vision. "The perspective offered by Nagarjuna may perhaps make it a little easier to think about the quantum world. If nothing exists by itself, everything exists only through dependence on something else, in relation to something else. The technical term used by Nagarjuna to describe the absence of independent existence is 'emptiness' (*sunyata*): things are 'empty' in the sense of having no autonomous existence. They exist thanks to, as a function of, with respect to, in the perspective of, everything else."[38]

PARTICIPATORY COMMUNICATION

Whatever progress I have been able to make in untangling quantum entanglement has been the result of the patient assistance of my friend and colleague William

Wootters. Bill is not only a superb theoretical physicist, who is regularly mentioned as being on the shortlist for the Nobel Prize in physics, but is also an outstanding teacher who has the rare ability to explain baffling concepts without incomprehensible mathematical formulas. He introduced me to Rovelli's work as well as the work of his own teacher, John Wheeler. Bill's primary contribution has been in an area known as quantum information theory. This esoteric field has very important "real" world applications ranging from quantum computing to cryptography. Wheeler has gone so far as to claim, "No one has done more than William Wootters toward opening up a pathway from information to quantum theory."[39]

Wheeler studied under Bohr and participated in the Manhattan Project. His pioneering work for which he is best known is in quantum information theory. This interdisciplinary field includes quantum mechanics, computer science, information theory, and cryptography. As we will see in the next chapter, information can be interpreted in many different ways. In this context, it is important to understand that information is not limited to mental activity or its simulation in machines but, like Hegelian *Geist*, is distributed throughout all natural and sociocultural processes. Never fixed, static, or stable, quantum information is always in-formation.

Information, as Bateson claimed, is "a difference that makes a difference." The differences that constitute information assume a variety of patterns in physical, chemical, biological, social, political, economic, and cultural processes. In quantum information theory, information is encoded in the physical properties of entangled events. The basic unit of information is called a quantum bit, which is abbreviated as qubit. This term was coined by Benjamin Schumacher in a 1995 conversation with Wootters. Qubits are variations of the on-off switches used in classical computers. In both cases, a bit is the basic unit of information. Classical bits represent different states separated by a spatiotemporal gap. This structure has two important consequences: first, there is a slight temporal delay when switching between on and off; and second, because the states are separate, the measurement of one does not influence the other. Qubits, by contrast, are in a state of coherent superposition and thus are inextricably interrelated. Consequently, the measurement of one state instantaneously yields information about the other state. This means that while bits carry one unit of information, qubits carry two.

The peculiar character of qubits has numerous practical applications, ranging from super high-speed computers to ultrasensitive sensors used in everything from medical devices to surveillance and communications systems. However, the most important application of quantum information theory is cryptography. The intricate interrelation of quantum particles at the microlevel means that they are extraordinarily sensitive to any disturbance or intervention, so much so that observation changes what is observed. It is important to recall that particles are events, and that not only human beings but any physical system can function as an observer. The constitutive relationality of quantum systems means that they cannot be observed without alteration and hence detection and cannot be cloned or copied. This is the basis for the cryptography that is essential for today's communications and surveillance systems as well as financial markets and digital currencies like Bitcoin.

Wheeler once summarized the course of his long and influential career that led to his most important insights in three stages: (1) everything is particles; (2) everything is fields; (3) everything is information.[40] He regards his final position as the third stage in the history of physics:

Era I   Motion with no explanation of motion—the parabola of Galileo and the ellipse of Kepler
Era II   Law with no explanation of Law: Newton's laws of motion, Maxwell's electrodynamics, Einstein's geometrodynamics
Era III   Information-based physics[41]

In an article entitled "World as Self-Synthesized by Quantum Networking" (1988), Wheeler describes the world as "inter-communicating systems . . . based on quantum-plus-information theory." This "system of shared experiences, which we call the world," he argues, "is viewed *as building itself* out of elementary quantum phenomena, elementary acts of 'observer-participancy.' In other words, the questions that the participants put—and the answers they get—by their observing devices, plus their communication of their findings, take part in creating the impressions, which we call the system: the whole great system which to a superficial look is time and space, particles, and fields. That system in turn gives rise to the observer-participants."[42] In a strange loop that by now should be familiar, the world gives rise to the participant-observers who give rise to the world.

Wheeler had a unique talent for coining cryptic phrases to summarize his basic points. Two of his most famous are "It from bit" and "Theory is reality." These two formulations are actually different ways of making the same point.

According to Wheeler, "It from bit symbolizes the idea that every item of the physical world has at bottom—at a very deep bottom, in most instances—an immaterial source and explanation; that what we call reality arises in the last analysis from the posing of yes-no questions and the registering of equipment-evoked responses; in short, that all things physical are information-theoretic in origin and this is a participatory universe."[43] Wheeler insists that his information-theoretic model represents the logical outworking of Bohr's interpretation of quantum mechanics.

> The quantum, strangest feature of this strange universe, cracks the armor that conceals the secret of existence. In contrast to the view of the universe as a machine governed by some magic equation here the view of the world as a *self-synthesizing system* of existences, built on *observer participancy* via a *network of elementary quantum phenomena*. The elementary quantum phenomenon in the sense of Bohr, the elementary act of observer-participancy, develops the definiteness out of indeterminism, secures a communicable reply in response to a well-defined question.[44]

Three phrases in this summary require clarification: self-synthesizing system of existences, observer-participation, and network of elementary quantum phenomena.

Wheeler's notion of a self-synthesizing system of existences recalls the self-organizing, autotelic, and autopoietic systems identified by Kant and elaborated by Hegel. Organization initially *emerges* from the interrelation of parts, which constitute the whole, and the whole, which constitutes the parts. Rather than being preprogrammed or programmed by an "external" agent, self-organizing systems involve "a plan without a plan." So understood, these self-synthesizing systems form networks of networks in which quantum phenomena are coemergent and codependent events. This network of networks forms what Rovelli describes as "the web of relations that weave reality." This leads to the third and most important characteristic of self-synthesizing systems—observer-participant. For many critics, Bohr and Heisenberg's theory marks a significant step down the slippery slope of solipsistic subjectivism. Wheeler's precise formulation suggests an alternative interpretation of quantum mechanics. In a very important clarification of observer-participant, Wheeler writes:

> Observer-participant is the central feature of the world of quantum. We used to think of the electron in the atom as having a position and a momentum

whether we observed it or not, as I thought the word already existed in the room whether I guessed it or not. But the world did not exist in the room ahead of time, and an electron in the atom does not have a position or a momentum until an experiment is conducted to determine one or the other quantity. The questions I asked had an irretrievable part in bringing about the word that I found *but I did not have the whole voice*. The determination of the word lay in part with my friends. They played the role that nature does in the typical experiment, where so often the outcome is uncertain, whether with electron or with photon. In brief, complementarity symbolizes the necessity to choose a question before we can expect an answer.... We once thought, with Einstein, that nature exists "out there," independent of us. Then we discovered—thanks to Bohr and Heisenberg—that it does not.[45]

The qualification "but I did not have the whole voice" is very important because it preserves a factor in the participatory universe that is not merely subjective. Wheeler's position is consistent with Bohr's repeated resistance to a subjectivist interpretation of quantum theory. Karen Barad argues persuasively that a pivotal point in Bohr's

> analysis is that the physical apparatus, embodying a particular concept to the exclusion of others, marks the subject-object distinction: the physical and conceptual apparatuses form a nondualistic whole marking the subject-object boundary. In other words, concepts obtain their meaning in relation to a particular physical apparatus, which marks the placement of a Bohrian cut between the object and the agencies of observation, resolving the semantic-ontic indeterminacy. This revolution of the semantic-ontic indeterminacy provides the condition for the possibility of objectivity.[46]

To appreciate the importance of this insight, it is necessary to remember that any system—physical or otherwise—can function as an observer-participant.

Wheeler describes his position as *participatory realism*. Obviously, participation requires at least two. Just as self-relation (identity) is impossible apart from relation-to-other (difference), so participation requires two or more participants. Wheeler's participatory universe offers a further twist of Rovelli's relational universe. In a relational world, all participation is coparticipation, and in a participatory universe, all relation is interrelation. Not one, not two, but one-in-two and two-in-one simultaneously joined and separated in and through the ceaseless oscillation and alternation of differentiating relation. Wheeler and Rovelli, following Bohr and Heisenberg, insist that nature does not exist "'out there' independent of us." We are part of nature, and nature is part of us. Nothing could be farther from Descartes's mechanical universe in which matter and mind are alienated from each other. *Neither* objective *nor* subjective but some-thing or no-thing in between.

Wheeler, like Rovelli, has an expansive vision that extends physical systems to biological life, and beyond to culture and technology.[47] Not only are all these systems networks, but they are all interrelated with each other. In entangled networks, the fundamental question once again comes down to the interrelation between time and the self.

> The heart of the matter is the word *self*. What is to be understood by the word self we are perhaps beginning to understand today as well as some of the ancients did. We know that in the last analysis there is no such thing as self. There is not a word we speak, a concept we use, a thought we think which does not arise, directly or indirectly, from our membership in the larger community. On that community the mind is as dependent as the computer. A computer with no programming is no computer. A mind with no programming is no mind. Impressive as is the greatest computer program that man has ever written and run, that program is nothing as compared to the programming by parents and the community that makes a mind a mind.
>
> The heart of mind is programming, and the heart of programming is communication. In no respect does the observer-participancy view of the world separate itself more sharply from the universe-as-machine than in its information on information transfer.[48]

Dynamic interrelated physical, chemical, biological, social, political, economic, and technological systems are communication networks whose coemergence and codependence weave the worldwide web in which we are all inescapably entangled.

CHAPTER 6

INFORMATION IN FORMATION

But maybe everything is information . . . Maybe that's what it's all about.

—Karl Ove Knausgaard

HOW THIS BOOK IS BEING WRITTEN

My study is in a refurbished barn with both too much space and too little space for books. My library overflows the barn and spills into multiple rooms in the house, and beyond to my apartment and office in New York City. While I have never counted the number of books I own, I would guess it is somewhere between eight and nine thousand. The first question people ask when visiting my barn is, "Have you read all these books?" My answer is always the same, "Yes, I decided many years ago not to keep any book I had not read." The books surrounding me tell the story of my life, which is why I find it so difficult to get rid of even a single one. In addition to my books, I have notebooks for every course I took as an undergraduate and graduate student and every course I have taught during the past half century. I also have file boxes with notes, outlines, and drafts for every book I have written. In addition to all this, there are various models, drawings, and designs for past and future art works. It gets crowded in here, but nowhere do I feel more at home. I have not counted the number of books and articles I've read in preparation for writing this book, but I am sure it is well over one hundred books plus countless articles. Since I have run out of shelf space, I have resorted to stacking books in the loft and on the floor.

As I ponder what I have been doing all these years, sitting at my desk in the barn, I often imagine myself *participating* in a conversation that was going on long before I arrived and hopefully will continue long after I die. Those who have come before me I call my ghosts, and I imagine myself ghosting my students and people in the future who might read the books I have written. Never one way, conversation is always a multidirectional dialogue of question or comment and response. Writers and artists are always haunted because creativity is never *de novo* but is always in conversation with ghosts both known and unknown. Every course I have taught and every book I have written *emerges* from and contributes to this ongoing conversation by developing a vision that becomes clear only as the end approaches. This emerging vision represents an *interpretative schema* through which I *filter the data* I have gathered from my reading and discussions and *process data to create information*. Over time, these schemata have become *increasingly complex* with *many intersecting and internally differentiated codependent parts*. Different schemata function as something like mini-networks within a more comprehensive web and interact with each other in unexpected ways. As they accumulate and diversify, these local networks form a soup that must simmer. When this soup reaches a certain density and diversity, a new schema *emerges*, which makes new *concepts* and *ideas* possible. This is not a conscious process and cannot be planned or programed; it is an unpredictable, *spontaneous event*. Often the best way to think about what you want to think about is, paradoxically, not to think about it.

The boundaries joining and separating interpretative schemata are porous, thereby allowing mutual modification through reciprocal exchange. What is commonly called "mind" is nothing more than the interaction of systems, subsystems, and schemata—there is no "I" apart from this dynamic relational process. Rather than the I operating as something like a transcendental subject, schemata *self-organize* in relation to each other as well as in conversation with the surrounding world. Just as the boundaries between different schemata are porous, so the boundary between "internal" subsystems and "external" systems is semipermeable. Quantum theory, relativity theory, information theory, cognition, bioinformatics, plant and animal intelligence, "artificial" "intelligence"—each is subject to multiple interpretations and, thus, requires interpretative schemata to filter and organize often conflicting data into intelligible information. Furthermore, if, as Hegel insisted, truth is the whole and, therefore, relative or relational, then none of these areas can be understood apart from the others.

The more I read, the more information I accumulate, and the more overwhelmed I often become. Eventually, *something clicks* and suddenly I "see" the book as a whole. This happens unexpectedly, and I have no control over if or when it occurs. I did not plan to write these words this morning; indeed, having been suffering from Covid for the past week, I was not sure if the brain fog would have lifted enough for me to read or write anything. But as so often happens, when I settled at my desk surrounded by books and unorganized notes, suddenly these words were *there*, wherever *there* might be, waiting for me. It has been happening this way for years, and I am still struggling to understand how "it" happens. As I reread these words, I hear not only the writers I've been "talking" with for the past year or so, but also the voices of ghosts from my distant past. Every time I hear these voices, they are different because their words are filtered through what I have learned since the last time we spoke. It is as if their works create my work, which, in turn, recreates their works in what Maurice Blanchot aptly calls an "infinite conversation."

IN(-)FORMATION

Few words are more important and more misunderstood today than "information." Information society, information revolution, information technology, information economy, information superhighway, information sciences, information networks, information security, information warfare, information overload, information theory. It is not an exaggeration to insist that information defines the contemporary era. But what *is* information? The *Oxford English Dictionary* defines "information" as "the action of informing; formation or molding of the mind or character, training, instruction, teaching, communication of instructive knowledge." According to this definition, "information" is a verb rather than a noun and, thus, is an activity instead of a state. This activity involves training, instructing, communicating knowledge. The association of information with knowledge raises further questions. What is knowledge, and how is it related to information? Is all information knowledge? Can there be information without knowledge? How does information become knowledge? Does knowledge provide information? What is the relation of information, knowledge, and mind? Does information form mind and constitute knowledge as the OED claims, or does mind form information, which issues in knowledge?

The history of the word provides clues to its meaning. "Information" derives from the Latin *informationem*, meaning outline, concept, idea, which, in turn, can be traced to the verb *informare*, to train, instruct, educate, shape, give form to. "Information" is best understood as the *process* of giving form or in-forming. More precisely, information is the transformation of data, which are given, into schemata that filter noise to create the concepts and ideas that *form* knowledge. Just as apparently stable entities are interrelated events, so information is neither static nor circumscribed but is an open-ended process that is always in formation. As such, information is related to but distinguished from noise, data, concept, idea, and meaning. Information processing is constitutive of all natural (physical, chemical, and biological), psychological, social, political, cultural, and technological systems and networks. With the recognition of the ubiquity of information processing, it becomes necessary to develop an expanded notion of cognition and related activities like perception, conception (present), memory (past), prediction (future), and adaptation-learning. The reconfiguration of this constellation of activities and processes leads to an expanded and more inclusive notion of mind.

Just as the scientific breakthroughs that led to quantum mechanics were inseparable from the invention of the atomic bomb, so the information revolution grew out of scientific and technological developments during World War II. In 1935 Alan Turing heard a lecture at Cambridge University in which M.H.A. Newman asked whether it is possible to create a mechanical process to determine whether a mathematical problem could be solved. Turing's effort to answer this question literally changed the world. In 1936 he developed what came to be known as the Turing Machine, which appears to be a deceptively simple device consisting of a long tape coded with abstract symbols that can be manipulated according to prescribed rules.

The Turing Machine is the prototype for today's computers. In a prescient article entitled "Intelligent Machinery" (1948), Turing explored the far-reaching implications of his revolutionary work. "I propose to investigate the question of whether it is possible for machinery to show intelligent behavior." He begins by arguing that the widespread resistance to considering the possibility of an intelligent machine is the result of the unjustifiable belief in human exceptionalism. "An unwillingness to admit the possibility that mankind can have any rivals in intellectual power. This occurs as much amongst intellectual people as amongst others: they have more to lose. Those who admit the possibility all agree that its realization would be very disagreeable. The same situation arises in connection

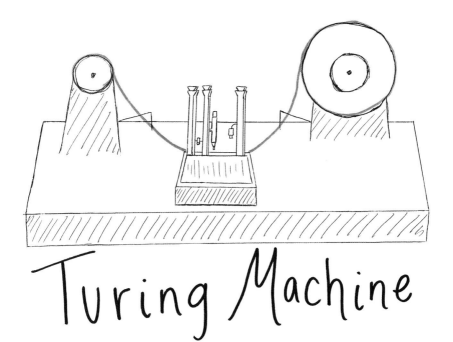

6.1 Turing Machine

with the possibility of our being superseded by other animal species. This is almost as disagreeable, and its theoretical possibility is indisputable." Turing was convinced that resistance to the development of machines that are intelligent is futile. Describing his machine, he wrote:

> In Turing (1937) a certain type of discrete machine was described. It had an unlimited memory capacity obtained in the form of an infinite tape marked out into squares, on each of which a symbol could be printed. At any moment there is one symbol in the machine; it is called the scanned symbol. The machine can alter the scanned symbol, and its behavior is in part determined by that symbol, but the symbols on the tape elsewhere do not affect the behavior of the machine. However, the tape can be moved back and forth through the machine, this being one of the elementary operations of the machine. Any symbol on the tape may therefore eventually have an innings.[1]

As the argument unfolds, it becomes clear that Turing's understanding of information, computation, and even intelligence is not limited to machines. Five years before James Watson and Francis Crick "cracked" the genetic code, Turing extended information processing and computation to biological and neurological processes. I will return to this issue in subsequent chapters; for the moment, it is necessary to understand that with the outbreak of the war, developing a functional Turing Machine became an urgent necessity. The decisive turning point in the research that eventually led to the computer was reached when Turing, working with John von Neumann, recognized the inherent limitations of mechanical devices and developed machines that performed logical calculations with electronic circuitry.

At the same time that Turing and von Neumann were creating computational machines, Norbert Wiener was developing a new area of research, named cybernetics, which was devoted to the study of regulatory processes in mechanical, electronic, and biological systems. As mechanical devices and systems proliferated during the Industrial Revolution, it became necessary to develop a way to maintain their stability by regulating their operation through negative feedback. During World War II, weapons systems became more complex and required more sophisticated control devices. The emergence of cybernetics in the 1930s and 1940s led to the use of information to regulate both mechanical and electronic machines. Servomechanisms are devices that detect errors in the operation of machines and automatically correct malfunctions in the determination of the position, speed, or altitude. In more elaborate applications, human beings and machines are joined in corrective feedback loops.

The common denominator in the work of Turing, von Neumann, and Wiener is the use of information in mechanical and electronic devices. After the war, a group of influential scientists met under the auspices of the Josiah Macy Foundation for more than a decade to develop a theory of information and communication that could be used in technologies that control both nonhuman and human systems. The same year that Turing published "Intelligent Machinery," Claude Shannon published a very important paper entitled "A Mathematical Theory of Communication." This essay included an introduction to Shannon's complex argument by Warren Weaver. Though Shannon formulated his theory to determine the requirements for the transmission of messages on telephone or telex lines, his analysis has played a decisive role in defining information and determining the conditions necessary for communication. The specific technical problems Shannon attempted to solve led him to interpret communication as the transmission of a message from an information source (sender) along a channel to a destination (receiver). Within this scheme, noise designates anything that interrupts the transfer of the signal.

> The fundamental problem of communication is that of reproducing at one point either exactly or approximately a message selected at another point. Frequently the messages have *meaning*; that is they refer to or are correlated according to some system within certain physical or conceptual entities. These *semantic aspects of communication are irrelevant* (emphasis added) to the engineering problem. The significant aspect is that the actual message is one *selected* from a set of possible messages. The system must be designed to operate for each possible selection, not just for the one which will actually be chosen since this is unknown at the time of design.[2]

Since Shannon brackets semantic issues, information does not involve meaning. Weaver explains, "The word *information*, in this theory, is used in a special sense that must not be confused with its ordinary usage. In particular, information must not be confused with meaning." It is also important to note that Shannon defines communication in terms of the *selection* from a set of *possible* messages. The most rudimentary form of information is the bit, which represents a choice between two alternatives (0 or 1). As possibilities for choice proliferate, the situation becomes more complex, and, thus, the amount of information communicated increases.

> This word information in communication theory relates not so much to what you *do* say as to what you *could* say. That is, information is a measure of one's freedom of choice when one selects a message. If one is confronted with a very elementary situation where he has to choose between two alternative messages, then it is arbitrarily said that the information associated with the situation is unity. Note that it is misleading (although often convenient) to say that one or the other message conveys unit information. The concept of information applies not to the individual messages (as the concept of meaning would), but rather to the situation as a whole, the unit of information indicating that in this situation one has an amount of freedom of choice, in selecting a message, which is convenient with regard as a standard or unit amount.[3]

Insofar as information is a measure of the freedom of choice within a set of possible messages, it is a function of probabilities. More precisely, in Shannon's theory, information is a quantitative measure of *improbability*.

Though at first glance the association of information with improbability seems puzzling, further reflection suggests that Shannon's theory is a mathematical formulation of a simple insight. As anyone who surfs the internet knows, information is news, and news must be new. When we read or hear what we expect or already know, little or no news is conveyed. Information involves what is unexpected and, thus, is related to improbability. In other words, *information is inversely proportional to probability: the more probable, the less information; the less probable, the more information.*

The definition of information in terms of improbability establishes its difference from redundancy. Redundancy functions as a constraint that increases the likelihood of the message's arrival at its destination. While such constraints can take different forms, they tend to function as rules, codes, or syntactic structures through which transmitted messages can be selected. These structures also serve as parameters of expectation for the reception of transmitted information. The probability at any stage of the generation of a message is related to previous decisions and events. Here, as in our consideration of Heidegger and quantum processes, it is important to note that decisions do not presuppose consciousness, self-consciousness, or intentionality. Patterns of previous decisions establish parameters of constraint that function as rules guiding but not determining present and future decisions. Possibilities among which decisions can be made are produced through a stochastic process. The more fixed the rule, the greater the redundancy and thus the less information is conveyed; the less fixed the rule, the less redundancy and the more information is transmitted. Insofar as it is inversely related to probability and directly related to improbability, information is inseparable from *uncertainty* or, more precisely, to the resolution of uncertainty. By reducing the number of ways in which various parts of a system can be arranged and thereby reducing the number of possible messages, redundancy increases certainty while at the same time decreasing information. For information to be conveyed, there must be *neither* too much *nor* too little redundancy. If everything is predictable, no information is conveyed; if there is no redundancy, which determines the parameters of possibility and probability, uncertainty cannot be resolved, and again no information is conveyed.

With information understood in terms of decision, stochastic processes, probability/improbability, and uncertainty, we return to quantum mechanics and quantum information theory. Before turning to the constellation of terms associated with information, it is necessary to understand an important assumption that Wiener, Shannon, and Weaver, though not von Neumann, share. As we have seen, Katherine Hayles argues that these three leading theorists presume "a conception of information as a (disembodied) entity that can flow between carbon-based organic components and silicon-based electronic components to make protein and silicon operate as a single system. When information loses its body, equating humans and computers is especially easy, for the materiality in which the thinking mind is instantiated appears incidental to its nature." Hayles sees in the most sophisticated versions of contemporary information theory a reinscription of classical binary oppositions like immaterial/material, form/matter, pattern/substance, and mind/body-brain that have defined the Western ontotheological tradition from Plato to Descartes. Furthermore, she takes this opposition to be definitive of posthumanism.

> What is posthumanism? . . . First, the posthuman view privileges informational pattern over material instantiation, so that embodiment in a biological substrate is seen as an accident of history rather than an inevitability of life. Second, the posthuman view considers consciousness regarded as the seat of human identity in the Western tradition long before Descartes thought he was

a mind thinking, as an epiphenomenon, as an evolutionary upstart trying to claim that it is the whole show when in actuality it is only a minor sideshow. Third, the posthuman view thinks of the body as the original prosthesis we all learn to manipulate, so that extending or replacing the body with other prostheses becomes a continuation of a process that began before we were born. Fourth, and most important, by these and other means, the posthuman view reconfigures human being so that it can be seamlessly articulated with intelligent machines.[4]

Hayles's concern is twofold: first, to reverse processes of dematerialization by recovering the materiality of the body, and second, to resist the integration of humans and machines. The difficulty with this line of argument is that it presupposes the dualism it intends to undo. The challenge is not to solicit the return of repressed materiality but to reconceive oppositions like immaterial/material, form/matter, and pattern/substance in a way that *neither* reduces the former to the latter *nor* reduces the latter to the former.

IN-FORMING DIFFERENCE

As we have seen, the foundational gesture of modern philosophy (Descartes) and science (Newton) is the disenchantment of nature through the elimination of mind from matter. Descartes translates the polarity of *res extensa* (body and world) and *res cogito* into the opposition between brain and mind. In an imaginary dialogue between Galileo and a Frenchman, who is a thinly disguised persona of Descartes, Giulio Tononi writes that for Descartes, "consciousness never expires, not even for a moment. What the brain does is immaterial; consciousness uses it to communicate with the body, and through the body with the world, but consciousness is a different substance and does not need the brain to exist.... The brain is an extended thing, a thing that occupies space and moves in it. And it is made of parts. But consciousness is not an extended thing—it is a thinking thing. And it's not made of parts but is a unity."[5] The internalization of dualistic opposition in terms of the brain and mind raises the question of how they are related. Here, as elsewhere, the way the question is framed precludes the possibility of a satisfactory answer. Once the mind/brain opposition is established, there are only three alternatives: reduce mind to brain (materialism), reduce brain to mind (idealism), or somehow cobble them together. While the first two positions are inverse forms of monism, the last position remains dualistic. Descartes attempted to solve the problem he created by arguing that the pineal gland, whose function still is not fully understood, mediates the mind-matter divide.

Descartes's way of posing the problem creates another difficulty that continues to plague philosophy—by reducing truth to self-certainty, he identifies mind with consciousness. While philosophers long avoided the question of consciousness, in recent decades the problem of the origin and function of consciousness has become a philosophical preoccupation, especially in the analytic tradition. David Chalmers, who is one of the leading figures in contemporary consciousness debates, has built his considerable reputation on renaming Descartes's dilemma "the hard problem." How can material processes give rise to the subjective experience of consciousness? The basic terms of the debate have not changed since the time of Descartes. With increasingly sophisticated technologies, neuroscientists are attempting to reduce mental activity to physiological processes (materialistic monism). Chalmers and his followers reject this position and insist that subjective experience (qualia) is irreducible to a material substrate. This position in effect reasserts a mind/matter dualism. More recently, there has been a surprising revival of the ancient tradition of panpsychism in which subjective experience and consciousness are extended to all entities and even the cosmos as a whole. This argument represents the inverse of material monism.[6]

None of these alternatives is satisfactory. First, the identification of the mind with consciousness extends Cartesian anthropocentrism, which, I have argued, is the philosophical foundation of the Anthropocene. While consciousness is undeniably a mental activity, we will see in the next section that the mind is more inclusive than consciousness. Second, the original opposition between brain and mind is, like every oppositional dualism, mistaken. Hegel's relationalism and Nietzsche's perspectivism demonstrate that dualistic opposition is self-negating because determinate identity is differential and, therefore, relative. Rovelli's relational interpretation of quantum mechanics and relativity theory combined with Wheeler's participatory physics, or, more precisely, metaphysics and quantum information theory, point to a third way that is *neither* monistic *nor* dualistic. Rovelli concisely summarizes these contrasting positions and points to a way out of the impasse between reductive monism and conflictual dualism:

> Ideas on the nature of the mind are often limited to just three alternatives: dualism, according to which the reality of the mind is completely different from that of inanimate things; idealism, according to which material reality

only exists in the mind; and naïve materialism, according to which all mental phenomena are reducible to the movement of matter. Dualism and idealism are incompatible with the discovery that we sentient beings are part of nature like any other, and with the overwhelming and ever-increasing evidence that nothing that we observe, including ourselves, violates the natural laws that we know. Naïve realism is intuitively difficult to reconcile with subjective experience.

But these are not the only alternatives. If the qualities of an object are born from the interaction with something else, then the distinction between the mental and physical phenomena fades considerably.... Subjectivity is not a qualitative leap with respect to physics: it requires growth in complexity ([Aleksandr] Bogdanov would say of "organization"), but always in a world that is made up of perspectives, already from the most elementary level.

Rovelli reconciles mind and matter by reconceiving both as interrelated information processes. "If we think in terms of processes, events, in terms of *relative* properties, of a world of relations, the hiatus between physical phenomena and mental phenomena is much less dramatic. It becomes possible to see both as natural phenomena generated by complex structures of interaction."[7]

With relative properties in a world of interrelations, we return to Bateson's definition of information as "a difference that makes a difference." Two points implicit in this definition must be rendered explicit. First, differences that are informative *make* a difference; that is to say, difference is not a static structure but is the dynamic activity of differentiating in which identity is established by simultaneously separating from and connecting with that which is different. Second, not all differences make a difference. Bateson explains, "There are differences between differences. Every *effective* [emphasis added] difference denotes a demarcation, a line of classification, and all classification is hierarchic. In other words, differences are themselves to be differentiated and classified." Though Bateson does not invoke the distinction between probability and improbability to classify differences, an effective way to understand the differences that make a difference is in terms of their improbability. Rather than redundant, the differences that *matter* are novel and, thus, unexpected and unpredictable. In the next section, we will see that such differences can be either creative or destructive.

Bateson formulates his definition of information in the context of developing a theory of mind. In contrast to Wiener and Shannon, for whom cybernetic mechanisms separated mind and matter, Bateson uses cybernetic theory to develop an expanded *ecology of mind* that effectively integrates mind and matter. "The mental world—the mind—the world of information processing is not limited by the skin." By identifying the mind with information processing, Bateson is able to extend mental functioning to natural, social processes, and, I would add, political, economic, cultural, and technological processes. "We get a picture, then, of the mind as synonymous with a cybernetic system—the relevant total information-processing, trial-and-error-completing unit. And we know that within Mind in the widest sense there will be a hierarchy of subsystems, any one of which we can call an individual mind.... I now localize something which I am calling 'Mind' immanent in the large biological system—the ecosystem. Or, if I draw the system boundaries at a different level, then the mind is immanent in the total evolutionary structure."[8] The structure of the mind is fractal—it is a porous system of subsystems, or network of networks in which interiority and exteriority are folded into each other in a way that subverts their opposition. When fully extended, interrelated systems form a global network that is a node in a cosmic web.

At this point in his argument, Bateson makes a move that appears to be completely unexpected—he shifts his argument to a theological register by identifying this cosmic web with God:

> The cybernetic epistemology which I have offered you would suggest a new approach. The individual mind is immanent but not only in the body. It is immanent also in pathways and messages outside the body; and there is a larger Mind of which the individual mind is only a subsystem. This larger Mind is comparable to God and is perhaps what some people mean by "God," but it is still immanent in the total interconnected social system and planetary ecology.
>
> Freudian psychology expanded the concept of mind inward to include the whole communication system within the body—the autonomic, the habitual, and the vast range of unconscious processes. What I am saying expands the mind outward. And both of these changes reduce the scope of the conscious self.[9]

Bateson was acutely aware of the practical and ethical implications of his expanded notion of mind; indeed, his entire effort in *Steps Toward an Ecology of Mind* is to develop a philosophical vision that provides an interpretive framework for taking actions that might delay or even prevent what he recognized to be the looming ecological disaster. The immanent God of his cybernetic cosmos is the polar opposite of the transcendent God whose followers, he argues, are destroying the planet. "If you put God outside and set him vis-à-vis his creation and if you have the idea that you are created in his image, you will logically and naturally see yourself as outside and against the things around you. And as you arrogate all mind to yourself, you will see the world around you as mindless and therefore not entitled to moral or ethical consideration. The environment will seem to be yours to exploit. Your survival unit will be you and your folks or conspecifics against the environment of other social units, other races, and the brutes and vegetables." These words could have been written by Lynn White, who confidently declared that Western Christianity "is the most anthropocentric religion the world has seen."[10]

While White's claim is not wrong, neither is it completely right. In the Judeo-Christian tradition, God appears primarily as a transcendent personal creator who is omniscient, omnipresent, and omnipotent and both imposes order on the world and guides individual lives. The crucial passage for White's argument is Genesis 1:26: "And God said, 'Let us make man in our image, after our likeness: and let them have dominion over the fish of the sea, and over the fowl of the air, and over cattle, and over all the earth, and over every creeping thing that creepeth upon the earth.'" As the appointed sovereign of creation, man exploits the world for his purposes. The ultimate goal of human existence is to escape the world and be united with the transcendent Sovereign. While the influence of this theological vision cannot be overestimated, it is important to realize that there is a countercurrent to this theistic image of God. In this alternative vision, the divine is immanent in nature and history, and human beings are the self-embodiment of God. The most prevalent version of this a-theistic theology is mystical and appears in writers as various as Nicholas of Cusa, Pseudo-Dionysus, Meister Eckhart, Friedrich Schleiermacher, Friedrich Schelling, and Ralph Waldo Emerson. Though differing in important ways, all these versions of mysticism tend to be ontologically monistic—the individual becomes lost in the divine totality. In other words, difference collapses into identity, or beings disappear in Being. By reabsorbing worldly appearances in an undifferentiated totality, monistic mysticism is as nihilistic as transcendent dualism, which denigrates the value of worldly existence.

This unity is, in Hegel's famous words criticizing his erstwhile roommate Schelling, "the night in which all cows are black." Hegel's system is a sustained effort to overcome all forms of opposition without losing difference in identity.

The theological doctrines of the Trinity and the Logos prefigure Hegel's dialectical logic, which, we have seen, constitutes the relational structure of nature and history. Father–Son–Holy Spirit (*Heiliger Geist*); Three-in-One, One-in-Three; Difference-in-Identity, Identity-in-Difference. It is important to recall that the dialectical structure of reason (*Vernunft*) overcomes the principle of noncontradiction, which is the foundation of understanding (*Verstand*). Hegelian logic is neither abstract nor otherworldly but is always already embodied or incarnate in nature (physical, chemical, biological) and history (society, politics, culture). In his *Philosophy of History*, Hegel anticipates Bateson's ecology of mind when he declares, "Reason, in its most concrete form, is God. God governs the world; the actual working of his government . . . is the history of the world."[11] The most complete realization of reason is *Geist*—spirit or mind. Hegelian *Geist* redeems the world by redeeming what once was called "man." For Hegel, as for Bateson, "the individual mind is immanent but not only in the body, but also in the pathways and messages outside the body." This is the "larger Mind of which the individual mind is only a subsystem." By reading Hegelian *Geist* through Rovelli's and Wheeler's interpretation of quantum mechanics and vice versa, it is possible to develop an understanding of mind that includes nature and extends to what is misleadingly called "artificial" "intelligence." The hinge on which this argument turns is a reconfigured understanding of cognition.

COGNITION

In Book 10 of the *Confessions*, Augustine's remarkable meditation on memory leads him to reflect on cognition.

> By the act of thought we are, as it were, collecting together things which the memory did contain, though in a disorganized and scattered way, and by giving them our close attention we are arranging for them to be as it were stored up ready to hand in that same memory where previously they lay hidden, neglected, and dispersed, so that now they will readily come forward to the mind that has become familiar with them. . . . In fact what one is doing is collecting them from their dispersal. Hence the derivation of the word "to cogitate." For *cogo* (collect) and *cogito* (recollect) are in the same relation to each other as *ago* and *agito*, *facio* and *factito*. But the mind has appropriated to itself

the word ("cogitation"), so that it is only correct to say "cogitated" of things which are "re-collected" in the mind, not of things re-collected elsewhere.[12]

In a manner similar to philosophers from Plato to Hegel, Augustine argues that cognition and memory are inseparable—collection presupposes recollection, which presupposes collection. There is, therefore, an unavoidable temporal dimension to cognition. The present shapes and is shaped by the past. Rovelli explains Augustine's point: "We are not a collection of independent processes in successive moments. Every moment of our existence is linked by a peculiar triple thread to our past—the most recent and the most distant—by memory. Our present swarms with traces of our past. We are *histories* of ourselves, narratives."

The temporality of cognition is not limited to the interplay of past and present. Though Augustine does not invoke the future in his account of cognition, his interpretation of time as consisting of the present of things present, the present of things past, and the present of things future implies that the future is integral to cognition. Rovelli's explanation once again is helpful:

> To a large extent, the brain is a mechanism for collecting memories of the past in order to use them continually to predict the future. This happens across a wide spectrum of time scales, from the very short to the very long. If someone throws something at us to catch, our hand moves skillfully to the place where the object will be in a few instants: the brain, using past impressions, has very rapidly calculated the future position of the object that is flying toward us.... The possibility of predicting something in the future obviously improves our chances of survival and, consequently, evolution has selected the neural structures that allow it. We are the result of this selection. This being *between past and future events* is central to our mental structure.[13]

Cognition is tensed—it involves collection (present), recollection (past), and anticipation, which predicts the probability and improbability of future events.

There is another important etymological nuance to "cognition" that Augustine overlooks. The Latin *cogito*, which means getting to know, acquaintance, knowledge, derives from *com*, together, + *gnoscere*, to know. Accordingly, cognition is coknowing and therefore is differential, reciprocal, and relative. This interpretation of cognition charts a middle way between the two dominant epistemological positions in Western philosophy—empiricism and idealism. For empiricists, the mind is, in Locke's famous phrase, a tabula rasa upon which sense impressions are inscribed. More complex concepts and ideas are formed by the association of simple perceptions. In this scheme, the mind is primarily passive, and all knowledge is derived from experience. In the alternative epistemological tradition, the mind is not a blank slate but is structured by innate ideas, which function like Plato's transcendent forms to create order out of disorder. These ideas are prior to and independent of experience, that is, a priori. For the idealist, the mind actively shapes, figures, and forms experience.

Kant attempts to mediate these extremes by arguing that the mind is both passive and active in cognition. He agrees with the empiricists that all knowledge derives from experience (a posteriori) but disagrees with their claim that the mind is a blank slate. As we saw in our discussion of space-time, Kant argues that the mind is structured by a priori *forms* of intuition (space and time) and categories of understanding (I. Quantity: Unity, Plurality, Totality; II. Quality: Reality, Negation, Limitation; III. Relation: Inherence and Subsistence, Causality and Dependence, Community; IV. Modality: Possibility-Impossibility, Existence-Nonexistence, Necessity-Contingency). The process of cognition, which produces knowledge, filters the "sensible manifold of intuition" first through the forms of intuition and then through the categories of understanding. Knowledge sensu stricto requires both sense experience and organizational structuring. To appreciate the importance and the limitations of Kant's argument, it is helpful to translate it into a contemporary idiom: *cognition is data processing*. The sensible manifold is the equivalent of data and the forms of intuition and categories of understanding function as a program or algorithm that transforms data into concepts, which are organized to form knowledge.

This line of analysis poses three critical problems. First, while Kant correctly argues that all knowledge is phenomenal, he incorrectly posits an unknowable thing-in-itself beneath or behind appearances. As Hegel insisted, the thing-in-itself is nothing more than a concept posited to determine the limitation of all concepts. Rather than establishing the conditions of the possibility of knowledge, as Kant claims to do, Hegel correctly insists that since the thing-in-itself is unknowable, he establishes the *impossibility* of knowledge. For Hegel, as for the Copenhagen School, Rovelli, and Wheeler's version of quantum mechanics, cognition is participatory or relational, and, thus, all knowledge is phenomenal—there is no thing-in-itself or objective reality. Bohm, by contrast, sees in Kant's notion of the thing-in-itself support for insistence that quantum mechanics is incomplete and requires an as-yet-unknown "hidden variable." He translates this unknown hidden variable into an unknowable "implicate order." "Intelligence and material process have thus a single origin, which is ultimately the unknown totality of the universal flux. In a certain sense, this implies that what have been commonly called mind and matter are abstractions from the universal flux, and that both are to be regarded as different and relatively autonomous orders within the one whole movement.... It is thought responding to intelligent perception which is capable of ringing about an overall harmony or fitting between mind and matter."[14] The question Bohm never answers is whether the world itself is as yet unknown or is forever unknowable.

Second, for Kant, the forms of intuition and categories of understanding are fixed and, thus, static; the mind, in other words, is hardwired. The schemata, which are the mind's ordering principles, are universal and do not change. Cognition is the machinic application of the same schemata, that is, program or algorithm, to different data. So-called learning devolves to little more than endless repetition, which produces the eternal return of the same. There is no mental innovation or evolution. Without the ability to learn and change, the adaptation necessary for survival is impossible. Third, Kant does not realize that cognition is always co-cognition or reciprocal. His epistemology is modeled on the commonsense understanding of the relation between a subject and an object.

In this relationship, the object supplies the data that the subject processes and, therefore, cognition is limited to subjects. Relational and participatory quantum mechanics exposes the limitations of this model. Objects are not separate things or entities but are *relational events*. Since relations are differential and information is a difference that makes a difference, subjects and objects are information processing events. In other words, subjects are simultaneously objects, and objects are simultaneously subjects.

The reciprocal relation between S-O_1 and O-S_2 creates a two-way feedback loop that modifies schemata, programs, and algorithms. Rather than pre-scribed pro-grams (*pro*—beforehand, in advance, + *graphein*, to write) organizing structures are shifting schemata that become by evolving patterns. The interrelation of systems issues in the intrarelation of schemata, programs, and algorithms that are in a constant process of self-organization that make learning, change, and adaptation possible.

In sum, cognition is reciprocal in-formation processing. As such, it involves *neither* the impression of data on a passive recipient *nor* the imposition of fixed forms on formless data. Cognition or, more precisely, co-cognition is an interactive self-organizing process of co-responsiveness in and through which correspondents communicate with and mutually adapt to each other. While effective

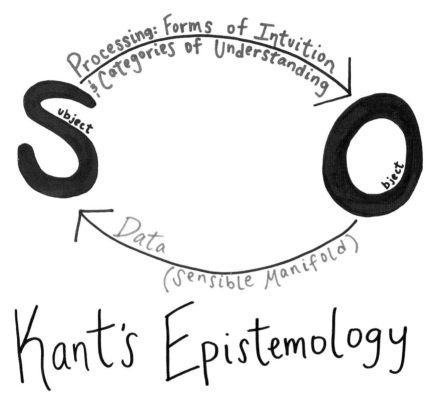

6.2 Kant's epistemology

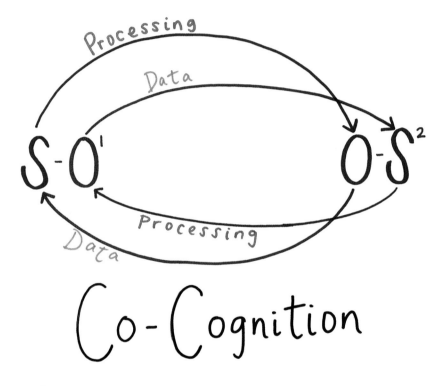

6.3 Co-cognition

adaptation issues in creative intervolution, the failure to adapt results in death, destruction, and extinction.

It is essential not to confuse cognition and consciousness—consciousness presupposes cognition, but cognition requires neither a brain nor consciousness. Therefore, cognition is not limited to human beings and higher primates; as we will see in the concluding section of this chapter and in subsequent chapters, *cognition is ubiquitous*—it is structurally and functionally isomorphic across all operational systems and networks. Consciousness emerges from the increasingly complex interplay of the subsystems that constitute the mind. To understand this expanded interpretation of cognition and its relation to consciousness, it is necessary to de-anthropomorphize cognition. Since cognition does not require a brain or nervous system, it can be deployed in a variety of ways in different nonhuman forms of life as well as in so-called artificial sensors and sensory apparatuses.[15]

To grasp the distinction and interrelation between cognition and consciousness, it is necessary to understand the role of cognition in the overall operation of the mind. Bateson, we have seen, argues that "within the Mind in the widest sense there will be a hierarchy of subsystems, any one of which we can call an individual mind." He proceeds to explain, "The mental world—the mind—world of information processing—is not limited by the skin."[16] Figure 6.4 is a pictorial representation of the metastructure of this system of systems.

Cognition is the activity through which information becomes effective. In his remarkably illuminating book, *The Self-Assembling Brain: How Neural Networks Grow Smarter*, Peter Hiesinger writes, "The information is in the connectivity."[17] If information is in the connectivity, then information, like everything else, is relational. There is no information until it is received and processed. Since relation is primary and not secondary, information processing is always a reciprocal or interactive process, which involves the following steps:

1. *Receiving data* from one or more sources
2. *Filtering data* through schemata, programs, or algorithms to discern differences that make a difference, that is, information
3. *Integrating new information* with previous information
4. *Predicting and responding* to co-relative responses
5. *Adjusting or adapting* schemata, programs, or algorithms based on co-response and the effectiveness of the interaction
6. *Remembering and storing* reconfigured schemata for future deployment

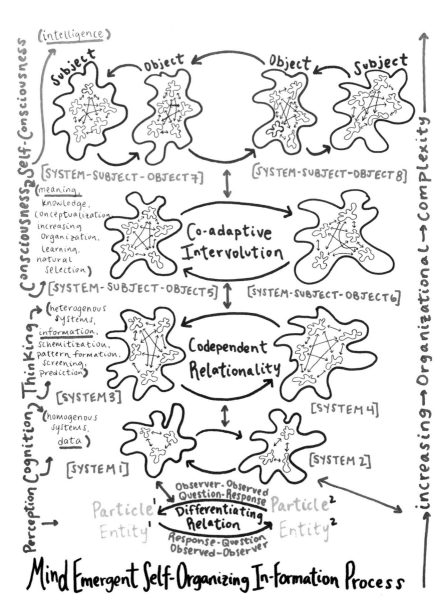

6.4 Mind: Emergent Self-Organizing In-Formation Process

As we have seen in our consideration of quantum mechanics, what had long been understood to be physical or material entities are actually events that are interactive information processes. Wheeler goes so far as to claim that "everything is information." In what he describes as "the participatory universe," everything is connected. "The system of shared experiences which we call the world is viewed as building itself out of elementary quantum phenomena, elementary acts of observer-participancy. In other words, the questions that the participants put—and the answers they get—by their observing devices—plus their communications of their findings take part in creating the impressions which we call the system: that the whole great system which to a superficial look is time and space, particles and fields. That system in turn gives birth to the observer-participants." Once again, a strange loop—observer-participants give rise to the world, which gives birth to observer-participants.

At the most rudimentary level, two eventful particles or entities interact in what is, in effect, a communication process with messages/data being sent and received. This is Rovelli's point when he claims that "it is possible to think of quantum physics as a theory of information. . . . that systems have about one another. The properties of an object can be thought of, as we have seen, as the establishment of correlation between two objects, or rather as *information* that one object has of another."[18] In figure 6.4, differentiating relation constitutes the neχus that simultaneously joins and separates Particle-Entity 1 and Particle-Entity 2. Each participant in this reciprocal relationship is simultaneously observer (subject) and observed (object). This mutual exchange constitutes the most fundamental form of perception. Wheeler clarifies his position in a comment on John D. Barrow and Frank J. Tipler's book, *The Anthropic Cosmological Principle*. The anthropic principle is a reformulation of the traditional argument from design in which the universe appears to be fine-tuned to create the conditions necessary for human life. Barrow and Tripler present a strong version of this argument in which the development of human conscious is inevitable. Obviously, the anthropic principle is a different version of the anthropocentrism that is the foundation of the Anthropocene. Wheeler makes his differences with this position perfectly clear:

> The contrast between the two views could hardly be greater: selection-from-an-ensemble and observer-participancy. The one not only adopts the concept of the universe, and this universe as machine, it also has to postulate, explicitly or implicitly, a super machine, a schema device, a miracle, which will turn

out universes in infinite variety and infinite number. The other takes as foundation notion a higgledy-piggledy multitude of existences, each characterized, directly or indirectly, by the soliciting and receiving of answers to yes-no questions, and linked by exchange of information.[19]

The observer/observed relationship between two eventful particles-entities forms a system when the two-particle events enter a reciprocal relationship with each other. When this occurs, each system becomes data to be processed by the other system. If they combine, they form a more encompassing *homogeneous system* made up of two isomorphic subsystems. Since identity is differential, not all systems and subsystems are the same. Furthermore, the interrelation among different systems and subsystems is not prescribed or programmed, that is, necessary, but is probabilistic, that is, aleatory. Through their interactions, different systems become subsystems in more inclusive *heterogenous systems*. The interplay among subsystems in heterogeneous systems forms schemata that function like changing programs or algorithms to *screen* data and filter noise. Screening both excludes (hides, conceals) and includes (shows, reveals) data. While constitutive relationality precludes any fixed or sharp opposition between inside and outside, stronger and weaker relations form distinctions by marking and remarking shifting margins of difference. This in no way interrupts or negates the reciprocity of the subsystems within heterogenous systems or the reciprocity between and among different systems. The interrelation of systems can, however, reconfigure the subsystems. At every level, subsystems and systems are coemergent and codependent.

As systems become more "internally" and "externally" heterogeneous, the process of screening data to identify differences that make a difference becomes more complex. Significant differences are a function of strong relations, and indifferent differences are symptoms of weak relations. The more variegated the schema, the more comprehensive and complex the information processed from the data become. In the initial phases of cognition, heterogeneous systems *emerge* from homogenous systems, which, in turn, *emerge* from simple systems. Emergence is a distinctive feature of complex systems. For emergence to occur, the interrelation of different systems must issue in increasingly diverse intrarelations among diverse subsystems. Systemic interrelations and intrarelations are not preprogrammed but occur by chance through probabilistic interactions. In his book *Emergence: From Chaos to Order*, John Holland, who was a student of John von Neumann, explains,

"Emergence ... occurs only when the activity of the parts [that is, subsystems] do *not* simply give rise to the activity of the whole."[20] Physicist Per Bak analyzes emergence in terms of what he describes as "self-organized criticality":

> Complex behavior in nature reflects the tendency of large systems with many components to evolve into a poised, "critical" state, way out of balance, where minor disturbances may lead to events called avalanches, of all sizes. Most of the changes take place through catastrophic events rather than by following a smooth gradual path. The evolution of this very delicate state occurs without design from any outside agent. The state is established solely because of the dynamical interactions among individual elements of the system: the critical state is *self-organized*.[21]

At a certain point, increasing interrelations among particles, entities, events, subsystems, and systems lead to a *phase shift* in which a new structure unexpectedly emerges. This is the "it" moment in which the book or anything else appears as *a whole* and reshapes gathered data, information, and knowledge. Since parts and whole are coemergent and codependent, neither can be reduced to the other. Four points in this description of self-organized criticality must be emphasized in the consideration of the emergence of consciousness from cognition. First, self-organized criticality results from interactions between systems and subsystems. Second, in this state, nonlinear events can have effects disproportionate to their causes. Third, the dynamic interactions among subsystems generate events that require a holistic description, which cannot be reduced to an account of the individual elements. Finally, at the tipping point, the effect of individual relations and events is unpredictable.

While none of these processes presupposes consciousness, consciousness is impossible without them. For schemata to be productive, they must anticipate the response of other systems and predict what response to this response will be most productive. As systems become more diverse, co-respondence becomes more unpredictable. Unexpected or novel data initially appear as noise, which can be distracting, destructive, or creative. This noise can be generated by emerging relationships with an expanding network of other systems. At a certain point, a system must either reconfigure its operative schema to accommodate disruptive data or disintegrate. When adaptation is successful, noise becomes information. This process through which schemata learn and evolve is similar to natural selection—effective schemata survive and ineffective schemata become extinct. Such adaptation results in "internal" complexification, which makes the higher cognitive processes of consciousness and self-consciousness possible. It is important to understand that there is no organizing agent—*the emergence of consciousness and self-consciousness from cognition is a self-organizing process.*

How, then, do cognition and consciousness differ, and how are they related? In recent philosophical debates, the primary preoccupation has been with the question of how the material processes of the brain can create immaterial consciousness. One of the most interesting and provocative participants in this discussion is Giulio Tononi, professor of sleep medicine and chair in consciousness science at the University of Wisconsin. Tononi argues that the traditional way of posing the question of the relation between the brain and consciousness makes finding an adequate answer impossible.

> As long as one starts from the brain and asks how it could possibly give rise to experience—in effect trying to "distill" mind out of matter, the problem may be not only hard, but almost impossible to solve. But things may be less hard if one takes the opposite approach: start from consciousness itself, by identifying its essential properties, and then ask what kinds of physical mechanism could possibly account for them. This is the approach taken by integrated information theory (IIT), an evolving formal and quantitative framework that provides a principled account for what it takes for consciousness to arise, offers a parsimonious explanation for the empirical evidence, makes testable predictions, and permits inference and extrapolations.[22]

Tononi develops his theory in a book entitled *Phi Φ: A Voyage from the Brain to the Soul*. The pivotal chapter in the book begins:

> Integrated Information:
> The Many and the One
> *In which is shown that consciousness lives*
> *Where information is integrated*
> *By a single entity above and beyond its parts*[23]

While I have profound disagreements with Tononi on some issues, his interpretation of consciousness in terms of integrated information is both very insightful and very useful. His primary interest is to develop a quantitative measure of integrated information to use in establishing the presence or absence of consciousness for diagnostic purposes. He also develops a graduated scale for measuring degrees of consciousness in humans, animals, and so-called nonnatural systems. "*Integrated information*," he explains, "*measures how much can be distinguished by the whole above and beyond its parts, and Φ is its symbol. A complex is where Φ reaches its maximum, and therein lives one consciousness—a single entity of experience.*"[24] He summarizes his argument by identifying five attributes of consciousness: intrinsic existence, composition, information, integration, and exclusion. In his most useful explanation of his theory, Tononi describes consciousness in terms of the interplay between differentiation (many) and integration (one).

> The central assumption of IIT is that these two fundamental properties of conscious experience—information and integration—rest on the coexistence of differentiation and integration, that is, on a unique balance between diversity and unity in the physical brain. This is a nontrivial assumption as, at first sight, it would seem that these two properties are extremely difficult to reconcile. They act as two forces working against each other. The more specialized the elements of a system are, the more difficult it is to make them interact and, consequently, integration will be extremely arduous. On the other hand, the stronger and more efficient the interaction between the elements, the more homogeneous their behavior will be and the overall level of integration will be reduced. Loosely speaking, the coexistence of differentiation and integration is no easy matter in any walk of life, from the personal to the political, from the biological to the organization of human society, let alone in physical systems.[25]

This interplay between differentiation and integration is operative at every level of consciousness. At one point, Tononi quotes Bateson's definition of information as a difference that makes a difference. In an imaginary dialogue between Galileo and Alturi (Alan Turning), Tononi uses this definition of information to interpret perception. "This is the trick that consciousness plays on you—your experience of pure dark is given immediately. But in reality, it can be dark only in relation to what it's not. And it is only if you have a brain that can distinguish darkness from untold other states, from each in a specific way, that you can actually seek dark."[26] Here Tononi approaches a dialectical interpretation of constitutive relationality; however, at other crucial points in his argument, he backs off from this important insight.

While Tononi's understanding of consciousness as integration is, I believe, correct, he makes four noteworthy mistakes. The first two are inextricably interrelated. "The axioms and postulates of IIT say that consciousness is a fundamental, *observer-independent* property that can be accounted for by the *intrinsic* cause-effect power of certain mechanisms in a state—how they give form to the space of possibilities in their past and future."[27] As the argument developed in the previous chapters makes clear, *nothing is observer-independent*. Tononi insists on observer-independence because he is convinced that "consciousness exists intrinsically." But nothing is intrinsic in the sense Tononi claims because everything is interrelated. The most telling point seems to be too obvious to miss—nothing can be known about anything that purports to be observer-independent.

What Tononi describes as "intrinsic existence" can, however, be reinterpreted in a way that is useful and suggests how to overcome his third mistake—the denial that consciousness is an emergent phenomenon. The crucial point in Tononi's argument about the intrinsicality of consciousness is the relation between part and whole. Consider once again his definition of integrated information. "The information generated *above and beyond its parts*" is "integrated information" (emphasis added). The whole (consciousness) is more than the sum of its parts. As we have seen in our discussion of inner teleology, part and whole are reciprocally or dialectically related—each becomes itself in and through the other, and neither can be itself apart from the other. Parts and whole are *coemergent and codependent*, and, thus, neither can be reduced to the other. As a system of subsystems, consciousness like everything else must have *both* a top-down *and* a bottom-up explanation.

The relation between system and subsystems is two-way—systems emerge from subsystems, which they constantly reconfigure, and reconfigured subsystems transform interrelated systems in adaptive and maladaptive ways. This relational process is *self-organizing*.

Before proceeding to my fourth and final criticism of Tononi, it is necessary to consider the implications of his integrated information theory for consciousness. Cognition, I have argued, gives rise to heterogeneous systems that interact with each other to create more inclusive systems. When the level of complexity, integration, and self-organization passes the tipping point, consciousness emerges and conceptualization becomes possible. Concepts and ideas organize percepts in ways that enable increasingly sophisticated discriminations and relations. Through the further integration of different concepts and ideas, meaning eventually emerges. With

Self-Consciousness
Consciousness
Cognition

Co-Emergence

the emergence of meaning, knowledge becomes possible. Meaning is never fixed or stable; furthermore, since becoming is both synchronically and diachronically relational, meaning is always contextual, and knowledge is inevitably perspectival.

In a final twist, consciousness turns back on itself to become self-consciousness. Self-consciousness is both "internally" and "externally" self-reflexive. The reflexivity of self-consciousness emerges in the interrelation of different self-consciousnesses. Like two entangled particles, I become myself in and through you, and you become yourself in and through me; neither of us can be ourself apart from the other. In Hegel's succinct phrase, *Geist*—mind or spirit—is "an I that is a we and a we that is an I." Only at this highest level of awareness and self-awareness does intelligence emerge. "Intelligence" is commonly understood as the capacity to acquire and apply knowledge. Once again, etymology enriches our understanding of term. "Intelligence" derives from the Latin *intelligentia*, which means knowledge, power of discerning; art, skill. The present participle of *intelligere*, to understand, comprehend, come to know, is formed by *inter*, between + *legere*, to choose, pick out, read.[28] Intelligence, then, does not just involve the gathering of knowledge but also requires two interrelated operations that we have seen in other mental activities—differentiating (selecting, choosing, picking out) and integrating (collecting, gathering). Catherine Malabou underscores a further aspect of intelligence. "Although philosophers do use what the Latin word *intelligentia* refers to, the 'faculty of understanding,' which the prefix *inter-* and root *legere* ('to choose, to pick') or *ligare* ('to relate') suggest that we interpret as the ability to establish relations among things, they more readily use the term 'intellect.'"[29] Intelligence is the gathering, differentiating, and selecting of information and knowledge that discerns and establishes effective relations among eventful things. The greater the relationality, the greater the intelligence. In contrast to current academic norms, which promote more and more knowledge about less and less, intelligence involves more and more knowledge about more and more. However, a person can know a lot but not be intelligent, and, as we will see in chapter 10, a machine can transform massive amounts of data into information but not be intelligent. Intelligence requires discerning connections between and among seemingly unrelated things and events.

As a system of different but integrated subsystems, *the mind is an emergent self-organizing information process*. Since system and subsystems are coemergent and codependent, each conditions and is conditioned by the other. While cognition is operative at every level, consciousness and self-consciousness emerge only at

Intelligence

6.6 Intelligence

"higher" levels of differentiation, integration, and organization. Once they have emerged, they shape but do not determine cognition. Mind is not identical with consciousness and/or self-consciousness. While much, perhaps most, mental activity is nonconscious or unconscious, all mental activity is cognitive. This means that mind extends beyond human beings and higher primates to all cognitive systems both "natural" and "artificial."

PANPSYCHISM—PANCOGNITIVISM

As I have previously noted, in recent years there has been a surprising revival of panpsychism in the most unlikely of circles—analytic philosophy. In a suggestive article, "The Intelligence of Swine," Evan Malmgren quotes British philosopher Philip Goff: "According to panpsychism, consciousness pervades the universe and is a *fundamental* feature of it."[30] To claim that consciousness is a fundamental feature of the universe is to insist that it is irreducible to anything other than itself. In his contribution to the *Stanford Encyclopedia of Philosophy*, Goff explains the reason for the resurgence of panpsychism.

> Panpsychism is the view that mentality is fundamental and ubiquitous in the natural world. The view has a long and venerable history in philosophical traditions of both East and West and has recently enjoyed a revival in analytic philosophy. For its proponents, panpsychism offers an attractive middle way between physicalism on the one hand and dualism on the other. The worry with dualism—the view that mind and matter are fundamentally different kinds of thing—is that it leaves us with a radically disunified picture of nature, and the deep difficulty of understanding how mind and brain interact. And whilst physicalism offers a simple and unified vision of the world, this is arguably at the cost of being unable to give a satisfactory account of the emergence of human and animal consciousness. Panpsychism, strange as it may sound on first hearing, promises a satisfying account of the human mind within a unified conception of nature.[31]

It is important to realize that this is not some fringe movement. Max Planck famously declared, "I regard matter as derivative from consciousness. We cannot get behind consciousness. Everything that we talk about, everything that we regard as existing, postulates consciousness."[32] In addition to panpsychism playing an important role in many spiritual and religious traditions, the list of philosophers drawn to it stretches from the Presocratics to leading modern and contemporary philosophers like Spinoza, Leibniz, Schopenhauer, Josiah Royce, William James, Pierre Teilhard de Chardin, Bertrand Russell, Alfred North Whitehead, Galen Strawson, and Thomas Nagel. The turn to panpsychism by analytic philosophers represents an effort to recast the mind-brain relationship in a way that is supposed to solve the problem of consciousness by redefining physical and biochemical processes as conscious. In his book *Mind and Cosmos: Why the Materialist Neo-Darwinian Conception of Nature Is Almost Certainly False*, Nagel writes, "The great cognitive shift is an expansion of consciousness from the perspectival form contained in the lives of particular creatures to an objective, world-encompassing form that exists both individually and intersubjectively. It was originally a biological evolutionary process, and in our species, it has become a collective cultural process as well. Each of our lives is a part of the lengthy process of the universe gradually waking up and becoming aware of itself."[33]

Others turn to Russell and Whitehead. In his book *Mindful Universe: Quantum Mechanics and the Participating Observer*, physicist Henry Stapp, who worked with Wheeler, develops what he describes as a "Whiteheadian Quantum Ontology":

> In the Whiteheadian ontologicalization of quantum theory, each quantum reduction event is identified with a Whiteheadian actual entity/occasion.
>
> Each Whiteheadian actual occasion/entity has a "mental pole" and a "physical pole."
>
> There are two kinds of actual occasions. Each actual occasion of the first kind is an intentional probing action that *partitions* a continuum into a collection of discrete experientially different possibilities. Each actual occasion of the second kind selects (actualizes) one of these discrete possibilities and obliterates the rest.
>
> According to this Whiteheadian quantum ontology, objective and absolute actuality consist of a sequence of psychophysical quantum reduction events, identified as Whitehead actual entities/occasions.

Stapp argues that the synthesis of the Copenhagen version of quantum mechanics and Whitehead's process theology results in a panpsychism that overcomes an anthropocentric interpretation of consciousness:

Science encompasses cosmology, and also our attempts to understand the evolutionary process that created our species. If we want to address the basic question of the nature of human beings, then we need more than merely a framework of practical rules that work for us. We need to be able to see this pragmatic anthropocentric theory as a useful distillation from an underlying non-anthropocentric ontological structure that places the evolution of our conscious species within the broader context of the structure of nature herself. We need a fundamentally non-anthropocentric ontology within which the anthropocentric pragmatic theory is naturally embedded.[34]

This is a very important point—overcoming the anthropocentrism of the Anthropocene *does* require de-anthropomorphizing both ontology and epistemology. Panpsychism cannot solve this problem because of its preoccupation with consciousness, which retains vestiges of anthropocentrism.

Tononi argues that panpsychism, which is intended to be a mean between the extremes of materialism and idealism, in the final analysis presupposes what it is supposed to explain:

> Materialism, or its modern offspring physicalism, has profited immensely from Galileo's pragmatic stance removing subjectivity (mind) from nature in order to describe and understand it objectively—from the extrinsic perspective of a manipular/observer. But it has done so at the cost of ignoring the central aspect of reality from the intrinsic perspective—experience itself. Unlike idealism, which does away with the physical world, or dualism, which accepts both in an uneasy marriage, panpsychism is elegantly unitary: there is only one substance, all the way up from the smallest entities to human consciousness and maybe to the World Soul (*anima mundi*). But panpsychism's beauty has been singularly barren. Besides claiming that matter and mind are one thing, it has little constructive to say and offers no positive laws explaining how the mind is organized and works.

The difficulty with Tononi's argument is that he continues to identify mental activity with consciousness. Consciousness, he claims, involves a "maximally conceptual structure also known as quale." Integrated information "*can be formulated quite simply: an experience is identical to a conceptual structure* that is *maximally irreducible intrinsically*."[35] The most important point in this claim is the following series of identifications: consciousness = integrated information = conceptual structure = quale = experience. Qualia, then, are necessarily experiential. While Tononi resists panpsychism's claim that every entity is conscious, his identification of mental activity with consciousness and consciousness with experiences reintroduces the anthropocentrism characteristic of panpsychism. Rather than overcoming human exceptionalism by inscribing human consciousness within a more encompassing mind, panpsychism reinforces human exceptionalism by projecting the experience characteristic of human beings onto the cosmos as a whole.

In my judgment, it *does not* make sense to claim that all events are experiential and, therefore, conscious. However, the investigation of Hegelian relationalism, relational quantum mechanics, relativity theory, and different aspects of information theory suggests that it *does* make sense to maintain that all entities and events are in some sense cognitive. Therefore, I propose substituting panpsychism with *pancognitivism*. All entities and events from the physical through the human to the so-called artificial are in some sense cognitive. This argument requires distinguishing cognition and consciousness as different though related activities of the mind. As I have argued, consciousness presupposes cognition, but cognition does not require consciousness. In an essay entitled "Cognition Is Not Exceptional," Simon McGregor, research fellow at the University of Sussex and associate at the Center for Cognitive Science and Center for Computational Neuroscience and Robotics, argues, "Humans have a long history of insisting that they are members of a special group with some uniquely privileged characteristic and eventually being proved wrong. This has played out in cognitive science as in disciplines such as history or ethology. . . . I spell out a position, here termed 'radical pancognitivism,' that constitutes the polar opposite of cognitive exceptionalism, in that it attributes cognition to literally every physical system in the universe."[36] Pancognitivism leads to an expanded notion of mind in which human consciousness is a complex subsystem within an all-inclusive cognitive process. In the following chapters, I will consider cognition in ecological systems (chapter 7), biological and biochemical systems (chapter 8), plant and animal systems (chapter 9), and technological systems (chapter 10).

CHAPTER 7

QUANTUM ECOLOGY

I now localize something which I call "Mind" immanent in the large biological system—the ecosystem. Or, if I draw the system boundaries at a different level, then mind is immanent in the total evolutionary structure. If this identity between mental and evolutionary is broadly right, then we face a number of shifts in our thinking.

—Gregory Bateson

ANTHROPAUSE

I live in the city and the country—New York City on the island of Manhattan and Williamstown in the Berkshire Mountains. They are divided and connected by a spatial and temporal gap of 158 miles and three hours. Every week for half the year I drive back and forth between the two. During this pause, interval, inter-mission, I sometimes listen to audio books or music, but often I drive in silence. I have traveled this road so many times in recent years—475 to be precise—that the auto-mobile virtually drives itself, and I drift into a meditative state. This is when some of my most creative thinking occurs. I do not exactly think; rather thinking without thinking happens *betwixt and between* city and country. A few days ago, the following words occurred to me. I am uncertain where they came from, but I have a pretty good sense where they are taking me.

Anthropause. Interruption. Interference. System failure because of an invisible virus, parasite that could be neither processed nor controlled. "The city rat invites the country rat onto the Persian rug. They gnaw and chew leftover bits of ortolan. Scraps, bits and pieces, leftovers: their royal feast is only a meal after a meal among dirty dishes of a table that has not been cleared. The city rat has produced nothing, and his dinner invitation costs him nothing." Do rats bring the plague to the city, or does the plague clear the way for the return of what has been repressed but not exterminated? Parasite of a parasite. City/Country. Host/Guest. Parasite/Host. Who is who? Which is which? Parasites feeding on leftovers, scraps, bits, pieces, crumbs. Philosophical crumbs—*Philosophiske Smuler eller En Smule Philosophi*. Crumbs swept away, under the (Persian) rug. "But we know that the feast was cut short. The two companions scurry off when they hear a noise at the door. It was only a noise, but it was also a message, a bit of information producing panic: an interruption, a corruption, a rupture of information. Was the noise really a message? Wasn't it, rather, static, a parasite? A parasite who has the last word, who produces disorder and who generates a different order." In his unsettling book *The Parasite*, Michel Serres explains that in French the word "parasite" means static—not immobility but interference, interruption, noise. Noise and information have an uneasy relation to each other. Too much "information" becomes noise, and sometimes what appears to be noise is really information. Information as noise, noise as information. What's the difference? If, as we have seen, information is a difference that makes a difference, then information is not the endless repetition of the same or the always already expected but is the interruption caused by the incursion of the unexpected or the new. News that is truly news is always unexpected and, therefore, is inevitably disruptive. This disruption can lead to system failure or it can be creative and productive. Since the new can be neither determined nor predicted, it always appears to be a chance variation. When it is not destructive, "Noise gives rise to a new system, an order that is more complex than the simple chain. The parasite interrupts at first glance, consolidates when you look again." Parasite/Host . . . Host/Parasite . . . coemergent and codependent. Bound in a reciprocal relationship, neither is precisely inside nor outside the other. "The whole question of the system now is to analyze what a point, a being, a station are. They are crossed by a network of relations; they are crossroads, interchanges, sorters. But is that not analysis itself: saying that this thing is at the intersection of several series. From then on, the thing is nothing else but a center of relations, crossroads, or passages."[1]

Within emergent self-organizing systems and networks, a thing is nothing other than a center of relations, crossroads, passages—in short a *nexus* of events and happenings. Parasite/Host—Host/Parasite. There is not one without the other—no system without parasites, no parasites without system. Sometimes noise *is* information. What message is the virus raging in our midst sending us? Will we listen? Will we hear it? Will the Anthropause mark the end of the Anthropocene and the unexpected beginning of something new, something unexpected, something so different that it can hardly be imagined?

"The ecology *of* mind is inseparable from the mind *of* ecology"—when I awoke this morning, this sentence was waiting for me. The message arrived unexpectedly and gave me pause. Where exactly did it come from, and what precisely does it mean? So much turns on so little—how is "of" to be understood? Is this a single or a double genitive? "Genitive" derives from *gignere*, which means "to beget." Does mind beget nature or does nature beget mind? What is ecology? What could an ecology of mind possibly mean?

By another strange coincidence, when I checked my email, I found a message from Jack in which he included an exchange of letters that did not make the final cut for our book.

April 4, 2020

Dear Jack:

I didn't immediately reply to your letter yesterday because I wanted to give myself some time to absorb all you said. Our roles were somewhat reversed, I am usually the one whose vision is dark, and you are the one who manages to see the crack that lets the light shine in. Perhaps the reason I am searching for a glimmer of hope is my constant thoughts about Aaron and Kirsten and their families, as well as the challenge of teaching young people in the time of the plague.

As day turned into night, your passing reference to the 1755 Lisbon earthquake stuck in my mind because of its relation to your concluding paragraph, where you wrote, "The word modernity comes from the Late Latin adverb *modo*, meaning 'just now.' So, 'now' was the Renaissance, 'then' was Classical Antiquity, and in between the *medio aevo*. But the Enlightenment was postmodern vis-à-vis the modernity of the Renaissance, and it's then, isn't it, that the melioristic notion of endless progress begins to take hold?"

Well, not precisely. The 1755 earthquake was one of the largest ever recorded at the time; between 10,000 and 100,000 people lost their lives. This unexpected event had a profound effect on European intellectuals. Voltaire was so distressed that he wrote a long poem, *Poème sur le désastre de Lisbonne ou Examen de cet axiom 'Tout est bien.'* A few days after the earthquake, he wrote to a friend, "You know the horrible event of Lisbon ... *voilà* a terrible argument against optimism." As you know, the Lisbon quake is important in *Candide*, where it is one of the primary reasons for Candide's disillusion with Panglossian optimism. So, I'm not so sure about when or how the belief in endless progress takes hold.

The more relevant association yesterday was not Voltaire, but Ephraim Lessing. Earlier in the semester I taught Lessing's *Education of the Human Race* in my undergraduate course. Lessing, like Voltaire, was totally rattled by the Lisbon earthquake. I was thinking of Lessing because I was thinking of rats. A few days ago, I mentioned an email Liane had sent me describing stars appearing in the newly dark sky over Manhattan. Liane is a former student of mine who is extraordinarily insightful and a fabulous writer. She wrote, "We're still in NYC. It's the strangest thing I've ever seen. There's no one there. My building is at 40 percent capacity. Anyone who can leave has left. The rats are retaking the city. We see scores of them when we're out walking Faraday (yes, Faraday is the name of her dog) in the evening, I think because there's no one there to scatter them. They're likely hungry as well." Rats are, as Camus makes clear, inseparable from the plague. Though rats do not carry the coronavirus, their appearance in New York City during the lockdown might be a harbinger of worse things to come. What do the rats know that we do not?

Perhaps the rats running around the deserted streets of Manhattan are telling us something. The disaster, it seems, will be environmental rather than nuclear. Is the plague the price we are paying for ravaging the earth, poisoning the seas, and polluting the air? Have I missed the gospel our rats have been preaching all these years? Will rats inherit the earth and have the last word?[2]

This exchange took me back to the early months of the plague when Jack reported that the sky in Los Angeles unexpectedly had become clear. Though there are conflicting explanations for this phenomenon, the popular press attributed it to the fact that during the pandemic automobile traffic decreased by

40 percent and air traffic by 75 percent. There was a corresponding decrease in travel and tourism. The immediate result of these developments was an improvement in quality. LA's legendary smog lifted and, as our Danish friends never tire of joking, U C L A.

Add to this a third weird coincidence and I begin to ask whether coincidences really are coincidental. As thoughts for this chapter were organizing themselves, I stumbled on an illuminating article by Emily Anthes entitled "Did Nature Heal During the Pandemic 'Anthropause'?"[3] The pandemic became a global experiment in the environmental effects of a decline in human activity. The preliminary expectation was that the decrease in human intervention would have a beneficial impact. Animals were retaking the territory that we had taken from them. It is undeniable that human beings have become an invasive species that upsets the ecological balance and puts the life of countless, perhaps all, species at risk. As humans were forced to pause their thoughtless frantic activity, not only rats, but many species that had been crowded out began reclaim their territory: dolphins in the Bosporus, wild boars in Haifa, pink flamingos in Albania, dungeons in Thailand, cougars in Santiago, Kashmiri goats in Wales, mountain lions in Santa Cruz. The list goes on and on. Animal behavior also began to change—birds literally changed their tune, deer roamed forests during all hours, sea turtles were able to spend more time in warm waters, which increased egg production. Deterritorialization . . . Reterritorialization.

And yet, and yet . . . There were also unexpected negative effects from the changing interactions of humans and animals and, correlatively, of animals with each other. Commenting on the benefits from decreased human activity, Anthers reports that "the return of animals might seem like a tidy parable about how nature recovers when people disappear from the landscape—if not for the fact that ecosystems are complex. A growing body of literature paints a more complex portrait of the slowdown of human activity that has become known as the 'anthropause.' Some species clearly benefited from our absence, consistent with early media narratives that nature, without people bumbling about, was finally healing. But other species struggled without human protection or resources."

Life is, indeed, complex. Loops within loops within loops. Humans adapt to the environment, which adapts to humans. Where does one end and the other begin? "The difference is part of the thing itself, and perhaps it even produces the thing. Maybe the radical origin of things is really that difference, even though classical rationalism damned it to hell. In the beginning was the noise."[4] Noise and information are not opposites; to the contrary, noise is information awaiting formation. Systems—all systems—think even as they are thought by processing noise into information. Ecosystems are smart, perhaps smarter than humans; they know what humans do not know or do not want to understand. The plague is the noise at our door, and processing the message it is sending is a matter of life or death.

CLASSICAL EVOLUTION

Though the word "ecology" was coined by the German zoologist Ernst Haeckel in 1866, the discipline of ecology, like so many of the trajectories we have been following, can be traced to the guests attending the Jena dinner party hosted by Goethe. Alexander von Humboldt is widely regarded as the "father of ecology." On his numerous trips throughout North and South America, he not only collected a vast array of flora and fauna but also studied the effect of climate on plant and animal life. The more differences he observed, the more connections he saw. "Gradually," he concluded, "the observer realizes that these organisms are connected to each other, not linearly, but in a network-like entangled fabric."[5]

More than quantum particles are entangled; life itself is an entangled network of networks. In the final paragraph of *The Origin of Species*, Darwin uses the image of the "tangled bank" to describe life:

> It is interesting to contemplate a tangled bank, clothed with many plants and many minds, with birds singing on the bushes, with various insects flitting about, and with worms crawling through the damp earth, and to reflect that these elaborately constructed forms, so different from each other in so complex a manner, have all been produced by laws acting around us.... There is a grandeur in this view of life, with its several powers, having been originally breathed by the Creator into a few forms or into one; and that, while the planet has gone circling on according to the fixed law of gravity, from so simple a beginning endless forms most beautiful and wonderful have be, and are being evolved.[6]

It took more than a century to appreciate the significance of entanglement and to recognize the far-reaching implications of Darwin's error of applying Newtonian mechanics to biological evolution.

In ways that are not immediately obvious, the presuppositions of Darwin's theory of evolution are as philosophical and theological as they are scientific. The position against which he formulated his argument was rational morphology, whose roots can be traced to the debates between medieval Realists and Nominalists. The chief proponent of the former position was Thomas Aquinas, and the leading Nominalist was William of Ockham, who taught at Oxford. The issue at the heart of the debate was the status of universal terms like "human being." Realism is a version of Platonism in which the universal (human being) is essential and individuals (human beings) exist only by their "participation" in the transcendent and eternal form of the human. Ockham rejected this claim and insisted that only individuals are real, and universals are nothing more than names (*nomen*), which are heuristic devices for organizing individuals. In other words, while realists claim that universals are ontologically and epistemologically prior to individuals, nominalists insist that individuals are ontologically and epistemologically antecedent to universals. The former position implies a deductive methodology and the latter an inductive methodology. By establishing the basic reality of individual beings and entities that can only be known empirically, Ockham laid the foundation not only for modern science but also for modern politics and economics.

Neither Darwin nor those he criticized recognized that the early purportedly scientific controversies about evolution replayed seemingly arcane medieval theological debates:

Rational Morphology	Darwinianism
Realism	Nominalism
Essentialist: Form-Structure ontologically prior to individual parts.	Antiessentialist: Individuals ontologically prior to Form-Structure.
Whole: structure of fixed relations irreducible to individual parts.	Whole: aggregate of and reducible to individual parts.
Relations internal.	Relations external.

While Darwin's lifelong commitment to empirical observation led to his suspicion of rational morphology, his appreciation for the design of nature led to a lingering fascination with a revised version of Aristotelian teleology. While a student at Cambridge University, he studied with William Paley, who at the time was the leading defender of the teleological argument for the existence of God. Arguing from effect to cause, Paley maintained that the order of the universe could only be explained as being the result of the grand designer named God. When Darwin set sail on the *Beagle* in 1831, he had committed Paley's influential book *Natural Theology, or Evidences of the Existence and Attributes of the Deity Collected from the Appearance of Nature* (1802) to memory and fully expected to become a country pastor when he returned to England. However, what he saw during his sojourn not only changed his mind but also transformed modern science and, by extension, modern society and culture.

Darwin was not the first to propose a theory of evolution. His grandfather Erasmus Darwin anticipated his grandson's argument in *Zoonomia* (1794–1796). What distinguishes the younger Darwin's theory is its grounding extensive empirical derived data from data derived from countless specimens he collected across near and distant continents. The intellectual climate in which Darwin began to develop his theory was dominated by the arguments of creationists like Paley and rational morphologists like Etienne Geoffroy Saint-Hilaire and Georges Cuvier. Disputes between uniformitarians and catastrophists about recent geological discoveries further complicated debates about evolution. One of the most influential lines of analysis grew out of the alliance between creationists and morphologists. Geoffroy Saint-Hilaire formulated the "Principle of Connections," according to which "certain patterns of relationship between structural elements in organisms remained constant even though the elements themselves changed."[7] In their seminal book *Darwinism Evolving: Systems Dynamics and the Genealogy of Natural Selection*, David Depew and Bruce Weber point out that Geoffroy's structuralist morphology had been influenced by a group of Romantic philosophers of biology who had been inspired by Friedrich Schelling's *Naturphilosophen* and Goethe's theory that all plants are transformations of "an 'original plant (*Urplanz*) in which the basic structure is revealed. . . . *Naturphilosophes*' vision can be seen, in fact, as Neoplatonic typological essentialism turned on its side and represented *as if* it were unfolding in time."[8]

The process of development is characterized by the progressive *unfolding* from a common origin. Geoffroy's theory sparked heated debates within and beyond natural philosophy or science. To counter the claim that all animal species are transformations of a single "ground plan," Georges Cuvier attempted to rationalize Aristotle's account of natural organisms by identifying four distinct types, or "embranchments," of animals. The cornerstone of his theory is his insistence that it is impossible for different types to develop from one to the other. Within this framework, God creates separate species, and when a species becomes extinct as the result of a natural catastrophe, God intervenes to re-create the old species or to create a new form of life. The theological appeal of this position is twofold: first, the creative role of God is preserved, and second, since the types remain distinct and cannot develop from one another, human exceptionalism is preserved and human difference from so-called lower animals is secured.

While Cuvier appealed to Aristotle, his conclusions are a version of Platonism. Different types of animals are in effect Platonic forms or archetypes, each of which is defined by certain inherent characteristics that are gradually revealed over the course of time. There is, however, no constitutive interrelation among different types of organisms. While some investigators like the Swedish botanist Carl Linnaeus (1707-1778) used such typologies for taxonomic purposes, rational morphologists considered the different forms to be the ontological ground for each determinate species.[9] Rational morphology is a latter-day version of philosophical realism in which the whole is prior to and a condition of the parts. Morphological analysis identifies the form or structure that constitutes the essence of the species and determines the traits of individual organisms. The "ground plan," or, in Heidegger's terms, the *Grundriss*, of the species is a structural whole of *fixed* relations that cannot be reduced to the sum of its individual parts. This structure establishes internal constraints on development within the species, but since the different forms remain strictly independent, it prevents any development between and among species. Instead of originating from a common ancestor in a linear evolutionary tree, each species is created independently of all the others.

Darwin overturned every assumption of rational morphology. Stephen Jay Gould explains the significance of Darwinian evolution. "Darwin's revolution should be epitomized as the substitution of variation for essence as the central category of natural reality.... What can be more discombobulating than a full inversion or 'grand flip' in our concept of reality: in Plato's world, variation is accidental, while essences record a higher reality; in Darwin's reversal, we value variation as a defining (and concrete earthly) reality, while averages (our closest operational approach to 'essences') become mere abstractions."[10] This reversal repeats nominalism's critique of realism. Darwin is consistently antiessentialist: the individual (organism within species as well as trait within organism) is ontologically prior to general forms and structures. The whole is the aggregate of the parts and is, therefore, subject to reductive analysis. Change, which is accidental and incremental, is directed by external circumstances (that is, independent environmental conditions). In developing his critique of rational morphology, Darwin draws on the insights of Newton, Thomas Malthus, and Adam Smith.

Darwin appropriates the principles of Newton's classical theory of physics to develop what is, in effect, a classical theory of evolution. He reportedly confessed to wanting to become "the Newton of biology."[11] Six basic principles inform his analysis: empiricism, individualism, efficient or external causality, equilibrium, continuity, and reductionism. Since true knowledge must be rooted in observation and sense experience, investigation must be inductive rather than deductive. Inquiry focuses on individual organisms and their particular traits. Darwin scrupulously followed these principles by gathering information about individual animals and plants and from these data built a general theory. Since causes in the Newtonian model are external rather than internal as they are for rational morphology, Depew and Weber argue that Darwin needed "to find an external force, like Newtonian gravity, rather than an internal drive, that impinged on the developmental and reproductive cycle with sufficient force to drive and shape evolutionary diversity. That force must, moreover, be a *vera causa*, a real cause that can be seen at work this very day."[12] To understand the implications of Darwin's theory, it is necessary to recognize several of his most important presuppositions. First, the organism and the environment are external to each other; second, the environment is fixed prior to the advent of any particular organism; third, to survive, organisms must adapt to niches constituted independently of them. In this theory, transformations occur when a particular mutation better equips an organism for survival. The core of Darwin's argument is *random* variation combined with selection by *external* environmental factors.

Within Darwin's schema, the Newtonian method of investigation and model of systems were necessary but not sufficient to explain the development of living organisms. The various strands of his theory did not come together until he had two moments of illumination separated by almost two decades. The first came when he was reading Malthus's *An Essay on the Principle of Population* (1798), and the second occurred eighteen years later when he recognized the importance of Smith's division of labor in *The Wealth of Nations* (1776). Darwin's mature theory of evolution results from the synthesis of his field observations with the population studies of Malthus and the economic theory of Smith. While many commentators and critics have argued that Darwinism has shaped social and economic theory, few have realized the extent to which Darwin's theory itself is based on population studies and economic speculation. As Darwin understood it, Malthus's claim that population increases geometrically and the food supply only arithmetically demonstrates that organisms are always subject to *external* pressures. The limitation of resources creates competition among both individuals within a species and different species. What Darwin needed to complete his theory was Smith's account of the division of labor. Survival in a ruthlessly competitive world depends on the ability of individuals and species to adapt to niches where they can best profit from their labors. Populations, like industries and companies, must diversify and specialize to remain competitive. Fitness is not a matter of relative strength but is the ability to adapt and fit into an available niche in the competitive landscape. When understood in this way, natural selection is the biological version of Smith's invisible hand. Darwinism presupposes a notion of individualism in which organisms are externally related to each other as well as to the environment. This understanding of individuality gives priority to isolation over relation and privileges competition over cooperation as the driving force of development.

No scientists or philosophers have more clearly understood the deleterious effects of Darwin's appropriation of Newtonian physics than the geneticist and evolutionary biologist Richard Lewontin and his Harvard colleague, geneticist and biomathematician Richard Levins. Describing the organism/environment relation, they write, "Classical biology, which is to say alienated biology, has always separated the internal and external forces operating in organisms, holding one constant while considering the other."[13] "Alienated biology" is the result of the combination of Cartesian dualism with Newtonian mechanism. In contrast to the oppositional either-or of classical biology, Lewontin and Levins develop what they label "dialectical biology," which presupposes the mediating logic of "both-and." As we will see in the following section, in their effort to correct Darwin's causal relation of the environment on the organism, they sometimes tend to overemphasize the constitutive relationship of the organism to the environment. Rovelli's relational interpretation of relativity theory and quantum mechanics suggests interplay between classical and quantum physics that creates the possibility of what can best be understood as quantum ecology. In quantum biology, organisms do not completely determine the environment, and the environment does not completely determine organisms. Rather, organisms and environments are coemergent and codependent.

For Lewontin and Levins, classical biology is alienated biology that is both a symptom and a cause of broader natural, social, political, and economic fragmentation.

> The dominant mode of analysis of how the physical and biological world as well as the social sciences has come into being, has been Cartesian reductionism. The Cartesian mode is characterized by four ontological commitments, which put their stamp on the process of creating knowledge:
>
> 1. There is a natural set of units or parts of which any whole system is made.
> 2. These units are homogeneous within themselves, at least insofar as they affect the whole of which they are the parts.
> 3. The parts are ontologically prior to the whole; that is the parts exist in isolation and come together to make wholes. The parts have intrinsic properties, which they possess in isolation and which they lend to the whole. In the simplest cases, the whole is nothing but the sum of its parts; more complex cases allow for interactions of the parts to produce added properties of the whole.
> 4. Causes are separate from effects, causes being the properties of subjects, and effects the properties of objects. While causes may respond to information coming from the effects (so-called 'feedback loops'), there is no ambiguity about which is causing subject, and which is caused object.[14]

It is noteworthy that Lewontin and Levins dedicate their book to "Friedrich Engels who got it wrong a lot of the time but who got it right where it counted." Giving a Marxist twist to Hegel's dialectical logic, they proceed to argue, "We characterize the world descried by these principles as the *alienated* world in which parts are separated from wholes and wholes reified as things in themselves, causes separated from effects, subjects separated from objects. It is a physical world that mirrors the structure of the alienated social world in which it was conceived."[15] This alienated world, they maintain, is the product of the capitalist economic system whose principles were formulated by Adam Smith.

Lewontin and Levins support their long and convoluted argument with ample data drawn from their analysis of biological organisms. Darwinism, they conclude, "places the organism at the nexus of internal and external forces, each of which has its own laws, *independent of each other and of the organism* that is their creation. In a curious way, the organism, the *object* of these forces, becomes irrelevant for the evolutionist because the evolution of the organism is only a transformation of the evolution of the environment. The organism is merely the medium by which the external forces of the environment confront the internal forces that produce variation."[16]

In addition to the difficulties posed by the externality of organisms and the environment, it is important to underscore three problematic implications of classical biology. Since organisms and the environment are externally related, they must be brought together by efficient causality, which is linear—causes are temporally prior to the effects, and effects are always proportionate to their causes. This leads to development that is gradual and continuous rather than accelerating and discontinuous. In emergent complex adaptive systems, by contrast, periods of relative stability are punctuated by aleatory disruptions, which issue in probabilistic but unpredictable phase shifts.[17] If organisms are *objects* alienated from the environment and acted upon by external forces, classical biology is a reformulation of the ontological and epistemological puzzles that we have encountered in our previous consideration of subject-object relations.

To overcome the impasse of traditional Darwinism, it is necessary to shift from the logic of either/or to dialectical logic of both/and. "The environment and the organism actively codetermine each other. The internal and external factors of genes and environment act upon each other through the medium of the organism. Just as the organism is the *nexus* of internal and external factors, it is also the locus of their interaction. The organism cannot be regarded as simply the passive *object* of autonomous internal and external forces; it is also the *subject* of its own evolution."[18]

Interruption

Parasite—parasite, interruption, static, noise.

> "Para" is a double antithetical prefix signifying at once proximity and distance, similarity and difference, interiority and exteriority, something inside a domestic economy and at the same time outside it, something simultaneously this side of a boundary line, threshold, or margin and also beyond it, equivalent in status and also secondary, or subsidiary, submissive as of guest to host, slave to master. A thing in "para," moreover is not only simultaneously on both sides of the boundary line between inside and out. It is also the

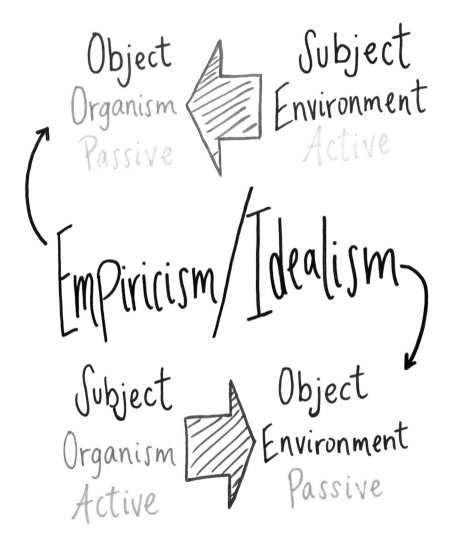

7.1 Empiricism-idealism

Object–Subject ⟶ Subject–Object
Organism   Environment
Passive–Active ⟵ Active–Passive

Quantum Ecology I

7.2 Quantum ecology I

> boundary itself, the screen, which is a permeable membrane connecting inside and outside.

"The boundary itself, the screen" which is *neither* inside *nor* outside. There is no parasite without a host and there is no host without parasites. Among many other things, "the word 'host' is of course the name for the consecrated bread or wafer of the Eucharist, from Middle English *oste,* from Latin *hostia,* sacrifice victim."[19] Turning everything inside out and outside in, parasites live from hosts that live from parasites.

Parasite. Noise is static that is never static but interrupts and often disrupts patterns of thinking and living long believed reliable. The interruption of interruptions is, of course, death. Even when expected, death comes like a thief in the night. The first sentence of the preface of this book is: "Philosophy is a prolonged mediation on death." Little did I know how true that claim would become. My writing has been disrupted at precisely this point by the death of my younger brother and only surviving sibling. I have long known but somehow momentarily forgot that every story is a ghost story. Host . . . Ghost . . . Hostel . . . Hotel . . . *Hospital.*

Beryl C. Taylor was a well-known equine veterinarian who specialized in thoroughbred and standardbred horses. In a manner similar to Gregor Mendel, who is widely regarded as the founder of modern science, Beryl selectively bred some of the finest racehorses in the country.[20] His illness arrived suddenly and took him far too soon. The night before I taught my last class, he was rushed to the hospital, and three weeks later he died. Though he and I had discussed the alternatives of burial or cremation, our conversations ended inconclusively. While no one in either the Taylor or the Cooper extended families had ever been cremated, his wife, Laura, and his sons, Mark and David, decided he would be cremated, and some of his ashes would be scattered in the natural park our father, Noel, had created in the town where we grew up. The remainder would be buried in the cemetery where Abraham Lincoln delivered his Gettysburg address. Beryl and Laura had both graduated from Gettysburg College, and our father was raised on a farm not far from the cemetery.

I have always known that the writing that really matters is directly or indirectly autobiographical. I did not, however, realize the autobiographical dimension of this book until I was forced to ponder Beryl's death and the scattering of his ashes. The Taylor homestead was only a few miles from where our ancestors, who were Mennonites and had fled persecution in Switzerland, originally settled.

My father's mother, Emma, was born in 1869 and his father, Calvin, was born in 1871, four and six years, respectively, after the end of the Civil War. Life on the farm of my father's youth had not changed much from the previous 150 years. There was no electricity, no running water, no indoor plumbing, no central heating. During the last year of his life, he wrote a 150-page memoir in which he described his life on the farm:

> I was out in the field plowing behind a mule at the age of eight. It was a long day—from sunrise to sunset. After we came in from plowing, harrowing, or cultivating, we had regular chores to do: water and feed all the livestock, which included chickens, hogs, cows, horses, guineas, turkeys, dogs and cats. Then milk the cows.... Before we went to school, we had to clean the stables of a dozen horses and mules and twenty cows.... After walking three miles to and from school, we had to return to the field and do the usual chores. That kept us busy until supper.

With no family support, he made his way from the farm to college, where he majored in biology. After graduating in 1929, he landed a job in a Pennsylvania coalmining town where he taught all the sciences and coached all the sports. There he met and married my mother, Thelma, who taught English. Somehow in the depths of the Depression they made their way to Duke University, where my mother earned a master's in English and my father received a master's in botany. Until his dying day, he could name every plant and flower he collected with Linnaeus's Latin binomial nomenclature. Sadly, he did not live long enough to know that his geochemist grandson, Aaron, who collected dirt to study climate change, named his half-Swedish daughter Linnaea after the great Swedish botanist.

When I was young, my father taught his first love, biology, in the suburban high school I attended. However, when the physics teacher retired in the early 1960s, he had to drop biology and teach physics. This was when the Soviet Union's launch of *Sputnik* had led to the radical revision of high school physics curricula from exclusively Newtonian classical physics to newer scientific theories that grew out of quantum mechanics. For some reason, he never returned to biology and never tried to reconcile physics and biology.

After retirement, my father returned to his lifelong interest in biology. By the 1980s, his growing ecological concerns led him to play a leading role in converting the last undeveloped area of our hometown into a natural park. Ever the teacher, he developed curricula to teach K–12 students about the environment. He identified all the flora and fauna in the park and wrote a booklet about its geological and natural history, which begins with the following epigraph: "A quiet sanctuary for all lovers of nature where one can relax and listen to the sounds of nature, enjoy its beauty or begin to understand the lesson that man depends on wildlife for survival. If one understands that man does depend on nature for his existence, one can also understand that man cannot continue to subdue the whole planet earth." Several years ago, I tried to reconcile my mother's interest in literature and my father's interest in biology and ecology in *Last Works: Lessons in Leaving*. My father, the scientist, prefaced the only book he ever wrote with Alfred Lord Tennyson's "Flower in the Crannied Wall."

> Flower in the crannied wall,
> I pluck you out of the crannies,
> I hold you here, root and all, in my hand,
> Little flower—but if I could understand
> What you are, root and all, all in all,
> I should know what God and man is.

"In these simple lines," my father wrote, "the poet Tennyson expressed the idea that even the humblest and most common form of life contains within it all the secrets and wonders of nature. That is not only a poetic but also a scientific truth. A knowledge of living things brings us a better understanding and appreciation of the world around us, of other people, of ourselves, and of the ultimate forces in the universe."[21] Not until the decision to scatter Beryl's ashes in Taylor Park and bury the remainder in Gettysburg did I realize that in what might well be my last work I am attempting to do what my father left undone—reconcile physics and biology in a way that will enable human beings to live with living things and thereby delay climate catastrophe at least for a while.

NEXUS

Quantum ecology is *radically relational*. As we have seen, even though quantum principles apply at the microlevel and not at the macrolevel, both micro- and macrolevels are characterized by the constitutive relationality of all events, entities, and beings. As Heisenberg explains,

> While in this way physics and chemistry have come to an almost complete union in their relations to the structure of matter, biology deals with structures of a more complicated and somewhat different type. It is true that in spite of the wholeness of the living organism a sharp distinction between animate and inanimate matter can certainly not be made.... [T]he kind of stability that is displayed by the living organism is of a nature somewhat different from the stability of atoms or crystals. It is a stability of process or function rather than a stability of form. There can be no doubt that the laws of quantum theory play a very important role in biological phenomena."

Heisenberg was acutely aware of the pernicious theoretical and practical effects of Cartesian dualism. By extending classical physics's Cartesian-Newtonian mechanism model to all reality, classical biology divides the world into irreconcilable fragments, which engage in endless conflict. This vision of world is both ontologically misguided and ethically suspect. Heisenberg correctly argues, "When we speak about the action of chemical forces, we mean a kind of connection, which is more complicated or in any case different from that expressed in Newtonian mechanics. The world thus appears as a complicated *tissue of events*, in which connections of different kinds alternate to overlap or combine and thereby determine the *texture of the whole*."[22]

What Heisenberg identifies as the "tissue of events," Merleau-Ponty labels the "tissue of things," which he describes as "flesh of the world." As we saw in the first chapter, the elemental is the matrix in which elements and all things mingle. "The flesh is not matter, is not mind, is not substance. To designate it, we should need the old term 'element,' in the sense it was used to speak of water, air, earth, and fire, that is, in the sense of a *general thing*, midway between the spatio-temporal individual and the idea, a sort of incarnate principle that brings a style of being wherever there is a fragment of being. The flesh is in this sense an 'element' of Being."[23] Neither immanent nor transcendent, neither mind nor body, neither information nor noise, flesh is the "intertwining" of the two. This is "the chiasm," symbolized by the Greek letter chi, χ. The intertwining of the chiasmus simultaneously brings together and holds apart the differences that constitute the identity of all things and all beings.

To understand this complicated tissue of events that determines the texture of the whole, it is necessary to return to the interplay of part and whole first formulated in Kant's notion of inner teleology and correlatively of self-organization, which forms the basis Hegel's dialectical logic. Lewontin and Levins follow Hegel by translating dialectical logic into dialectical biology:

> One way to break out of the grip of Cartesians is to look again at the concepts of part and whole.... What constitutes the parts is defined by the whole that is being considered. Moreover, parts acquire by virtue of being parts of a particular whole, properties they do not have in isolation or as parts of another whole.... [A]s the parts acquire properties by being together, they impart to the whole new properties, which are reflected in changes in the parts, and so on. Parts and wholes evolve in consequence of their relationship, and the relationship itself evolves. These are the properties of things that we call dialectical: that one thing cannot exist without the other, that one acquires properties from its relation to the other, that properties of both evolve as a consequence of their interpenetration.[24]

As we have seen, part and whole are reciprocally related in such a way that neither is what it is apart from the other. Since the whole *emerges* from but cannot be reduced to the parts, reductive analysis is both inaccurate and misleading. Rather than separate and externally related, parts within organisms, parts of the environment, and the relation between organisms and the environment are wholes that are coemergent and codependent. Joined in positive feedback loops, the environment shapes the organism, which shapes the environment.

Rovelli's relational interpretation of quantum mechanics suggests a way to develop a more effective understanding of ecology through a reconciliation of physics and biology. Extending quantum relations from the micro to the macro levels, Rovelli argues, life "is made up of individuals who interact with what surrounds them, formed by structures and processes that are self-regulating, maintaining a dynamic equilibrium that persists over time."[25] As in the quantum world, biological organisms and ecological systems are bound in a relational ontology that entails relativistic epistemology. The relational matrix that mediates organisms and environment is "*relative information*" that enables both cognition and communication. Rovelli argues that it is not sufficient to think of the world merely in terms of simple matter with variable properties. "Quantum physics is the discovery that *the physical world is a web of correlations: relative information*. The things of nature are not collections of isolated elements in haughty individualism. Meaning and intentionality are only particular cases of the ubiquity of correlations. There is a continuity between the world of meaning in our mental life and the physical world. Both are relations."

For organisms and environment, as for every other relationship, each member is simultaneously active and passive, observer and observed, subject and object. "This observation," Rovelli explains, "clarifies why we can only speak of meaning in the context of biological processes or processes rooted in biology. But it also grounds the notion of meaning in the physical world. *Meaning is not external to the natural world*. We can speak of intentionality without leaving the realm of naturalism. Meaning connects something with something else, *it is a physical link* that *plays a biological role*.... The distance between the way we think about the physical world and the way we think about our mental world diminishes."[26] Information, we have discovered, is a difference that makes a difference. For data to become information, they must be processed through a schema that filters out noise to determine differences that make a difference. Lewontin explains this complicated process:

> Organisms determine by their biology the actual physical nature of signals from the outside. They transduce one physical into quite a different one, and it is the result of the transduction that is perceived by the organism's functions as an environmental variable.... These are simple and obvious examples of the generality that it is the biology, indeed the genes of an organism that determine its effective environment by establishing the way in which external physical signals become incorporated into its reaction. The common external phenomena of the physical and biotic world pass through a transforming filter created by the peculiar biology of each species, and it is the output of this transformation that reaches the organism and is relevant to it.[27]

The activity of screening data to discern emergent patterns that constitute information is a cognitive process. As we discovered in chapter 6, cognition is not limited to the mental activity that occurs in the human brain but is distributed throughout physical and biological systems, as well as technological devices and networks. Since cognition is ubiquitous, "meaning is not external to the natural world." Not only are ecological systems mindful, but the mind is also ecological.

At this point in the argument, the most important insight afforded by radical relationality is that the interaction of organism and environment is *mutually constitutive*. The environment to which the organism adapts is not fixed, and niches are not predetermined. To the contrary, the environment processes data from organisms and changes with them. For organisms, genetic constraints developed over the long course of evolution are not fixed but adjust as necessary by the changing environment. For the environment, constraints and affordances developed throughout history shift with changing interrelations among organisms. In this way, organisms and environment form positive feedback loops in which organisms construct the environment that constructs organisms.

It is important to realize that the ancient philosophical and theological debates that shaped classical physics and classical biology continue to influence current scientific theories in ways that usually go unnoticed. Theories are never pure but always reflect the sociocultural context in which they emerge and that they contribute to shaping. Metaphors affect theories as much as theories inform metaphors. It was no more accidental that Newton formulated his mechanistic vision of the universe at the dawn of the industrial revolution than it was that Watson and Crick formed their theory of the genetic code at the dawn of the computer age. Nor is it an accident that critiques and revisions of what can only be described as genetic formalism or fundamentalism coincide with the network revolution brought about by the shift from centralized mainframe computers to personal computers and handheld devices connected in distributed networks.

One of the persistent debates in modern biology represents an alternative version of the perennial problem about the relationship between part and whole. Can biological organisms be understood better by studying the molecular processes that make them possible, or is a more holistic approach that considers the entire organisms and its relationship to its surroundings more effective? If parts are separate from each other and the whole is nothing more than the sum of the parts, a reductive analytic method is necessary. If, however, parts and whole are reciprocally related in such a way that each emerges from but cannot be reduced to the other, a more holistic dialectic method is required. While the former approach is characteristic of classic physics, the latter approach is more compatible with quantum physics.

Heisenberg suggested that quantum mechanics had much to teach biology, but it was left for Bohr and Schrödinger to explain what those lessons might be. In the essay entitled "Light and Life" (1933), Bohr argues that the principles of quantum mechanics not only require a revision of classical atomic theory but also create the conditions for a reconsideration of living systems that does not lead to the conclusion that biological organisms and processes can be reduced to mechanical processes. Schrödinger elaborated Bohr's suggestion in a popular book entitled *What Is Life?* (1944). He expanded his speculation by recasting the question: "How can events in space and time which take place within the spatial boundary of a living organism be accounted for by physics and chemistry?"[28] To answer this question, he attempts to demonstrate the applicability of quantum theory to biological organisms. Drawing on the work of the German physicist Max Delbruck and the Russian geneticist Timofeeff-Ressovsky's theory that genes are large molecules consisting of bonded atoms, Schrödinger argues that "life could be thought of in terms of storing and passing on biological information." This information "had to be packed into every cell" in what he described as "a 'hereditary code-script' embedded in the molecular fabric of chromosomes." To understand life, then, it was necessary to identify these molecules and "crack their code." James Watson acknowledged that Schrödinger's theory had inspired his research that led to the deciphering of DNA's structure. When Watson and Crick received the Nobel Prize in Medicine (1962), they were joined by a third person who contributed to the integration of physics and biology—the biophysicist Maurice Wilkins.

None of this would have been possible without major advances in molecular biology during the 1930s and 1940s. While Warren Weaver coined the term "molecular biology" in 1938, Max Delbruck had done much of the foundational work during the previous decade. As the field of molecular biology developed, two types of macromolecules became the primary focus of attention—nucleic acids—especially deoxyribonucleic acid (DNA), which constitutes genes, and proteins. The most pressing work during the following decades involved the effort to decipher the precise structure and function of DNA and figure out the relation between DNA and proteins that enables the development of the organism.

In their effort to offer a corrective to what they describe as "alienated biology" with an understanding of the organism as a neχus between interior and exterior forces, Lewontin and Levins confronted a latter-day version of the "preformatist" view of organisms characteristic of the rational morphology Darwin criticized. In his important book, *The Triple Helix: Gene, Organism, and Environment*, Lewontin writes, "Physicists speak of 'waves' and 'particles' even though there is no medium in which those 'waves' move and no solidity to those 'particles.' Biologists speak of genes as 'blueprints' and DNA as 'information.' Indeed, the entire body of modern science rests on Descartes's metaphor of the world as a machine, which he introduced in Part V of the *Discourse on Method* as a way of understanding organism but then generalized as a way of thinking about the entire universe."[29] A clue to Lewontin's argument can be found in the etymology of the word "evolution," which derives from the Latin *evolvere*, meaning to unroll. The biological theory Lewontin criticizes relies on what he regards as two mistaken metaphors—code or program and development. When the genetic code is interpreted as a pro-gram, evolution is seen as a process of unfolding. "Development, *Entwicklung* in German, is literally an unfolding or unrolling of something that is already present and, in some way, preformed. . . . Modern developmental biology is framed entirely in terms of genes and cell organelles, while environment plays only the role of a background factor. The genes in the fertilized egg are said to determine the final state of the organism, while the environment in which development takes place is simply a set of enabling conditions that allow genes to express themselves."[30] In molecular and developmental biology, the genetic code or program functions like the morphological type in rational morphology. The mature organism is the result of the unfolding of an essence that was present *ab initio*. While Darwin's criticism of the formal determinism of rational morphology errs in the direction of giving too much agency to the environment and too little to organisms, Lewontin and Levins's Marxist proclivities lead them to propose a "constructivist" alternative to genocentrism that sometimes tends to overemphasize organismic agency at the expense of environmental activity.

With the discovery of the genetic code, many scientists believed it would be possible to fully understand the structure and development of biological organisms. According to the programmatic developmental model, "molecules that reproduce themselves and that have the power to make the substances of which the organism is composed contain all the information necessary to specify the complete organism. The development of an individual is explained in standard biology as an unfolding sequence of events already set by a genetic program. The general schema of developmental explanation is then to find all the genes that provide instructions for this protein and to draw the network signaling the connections between them." The concurrent development of high-speed computers led to the Human Genome Project (1990–2003), which held out the promise of making this dream a reality. With the complete sequencing of the genome, it seemed that the Book of Life would finally be opened for all to read. Nobel laurate Walter Gilbert went so far as to declare that with the complete deciphering the genome, "we will know what it is to be human."[31]

While the sequencing of the genome was a major breakthrough, the dream of reducing trait or condition to gene proved to be a delusion because the project rested on several misleading assumptions: First, the recurrent presupposition that the whole is the sum of its parts and thus organisms can be understood through reductive analysis. Second, the belief that all the information necessary to produce proteins and enzymes necessary for life is contained in the genome. This implies that if the gene or sequence of genes can be isolated, it would be possible to establish a one-way causal link between gene and trait. In other words, the genotype would determine the phenotype. If this were to prove to be true, it would in principle be possible to develop therapies to prevent or cure disease as well as a more reliable form of genetics. Finally, as I have suggested, evolution involves the unfolding of what is implicit rather than an *intervolutionary process* in which different organisms as well as organisms and environment are entangled in mutually constitutive emergent complex adaptive networks.[32]

Interrelations both within organisms and between organisms and the environment are considerably more complex than this reductive analysis suggests. Contrary to the assumptions of the Human Genome Project, "not all the information about protein structure is stored in the DNA sequence, because the folding of polypeptides [that is, polymers consisting of a chain of amino acids] into proteins is not completely specified by their amino acid sequences. We do not, in fact, know what the rules of protein folding are, so no one has ever succeeded in writing a computer program that will take the sequence of amino acids in a polypeptide and predict the folding of the molecule."[33] Neither the precise relation between genes and proteins nor the exact functions of proteins is completely understood. Furthermore, another aspect to the internal structure of organisms complicates genocentrism. Rather than genes providing something like a prescribed program or algorithm for epigenesis, the ontogeny of an organism is the consequence of a unique interaction between the genes it carries, the temporal sequence of external environments through which it passes during its life, and random molecular interactions within individual cells." In other words, "the organism is not specified by its genes, but is a unique outcome of an ontogenetic process that is contingent on the sequence of environments in which it occurs."[34] This implies that the phenotype is not completely determined by the genotype but also conditions the genotype through interactions with the environment. Insofar as different phenotypes have different survival rates, the phenotype can modify the genotype. The genotype carries the memory of the organism, which provides the preliminary schema for noise and processing information. The flow of information is both from genes to organism and from organism to genes. As Depew explains, "The central dogma of molecular biology, according to which information flows solely from DNA to RNA to proteins, seemed to underwrite

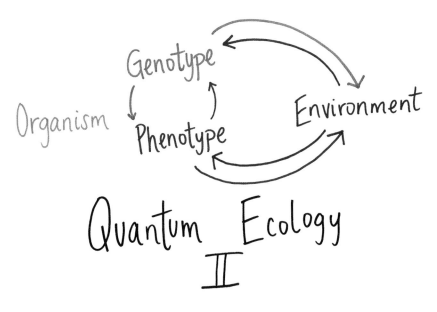

7.3 Quantum ecology II

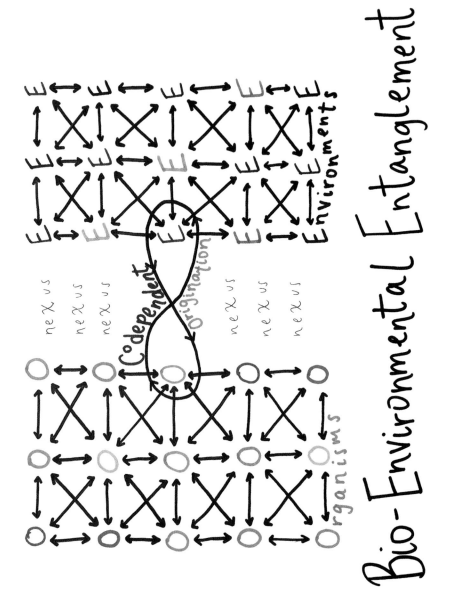

7.4 Bio-environmental entanglement

the isolation of evolutionary from developmental biology. It has become clear, however, that 'reverse information flow' does occur. Moreover, genomic DNA, once thought to be static and unchanging information store, turns out to be extremely fluid. Amplifications, deletions, rearrangements, and mutations occur frequently during development and in response to environmental stimuli."[35] It is, therefore, necessary to further modify the diagram of the interrelation between organism and environment.

The boundary between organism and environment is neither fixed nor stable but is constantly shifting in a way that folds interiority into exteriority and exteriority into interiority. Since nothing is itself by itself, every organism is the environment for other organisms. This complicates the parasite/host relationship. Consider, for example, the human microbiome, which consists of an estimated thirty-nine trillion microbial cells, including bacteria, viruses, and fungi. Which is parasite and which is host? What is noise and what is information? What is destructive and what is creative?

These questions suggest a final twist to the threads forming the tissue of life. As a result of the codetermination of organisms and the environment, each is constantly adapting to an other that is always changing in relation to it. Furthermore, the organism-environment relation is not one-to-one because both individual organisms and different species constantly influence each other. The relations that constitute both organisms and environments are both synchronic, that is, they are formed by connections determined at a particular moment (synchronic) and change over the course of time (diachronic). Like quantum particles, organisms and environments are not fixed things but are relational events whose occurrence is probabilistic and therefore uncertain.

To overcome the errors of Descartes, Galileo, Newton, and Darwin, it is necessary to return mind to nature and nature to mind. Since information processing is the matrix of communication among organisms and as well as between organisms and the environments, everything is joined in a quasi-cognitive network that forms an ecology *of* mind. As we have seen, cognition is not limited to the activity of the human brain but extends to bodies, plants, animals, and even technologies misleadingly labeled "artificial intelligence." By expanding the mind in this way, it becomes possible to imagine a superintelligence that returns human beings to earthly bodies rather than launching them into outer space or vaporizing them in screens and the ethereal clouds of virtual reality.

CHAPTER 8

MINDING THE BODY

That bacteria are simply machines, with no sensation or consciousness, seems no more likely than Descartes's claim that dogs suffer no pain.

—Lynn Margulis

DISEASE

When the phone rings in the middle of the night it's never good news. On the night before my last class after fifty years of teaching, my restless sleep was shattered by a call informing me that my brother had been rushed to the hospital suffering from sepsis. Having myself almost died of sepsis seventeen years earlier, dark memories rushed into my mind. Doctors had given me only a 20 percent chance of living, and my brother's case was much worse than mine had been, so it was hard to be optimistic.[1] By the time I arrived at the hospital, he was unconscious and on life support—a ventilator (lungs), an external pacemaker (heart), and dialysis (kidneys). But it was the eight minicomputers with tubes connecting them to his body that I found most overwhelming. All these devices transmitted data to larger computers, which projected it on screens in his room and at a central observational station. A team of dedicated nurses monitored his condition 24/7. This was not intensive care; it was acute care. Some memories you can never erase from your mind—they linger and change the way you see the world and understand life.

The following weeks were an excruciating rollercoaster ride. When I first saw him, I thought he had no chance of surviving and wondered what the quality of his life would be if he beat the odds. There was no way to know if he was aware of what had happened or if he could hear the words we spoke to him. So many of the questions I had been discussing in my classes and was preparing to write about in this book suddenly became frightfully real. Where *does* the human end and the machine begin? What *is* the relation between mind and body? *Does* thinking require consciousness? *Does* communication require consciousness? *Can* bacteria think? What *is* the relationship between parasite and host?

Disease disrupts quotidian activities and reminds you just how fragile the human body is. As I watched the doctors and nurses run endless tests of blood, urine, enzymes, and body gases, monitor their machines and, when necessary, making minute adjustments in his medications, Heidegger's suspicion of modern science's quantification and his critique of modern technology as an expression of the will to mastery seemed far less persuasive. What most impressed me was the intricate interrelation of bodily organs and processes. The heart, lungs, and kidneys function as a unified system—when one organ is compromised, the others suffer. It was not just the interplay of the organs, but also the incredibly delicate balance of the medications and process they facilitate. Nurses were constantly gathering information, making calculations, and recalibrating doses, but no matter how hard they tried, they never got it right. Pressors to raise his blood pressure increased his heart rate, and when the medications for his kidneys were right, his blood pressure was wrong. While all this was going on, his immune system was struggling to control the sepsis that was systematically shutting down his organs. Calculations of the most intelligent scientists and doctors as well as the most sophisticated computers were no substitute for the constant calculations our bodies are always performing.

This was not a new lesson for me. Half a lifetime ago, I developed type I diabetes.[2] Chronic illnesses are very difficult to manage, but if you are attentive, they have much to teach. What makes diabetes so challenging is that it never takes a holiday; it is necessary to monitor your body 24/7/365. Therapeutic technologies have changed significantly in the four decades I have been struggling with the disease. When I started, I had to test my blood seven or eight times a day, estimate my carbohydrate intake, calculate the amount of insulin I needed, and give myself injections. All of these calculations are now done and the insulin is delivered by a small computer I always wear on my belt. While I still must input the number of carbohydrates I consume and occasionally supplement my insulin dosage, most of the time my insulin pump is on autopilot. It calculates tests my blood glucose level every five minutes, decides whether it is rising or falling, and makes dosage modifications based on its prediction the future direction of my blood glucose. When everything is interrelated, minor adjustments can have disproportionate effects. Mistakes have consequential short-term and long-term effects and can even prove fatal. What we call normalcy, I have learned, is a very narrow bandwidth. My device is smart but not as smart as my body. Nearly forty years of minding my body has taught me to appreciate the extraordinary intelligence of the body's mind.

AUTOPOIESIS

Borrowing her title from Schrödinger, evolutionary biologist Lynn Margulis begins her book *What Is Life?* by quoting Thomas Mann's novel *The Magic Mountain*.

> What was life? No one knew. It was undoubtedly aware of itself, so soon as it was life; but it did not know what it was . . . it was not matter and it was not spirit, but something between the two, a phenomenon conveyed by matter, like the rainbow on the waterfall, and like the flame. Yet why not material?—it was sentient to the point of desire and disgust, the shamelessness of matter become sensible of itself, the incontinent form of being. . . . It was a stolen and voluptuous impurity of sucking and secreting; an exhalation of carbonic gas and material impurities of mysterious origin and composition.

Life: neither matter nor spirit, but something between the two. What lies between?
"Life," Margulis proceeds to explain, "is not distinguished by its chemical constituents but by the behavior of its chemicals. The question 'What is life?' is thus a linguistic trap. To answer according to the rules of grammar, we must supply a noun, a thing. But life on earth is more like a verb. It repairs, maintains, re-creates, and outdoes itself."[3] Life, then, is an event, or, more precisely, the self-organizing interplay of inextricably entangled events. As we have seen, the interpretation of life as a self-organizing process began with Kant's notion of inner teleology. In contrast to mechanisms in which separate parts are externally related, in the organism,

> the parts of the thing combined of themselves into the unity of a whole by being reciprocally cause and effect of their form. . . . Its parts must in their

collective unity reciprocally produce one another alike as to form and combination, and thus by their own causality produce a whole, the conception of which, conversely,—in a being posing the causality according to conceptions that is adequate for such a product—could in turn be the cause of the whole according to a principle, so that, consequently, the *nexus* of *efficient causes* might be no less estimated as an operation brought about by *final causes*.[4]

It is helpful to recall that Kant's interpretation of inner teleology or intrinsic purposefulness becomes the basis of modern art and literature as well as Marx's understanding of capital. In this context, the most interesting use of Kant's theory of self-organization is Hegel's appropriation of it to form the structure of the idea, which is the foundation of his ontological and epistemological speculative idealism. In his *Science of Logic*, the notion of life lies *between* the chapter entitled "Teleology" and the chapter entitled "The Idea of Cognition." Life, Hegel argues, is a *process*.

> In so far as the object confronts the living being in the first instance as an indifferent externality, it can act upon it mechanically; but in so doing it is not acting as on a living being; where it enters into relationship with a living being it does not act on it as a cause but *excites* it. Because the living being is an urge, externality cannot approach or enter it except in so far as it is in its own very nature already *in the living being*; therefore the action on the subject consists merely in the latter finding the externality presented to it *corresponding* [*entsprechend*]. This externality may not be conformable to the subject's totality, but at least it must correspond [*entsprechen*] to a particular side of it, and this possibility resides simply in the fact that the subject in its external relationship is a particular.[5]

This onto-noetic process is the minding of the body that returns mind to nature by rendering nature mindful.

The self-organizing structure of biological organisms Kant identified and Hegel elaborated is the basis of the influential notion of autopoiesis developed by Chilean biologists Humberto Maturana and Francisco Varela. "Autopoiesis" derives from the Greek words *auto* (self) and *poiein* (making). Maturana and Varela go so far as to claim that "the notion of autopoiesis is necessary and sufficient to characterize the organization of living systems."[6] Living organisms at every level from cells to mature humans are autopoietic in their origin, structure, and operational logic. In the absence of any external creator or designer, life must emerge through what theoretical biologist Stuart Kauffman defines as "autocatalytic" process. In explaining his criticism of genocentrism in terms of autocatalytic processes, Kauffman explicitly draws on Kant's interpretation of organisms as structured by intrinsic purpose or inner teleology.

> Immanuel Kant, writing more than two centuries ago, saw organisms as wholes. The whole existed by means of the parts; the part existed because of and in order to sustain the whole. This holism has been stripped of a natural role in biology and replaced by the image of the genome as the central directing agency that commands the molecular dance. Yet an autocatalytic set of molecules is perhaps the simplest image we can have of Kant's holism. Catalytic closure ensures that the whole exists by means of the parts, and they are present both because of and in order to sustain the whole. Autocatalytic sets exhibit the emergent property of holism.

Kauffman intends this analysis to describe very specific chemical processes. From its simplest to its most complex forms, life emerges in networks composed of webs of interconnected webs. For life to originate, the chemicals in the prebiotic "soup" must become sufficiently diverse and adequately, but not overly, interconnected. As the number of chemicals or nodes proliferates and interconnections grow, networks "evolve to a natural state *between* order and chaos, a grand compromise *between* structure and surprise." In other words, the emergence of life requires *neither* too little connectivity, which freezes systems, *nor* too much connectivity, which renders them chaotic. At the moment of self-organized criticality, *more becomes different*. This is the tipping point where order emerges from disorder and patterns develop from noise.

> At its heart, a living organism is a system of chemicals that has a capacity to catalyze its own reproduction. Catalysts such as enzymes speed up chemical reactions that might otherwise occur, but only extremely slowly. What I call a collectively autocatalytic system is one in which the molecules speed up the very reactions by which they themselves are formed: A makes B; B makes C; C makes A again. Now imagine a whole network of these self-producing loops. Given a supply of food molecules, the network will be able constantly to re-create itself

like the metabolic networks that inhabit every living cell, it will be alive. What I aim to show is that if a sufficiently diverse mix of molecules accumulates somewhere, the chances that an autocatalytic system—a self-maintaining and self-reproducing metabolism—will spring up become a near certainty.[7]

The structure and operation of the autocatalytic process Kauffman describes in the origin of life is isomorphic with the structure and function of living beings ranging from individual cells to mature organisms. In contrast to the critique of machinic interpretations of biological organisms developed by late eighteenth- and early nineteenth-century idealistic philosophers and romantic poets, Varela and Maturana attempt to synthesize the notion of machines and organisms. "We maintain that living systems are machines."

> An autopoietic machine is organized (defined as a unity) as a network of processes of production (transformation and destruction) of components that produces the components which: (i) through their interactions and transformations continuously regenerate and realize the network of processes (relations) that produce them; (ii) constitute it (the machine) as a concrete unity in space in which they (the components) exist by specifying the topological domain of its realization as such a network. It follows that an autopoietic machine continuously generates and specifies its own organization through its operation as a system of production of its own components, and does this in an endless turnover of components under the conditions of continuous perturbations and compensation of perturbations.

By identifying machines with their organization rather than their organized components, Varela and Maturana risk reinscribing the dualism between whole and parts or pattern and stuff that plagues classical biology. "Our approach" they explain, "will be mechanistic: no forces or principles will be adduced which are not found in the physical universe. Yet, our problem is that the living organization and therefore our interest will not be in properties of components, but in processes and relations between processes realized through components."[8] So understood, the organization/components relation resembles the one sided, top-down structure of both the Platonic form/matter and the posthumanist pattern/substance distinction. As our consideration of Hegel's dialectical logic and Lewontin and Levin's dialectical biology has made clear, parts and wholes or components and organization are inseparably interrelated. The identity of components is constituted by their differential relations within a whole, and there can be no organization without constitutive elements. The reciprocal relational structure among parts as well as between parts and whole results in a two-way process (that is, bottom-up and top-down) of coemergence.

While "autopoietic machines are autonomous; that is, they subordinate all changes to the maintenance independently of how profoundly they may otherwise be transformed in the process," "allopoietic machines have as their product something different from themselves."[9] This autonomous autopoietic structure first emerges at the cellular level. Kauffman's analysis of the chemical origin of the cell in terms of reciprocal autocatalytic processes suggests a corrective to the one-way causality implied in Varela and Maturana's account of autopoiesis. A cell is an internally heterogeneous network of recursive chemical reactions that adjusts and maintains its structure in relation to its environment. The cell initially emerges from the prebiotic chemical soup through a process of differentiation that creates a semipermeable membrane, which is the *nexus* simultaneously separating and connecting interiority with exteriority. Neither organism nor environment, this membrane is the margin of difference that is the condition of the possibility of the two. In his important book *Mind in Life: Biology, Phenomenology and the Sciences of Mind*, Evan Thompson explains,

> A cell is spatially formed by a semipermeable membrane, which establishes a boundary between inside and outside of the cell and the outside environment. The membrane serves as a barrier to free diffusion between the cell and the environment, but also permits the exchange of matter and energy across the boundary. Within this boundary, the cell comprises a metabolic network. Based in part on the nutrients entering from the outside, the cell sustains itself by a network of chemical transformations. But—and this is the key point—the metabolic network is able to regenerate its own components, including the components that make up the membrane boundary. Furthermore—and this is the second key point—without the boundary containment provided by the membrane, the chemical network would be dispersed and drowned in the surrounding medium. Thus the cell embodies a circular process of self-generation: thanks to its metabolic network, it continually replaces the components that are being destroyed, including the membrane, and thus continually recreates the difference between itself and everything else.[10]

The term "metabolism" specifies the chemical reactions necessary to sustain life. This involves three interrelated processes:

1. The conversion of the energy in food into the energy required for cellular functioning
2. The transformation of food into proteins, lipids, nucleic acids, and carbohydrates
3. The elimination of waste

The boundary marking the margin of difference that makes the identity of living organisms possible can be *neither* totally closed *nor* completely open. The cellular membrane functions as a filter that *screens* potentially productive and destructive molecules and chemicals. To explain the operation of membranes, Varela and Maturana draw a distinction between organizational closure and operational closure. Organizational closure refers to the self-reflexive network of relations *within* an organism, and operational closure denotes the "structural coupling" *between* organism and environment or *between* and among different organisms. The same principle we discovered in quantum ecology also operates at the cellular level—there can be no organism without environment, and no environment without organisms. The autonomy of organisms is "the capacity of a system to manage the flow of matter and energy through it so that it can at the same time regulate, modify, and control: (i) internal self-constructive processes and (ii) processes of exchange with the environment."[11] In their analyses of the cell/environment relationship, Maturana and Varela tend to be more interested in the way in which the environment reconfigures the relationship among internal components of the cell than in the correlative influence of cells on the structure and operation of its surroundings. I will return to this issue in the next section when considering Maturana and Varela's interpretation of the immune system in terms of autopoiesis.

This account of the cell-environment relationship points to one of Maturana and Varela's most suggestive and provocative claims: "A cognitive system is a system whose organization defines a domain of interactions in which it can act with relevance to the maintenance of itself, and the process of cognition is the actual (inductive) acting or behaving in this domain. *Living systems are cognitive systems and living as a process is a process of cognition.* This statement is valid for all organisms, with and without a nervous system."[12] In contrast to Descartes's and Newton's effort to take mind out of nature, Maturana and Varela follow Hegel and Rovelli by attempting to put mind back into nature. In chapters 5–6 I argued that, taken together, quantum theory and information theory lead to a pancognitivism in which physical, biological, cultural, and technological networks involve information processing that make unconscious and conscious communication possible. The communication necessary for life begins at the cellular level. In what appears to be a biological version of the ontological argument for the existence of God, Margulis maintains, "The mind and the body are not separate but are part of the unified process of life. Life, sensitive from the onset, is capable of thinking. The 'thoughts,' both vague and clear, are physical in our bodies' cells and those of other animals." Thought, like life, is matter and energy in flux; the body is the 'other side.' *Thinking and being are the same thing.*"[13]

To argue that living organisms, like quantum events, are quasi-cognitive or even cognitive requires an expanded interpretation of cognition. It is important to recall my insistence that consciousness presupposes cognition but cognition does not necessarily entail consciousness. Thompson condenses Maturana and Varela's argument in their widely influential book *Autopoiesis and Cognition* to a series of concise formulas:

1. Life = autopoiesis and cognition.
2. Autopoiesis entails the emergence of a bodily self.
3. Emergence of a self entails emergence of a world.
4. Emergence of a self and world = sense-making.

The most important point for cellular and organismic cognition is the claim that the emergence of self and world, and, I would add, organism and environment as well as cell and surroundings, is sense-making. Cognition is a process of sense-*making* rather than sense-receiving. Never antecedently given, "the environment is the sense [the organism] makes of the world. This environment is a place of significance and valence, as the result of the global action of the organism." While Maturana and Varela are concerned with the way in which the environment reconfigures the interrelation of internal components of purportedly autonomous systems, Thompson develops what he labels an "enactive" theory that emphasizes the interaction of cells and their biochemical environment:

5. Sense-making = enaction. Sense-making is viable conduct. Such conduct is oriented toward and subject to the environment's significance and valence. Significance and valence do not exist "out there," but are

enacted, brought forth, and constituted by living beings. Living entails sense-making, which equals enaction.[14]

This understanding of biological sense-*making* reinterprets adaptation as a cognitive process. As we saw in the last chapter, organisms do not adapt to a fixed environment with predetermined niches but adapt to a changing environment that is adapting to them.

In autopoietic systems, the relation between cell and milieu (*mi-*, middle, *lieu*, place) is *reciprocal* and, therefore, *relational*. Since nothing is itself by itself, every organism is an organism of other organisms. For example, in the human biome the 39 trillion microbial cells make up multiple interrelated ecosystems in which individual cells form and are formed by other intestinal microbial networks, which, in turn, form the biome that interacts with other organs and systems both within and beyond the supposedly individual body. Where is the boundary between cell and matrix, organism and environment, self and world to be drawn? Which is parasite and which is host? Bacteria cannot survive without the organism, and the organism cannot survive without bacteria. One of the gravest threats to human existence is the destruction of bacteria in the biome as the result of the overuse of antibiotics and the ingestion of antibiotics from eating meat of animals that have been fed excessive amounts of antibiotics and drinking water contaminated by antibiotics from animal waste and agricultural fertilizer. Sustaining the life of bacteria in the biome is a matter of life and death for human beings.[15]

The seeds for this understanding of the relationship between cells and milieu can be traced to questions about the entanglement between observer and observed in scientific inquiry. In a paper entitled "On Self-Organizing Systems and Their Environments," first presented in 1960, Heinz von Forester, who at the time was conducting research at the University of Illinois's Biological Computer Laboratory, describes the paradox of "observing systems." On the one hand, if the observer remains outside the system, it is not clear whether knowledge of its inner workings is possible; and on the other hand, if the observer is part of the system, self-observation leads to an infinite regress that makes complete knowledge of the system impossible. Von Foerster eventually became convinced that Maturana and Varela's notion of autopoiesis pointed to a way to resolve the paradoxes of reflexivity in a paper Maturana and his colleagues W. S. McCulloch and W. H. Pitts published two decades earlier, entitled "What the Frog's Eye Tells the Frog's Brain." By implanting electrodes in frogs' brains, Maturana and his colleagues were able to detect the way in which sensory receptors process the data the brain receives.

They report demonstrating that the frog's eye "speaks to the brain in a language already highly organized and interpreted instead of transmitting some more or less accurate copy of the distribution of light and the receptors." This insight leads them to conclude:

> Our results show that for the most part within that area, it is not the light intensity itself but rather the pattern of local variation of intensity that is the exciting factor. There are four types of fibers, each type concerned with a different sort of pattern. Each type is uniformly distributed over the whole retina of the frog. Thus, there are four distinct parallel distributed channels whereby the frog's eye informs his brain about the visual image in terms of local pattern independent of average illumination. We describe the patterns and show the functional and anatomical separation of the channels.[16]

The perceived object is not the result of light being imprinted on the retina as if it were a tabula rasa; to the contrary, perception, which is the basis of knowledge, is the result of subconscious processing registering patterns that are neither merely subjective nor merely objective.

Though Maturana and Varela never mention Bohr or Heisenberg, Thompson's account of the autopoietic interpretation of cognition makes its striking resemblance to the Copenhagen theory of quantum mechanics unmistakable:

> Cognitivists conceive of these representations as symbols in a computational "language of thought," and connectionists as constrained patterns of network activity corresponding to phase space "attractors" (regions of phase space toward which all nearby trajectories converge). In either case there is a strong tendency to adopt an objectivist conception of representation: representations are internal structures that encode context-independent information about the world, and cognition is the processing of such information.
>
> This objectivist notion of information presupposes a heteronomy perspective in which the observer or designer stands outside the system and states what is to count as information (and hence what is to count as error or success in representation). Information looks different from an autonomy perspective. Here the system, on the basis of its operationally closed dynamics and mode of structural coupling with the environment, helps determine what information is or can be.[17]

Whether at the quantum or the cellular level, information is both context dependent and relative to the system processing data. Since not all differences make a difference, systems must be able to filter noise by differentiating relations that are life enhancing from relations that are inconsequential or destructive. There can be no life without cognition and no cognition without life.

SMART CELLS

The staunch resistance to extending cognition and consciousness from humans to animals, plants, and even cells is the result of the residual anthropocentrism that continues to infect not only philosophy but also much modern science. In his insightful book *The First Minds: Caterpillars, 'Karyotes, and Consciousness*, cognitive psychologist Arthur S. Reber observes, "While there are enormous empirical, conceptual, and theoretical difficulties that confront anyone venturing into this epistemic domain, the real culprit is simple and we spotted it early on: that inordinate fascination with our own species and the narcissistic embracing of the specialness of human consciousness."[18] In recent years, a growing number of respected scientists have become more favorably inclined to embrace a pancognitivist positon.

While Maturana, Varela, and Thompson base their argument on a structural analysis of autopoiesis, other scientists take an evolutionary approach to prove cognition in biological organisms even at the cellular level. In her important article "The Conscious Cell," Lynn Margulis argues, "The evolutionary antecedent of the nervous system is 'microbial consciousness.' . . . The eukaryote fossil record begins about 2,000 million years ago. The first hard-shelled marine animals appeared 541 million years ago. This dates the evolution of hard parts. Brains appear later, but consciousness, awareness of the surrounding environment, starts with the beginning of life itself."[19]

No one has done more to explain the complexities of cellular cognition or what he labels "basal cognition," than Michael Levin, who is professor of biology at Tufts University and the director of the Allen Discovery Center at the Tufts Center for Regenerative and Developmental Biology. Prior to entering the field of biology, Levin was a software engineer and studied algorithms with which biological organisms implement complex adaptive behavior.[20] This research has important implications for new directions in artificial intelligence, which I will consider in chapter 10. Basal cognition, Levin maintains, does not require either a nervous system or a brain. "The emerging field of Basal Cognition focuses on the phylogenetic origins of learning and goal-directed activity, drawing a continuum between the humble origins of information in the metabolic homeostatic mechanisms of ancient cells and more complex learning, representation, and goal-directivity. Taken together, work on behavioral capacities of non-neural systems, and recent results of the molecular genomics pathways involved in learning and memory in brains, reveal a key insight necessary for broadening our understanding of the substrates of cognition."[21] Levin proposes a "cognitive approach," which, he believes, will lead to a "seismic shift" that will be nothing less than a new "Copernican revolution." From the most rudimentary to the most sophisticated systems, cognition involves perception, information processing, memory, learning, and decision. For any organism to survive, it must meet the following minimal requirements.

1. Identify and move toward what supports its existence (necessary chemicals, food, nurturing organisms) and avoid what threatens its existence (toxic chemicals, predatory organisms).
2. Process information about environmental light, temperature, and geomagnetic forces.
3. Make appropriate decisions based on relevant information.
4. Communicate information to other organisms.[22]

As we have seen in our consideration of autopoietic systems, cognition involves the interaction between organism and environment. Levin realizes that this boundary between the two is considerably more complex than is generally understood. "'Organisms' are not well-defined structures with clear boundaries. It is now well-established that most organisms are a patchwork of genomes, containing microbiota in addition to their metazoan host, and numerous complex cases of colonial organisms exist that make it very hard to draw specific lines about what exactly an organism is. Since microbes and parasites living on human and animal bodies contribute to personality traits and cognitive function, it is clear that the difficulties that evolutionary biologists have had in defining an 'organism' extends to cognitive science, with aspects of basal cognition of an agent receiving contributions from multiple co-existing genomes."[23] Since every organism is an organism of organisms, interiority (organism) and exteriority (environment) are folded into each other, forming a convolutional network. Communication

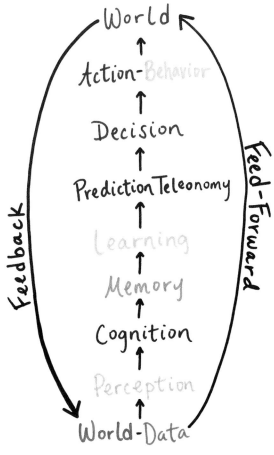

8.1 Biological cognition-decision

between and among these networks requires multiple layers of interpenetrating cognition. None of this activity presupposes consciousness or requires a nervous system or brain.

In an article provocatively entitled "Do Cells Think?," Harvard professor of neuroscience, molecular biology and applied physics Sharad Ramanathan argues that coping with a fluctuating environment "requires that the cell possess a sensory mechanism for detecting the changing condition and a signal transduction appropriate for converting that perception into an appropriate response. For instance e coli [which causes sepsis] is a robust example of cells sensing a specific signal and responding by simply changing their rate of tumbling."[24] After data have been screened, the resulting information must be processed and stored in memory. Thinkers as far back as Plato have recognized the necessary relation between cognition and memory. In his mediation on thinking and memory in his *Confessions*, Augustine writes, "In fact what one is doing is collecting them [that is, memories] from their dispersal. Hence the derivation of the word 'to cogitate.' From *cogo* (collect) and *cogito* (recollect).... But the mind has appropriated to itself this word ('cogitation'), so that it is only correct to say 'cogitated' of things which are 're-collected' in the mind, not things re-collected elsewhere."[25] The information collected and arranged through the process of cognition is stored in the memory, which Augustine famously described as "the belly of the mind." Modern science has translated Augustine's biological metaphor for memory into something like a multilayered "bioelectrically mediated" digestive process. Levin explains, "Memory does not exist at one level—that of neural networks—it can span levels of organization."[26] At the cellular level, interactions with the environment are encoded and stored to form the cell's memory. The microbial mind could not function without cellular memory. Ramanathan concludes his account of e coli activity by noting that "even this simple process requires that the cell be able to compare the level of the stimulating agent at two different times and to do so with equivalent sensitivity over multiple orders of magnitude of stimulant concentration. The ability of the cell to perform the former process means that the cell has the capability to remember a prior condition long enough to compare it to a current condition."[27]

Columbia University neuroscientist Eric Kandel received the 2000 Nobel Prize in Physiology and Medicine for his research on the biochemical basis of neuronal changes accompanying memory and learning. One of his most significant contributions has been the identification of the proteins that must be synthesized to convert short-term into long-term memories by means of synaptic modification.

Singular cellular organisms that collect data and process, store, and recollect information have the ability to learn. Since synapses are constantly being modified as a result of both repeated and new experiences, memories are constantly being revised. This is the biochemical basis of the probabilistic notion of time I discussed in chapter 4. Stored information that is constantly reconfigured forms the fluid template or genetic algorithm that results in the learned behavior necessary for effective adaptation to perpetually changing environments. Memory is not just about the past but is also what makes the anticipation of the future possible. Without informed predictive anticipation there would be no decisions.

Single-celled organisms not only think, they also choose. According to Berkeley biochemist Daniel Koshland,

> "Choice," "discrimination," "memory," "learning," "instinct," "judgment," and "adaptation" are words we normally identify with higher neural processes. Yet, in a sense, a bacterium can be said to have each of these properties.... It would be unwise to conclude that the analogies are only semantic since there seem to be underlying relationships in molecular mechanism and biological function. For example, learning in higher species involves long-term events and complex interactions, but certainly induced enzyme formation must be considered one of the more likely molecular devices for fixing some neuronal connections and eliminating others. The difference between instinct and learning then becomes a matter of scale, not of principle.[28]

To understand cellular decision making, it is important to distinguish teleology from teleonomy. I have argued that one of the distinguishing features of modern science is the shift from Aristotelian final causality to efficient causality. Final causality is purposeful or end-directed activity that presupposes divine or human consciousness and self-consciousness. Colin Pittendrigh coined the term "teleonomy" in 1958 to specify the purposeful activity of structures and living organisms brought about by processes like natural selection. Teleonomy, unlike teleology, does not require divine or human consciousness or self-consciousness. Cellular cognition is the condition of the possibility of cellular teleonomic decision-making. Reber summarizes the details of this process:

> To be able to respond effectively to shifts in the molecular concentrations in the surrounds, the individual organism needs to be able to hold onto a record

of the circumstances of its past. If a bacterium detects a particular concentration of lactose molecules in the water it just swam into, it is highly adaptive to be able to compare these conditions with a memory of the concentration in the environment just left the comparison will determine its subsequent movements. If a bacterium remembers that the kinds of ambient temperatures it currently is experiencing have, in the past, been routinely followed by a sudden rise in calefaction, it will be highly adaptive to be able to make an *anticipatory adjustment* in metabolism to accommodate these future circumstances.[29]

Memory and anticipation meet in decision. "Since the percentage of the cells choosing one response over the other is in proportion to the relative strengths of the signals, the cells are essentially 'weighing the odds' that one response or the other is the more favorable solution to the complex situation."[30] Cognition and decision are not limited to human beings but occur at every level of life. The basal cognition of smart cells is the most rudimentary form of the mind's activity that extends to the body as a whole.

SMART BODIES

Nothing shows bodily cognition more clearly than the immune system. I have previously noted the inextricable entanglement of metaphors and scientific concepts with theories. While scientists tend to insist that their theories and experiments disclose facts, one generation's concepts invariably become the next generation's metaphors. The immune system has been especially susceptible to metaphorization. Over the years, at least eight different metaphors or combinations of metaphors have oriented research and guided treatment: language and semiotics, illegal alien invasion, espionage, military warfare, cybernetics, code, information, and networks. One of the reasons the immune system is so important both physiologically and philosophically is that immune response not only protects the health of the body but also constitutes the biological identity of the self. One of the leading texts bears the suggestive title, *Immunology: The Science of Self-Nonself Discrimination*.[31] Identity is not merely a psycho-social construction but is also grounded in biochemical and biological process beginning at the cellular level. Rather than stable subjects, organisms and bodies are ongoing *relational events*. At the most rudimentary level, the immune self establishes and maintains the identity of the organism by differentiating it from and relating it to other organs and organisms. There have been three primary ways the immune system has been understood: clonal selection theory, autonomous network theory, and information theory. Each of these theories illuminates important aspects of the structure and function of the immune system, but none is fully satisfactory.

The field of immunology began in the late nineteenth century when the Russian zoologist Elias Metchnikff identified phagocytic cells that destroy invading pathogens. From the early years of investigation through the era of the Cold War and down to the present-day, the preoccupation with migration and illegal aliens has fostered immunological rhetoric that combines metaphors of alien invasion, espionage, and military invasion. In her informative book *Flexible Bodies: Tracking Immunity in American Culture from the Days of Polio to the Age of AIDS*, Emily Martin explores way the military and espionage language used by scientists to describe the immune system spills over into the media and popular culture.[32] She reports that the cover of *Time* magazine on May 23, 1988, carried the headline, "The Battle Inside the Body: New Discoveries Show How the Immune System Fights Disease." Two years later the cover story in *U.S. News and World Report* was entitled "The Body at War: New Breakthroughs in How We Fight Disease."

All these theories are based on the oppositional logic of Either-Or. In the body's struggle with disease, "invaders" are antigens, and the defense system is formed by an "army" of lymphocytes, some of which secrete antibodies and some of which attack cellular antigens directly. The cells involved in the immune system are lymphocytes, which are a class of white blood cells. Though all lymphocytes are produced in bone marrow, some migrate to lymphatic tissues and organs, and others pass through the thymus gland, where they undergo further specialization. The former are called B-lymphocytes, which produce the antibodies that are responsible for humoral response, and the latter are T-lymphocytes, which are responsible for cell-mediated response and are known as killer T-cells. Millions of foreign agents attempt to sneak across the body's borders. These hostile invaders are marked by antigens that bear a unique identity formed by specific molecules on their surface. If the organism is to survive, this protein structure is the code that must be deciphered by the body's defense system.

A major breakthrough in immunological theory occurred in 1954 when Niels Jerne discovered that the human body is programmed to detect and respond to an astonishing one hundred million or more antigens. Jerne credits his discovery to his understanding of Kierkegaard's account of the Platonism implicit in Hegel's onto-noetic system.

> "Can the truth (*the capability to synthesize an antibody*) be learned? If so, it must be assumed not to pre-exist; to be learned, it must be acquired. We are thus confronted with the difficulty to which Socrates calls attention in the *Meno*, namely that it makes as little sense to search for what one does not know as to search for what one knows; what one knows one cannot search for, since one knows it already, and what one does not know one cannot search for, since one does not even know what to search for. Socrates resolves this difficulty by postulating that learning is nothing but recollection. The truth (*the capability to synthesize an antibody*) cannot be brought in, but was already immanent."
>
> The above paragraph is a translation of the first lines of Søren Kierkegaard's *Philosophical Fragments*. By replacing the word "truth" with the italicized words, the statement can be made to represent the logical basis of selective theories of antibody formation. Or, in the parlance of Molecular Biology, synthetic potentialities cannot be imposed upon nucleic acid, but must pre-exist.[33]

According to Jerne's natural selection theory, the body randomly produces millions of antibodies, each of which is capable of recognizing a single antigen that might be encountered. Everybody's set of antibodies differs because the underlying gene arrangement that gives rise to so many different antibody genes is random. The revolutionary aspect of Jerne's theory is his contention that antibodies are produced in the absence of antigens. In other words, the body is *preprogrammed* for possible future battles.

In an article entitled "A Modification of Jerne's Theory of Antibody Production Using the Concept of Clonal Selection" (1957), Frank Macfarlane Burnet specified the precise mechanisms at work in the immune response in a way that makes the cognitive function of the immune system perfectly clear.[34] The strategies for recognizing antigens are different in humoral and cell-mediated response. In humoral response, a specific antibody produced by a B-lymphocyte and present as a receptor on the surface of the lymphocyte recognizes a limited range of antigens and responds by producing more antibodies, which are released into the bloodstream to combat the infection. The recognition mechanism in

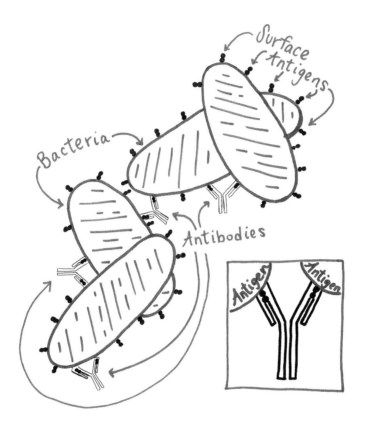

8.2 Antigen-antibody structure

B-lymphocytes is similar to the digital function in which the decoding of a message turns on a switch that sets in motion antibody production. The antigen-antibody dyad forms something like a lock-key structure with part of the surface structure of the antigen, the epitope, serving as a lock that can be opened by a customized key. The keylike part of the antibody is a protein structure that fits the antigen epitope. Since there are at least one hundred million antibodies in the human body, the sorting process necessary for an effective response is extraordinarily complex. When the key fits the lock, the immune response begins.

Once the antibody recognizes the antigen and understands what it is saying, the antibody knows how to respond. The response is twofold. First, the antibody's recognition of the antigen triggers the production of more antibodies by B-lymphocytes. According to Burnet's clonal selection theory, the lymphocyte clones itself, thereby producing additional lymphocytes, which make the same antibody. These antibodies are released into the bloodstream and circulate to destroy antigens. One of the most remarkable features of the immune system is that some of the cloned lymphocytes are brought to the brink of response and then deactivated. These antibodies are held in reserve and constitute a critical part of the organism's cognitive apparatus—the *immunological memory*. If the body is attacked again by the same antigen, these reserves can be activated to respond quickly. This mechanism explains why, in some cases, a person who has had a disease once is immunized against further occurrences of the illness.

T-lymphocytes are more complex than B-lymphocytes. There are three basic types of T-cells: regulatory T-cells, which suppress or turn on antibody production; suppressor T-cells, which turn off antibody production when there is an adequate humoral response; and killer T-cells, which directly attack infected cells. The killer T-cells need the assistance of cells in the blood known as macrophages, which consume the foreign microbes and present a portion of the protein structure to the T-cells. This enables T-cells to distinguish self from nonself and, using the lock-key pattern, pierce the membrane of the alien cell and release fluids that eventually kill it. It is important to note that the immune system does not always function properly and sometimes turns on the body it is supposed to correct. This results in autoimmune diseases like rheumatoid arthritis, lupus, multiple sclerosis, and, of course, type-I diabetes.

While clonal selection theory remains widely influential, it has not gone unquestioned. Two lines of criticism are especially relevant in this context—Irun Cohen's cognitive network theory and Varela's autonomous network theory. Cohen, who is an immunologist at the Weizmann Institute of Science in Israel, presents his theory in a brief paper entitled, "The Cognitive Principle Challenges Clonal Selection."[35] The designation of this theory as "cognitive" is somewhat misleading because all theories of the immune system demonstrate that immune response is essentially cognitive. Though he does not invoke philosophical ideas, Cohen's argument presents a biological version of the age-old debates between a priori and a posteriori epistemologies. As we have seen, the combination of Jerne and Burnet's clonal selection theory can be understood as involving something like an a priori epistemology. The proliferation of antibodies creates an *internal* program that filters disruptive noise by targeting specific antigens. Cohen, by contrast, argues that it is antigens, which are *external* to the immune system, that are responsible for organizing the immune system itself. "The paradigm, which for over three decades, has organized immunological thinking is clonal selection. The clonal selection paradigm holds the antigens responsible for organizing the immune system; only those lymphocytes bearing receptors that match the antigens encountered by the individual flourish." Since the identity of the immune self initially is constituted by a reaction to a foreign agent that is different or other, Cohen argues that this identity is negative. In other words, the immune self is defined through a process of double negation in which the self is not the nonself. Cohen attempts to replace this negative identity with a positive identity in which the immune self establishes its identity independent of and prior to the encounter with other cells that would destroy it.

> Cognitive paradigms are founded on the idea that any system which collects and processes information will do its job most efficiently by having an *internal representation of its subject*. Simply put, a cognitive system is a system that extracts information and fashions experience out of raw input by deploying *information already built into the system*; in a sense, a cognitive system is one that knows what it should be looking for. This *internal information*, which precedes and imposes order on experience can be seen conceptually as a *blueprint* for dealing with the world. In the abstract, cognitive systems can be said to behave with a sense of direction; their *internal organization* endows them with a kind of intentionality. Cognitive systems, then, are not passive processors or recorders of information; they are designed to seek very particular information from the domain in which they operate. (Emphasis added.)

According to Cohen's theory, the immunological self harbors an internal organization that is not dependent on external factors. The immune system processes the data from its surroundings and forms comprehensible information. Cohen describes the internal organization that makes information processing possible in different ways—blueprint, program, internal images, preformed images. He gathers these terms together with the idea of a "neurological homunculus," which is "encoded in the brain function not only to organize the information entering from the external environment." It is important to understand that this homunculus is not is a stable entity but a constantly changing process. Attempting to explain this odd construction, Alfred Tauber suggests, "Perhaps we require a new grammar, for the self is *neither* subject, *nor* object, but is actualized in action. The self becomes, on this view, a subject-less verb or perhaps predicate."[36] From this point of view, the immunological self is an *ongoing event* rather than a discrete entity.

While there are several insightful aspects of Cohen's argument, there are problems that must be addressed. By concentrating on the *internal* organization of the immune system, Cohen establishes an original separation between self (that is, subject, immune system) and world (that is, object, antigen). The process of immunological cognition involves the *representation* of a purported separate object to a supposedly autonomous subject. Without recognizing or acknowledging the genealogy of his epistemology, Cohen translates Kant's categorial processing of the sensible manifold of intuition into biological terms: "Inherent in any immune response is noise, a paralyzing degree of polyclonal activation. Indeed, some parasites neutralize the immune system by activating it polyclonally. The large numbers of potential epitopes present on every antigen could never be allowed a free hand in selecting all potentially reactive lymphocyte clones. The system must have discriminating filters. To make sense out of the input is to filter out the inessential noise and to concentrate on a manageably small part of reality. Focusing creates signal." However, just as Kant never explains the origin of forms of intuition and categories of understanding, so Cohen never explains the origin of his "conceptual blueprint for dealing with the world." Presumably he would attribute the internal *Grundriss* to the genetic code or program. But this begs the question by pushing the problem of origin farther back. Moreover, this solution would exacerbate the unidirectional relation between genotype and phenotypes that we have previously considered. Furthermore, Cohen does not realize the ontological implications of his immunological epistemology. In a manner similar to the way Kant's transcendental analysis of the structure of the mind establishes an unbridgeable gap between subject and object, Cohen's focus on the internal structure of the cognitive network creates an oppositional dualism characteristic of Cartesian philosophy and classical physics. There is, however, one point where he indirectly suggests a way to overcome this impasse. In a supplementary "digression," he suggests that the immune system and antibodies form a host/parasite relationship.

> It is worth digressing to note that, just as the immune system is adapted to our parasites, our parasites are adapted to the immune system. Their life depends on it. The open-ended antigenic variation of parasites is the parasitic internal image of the effectively open-ended repertoire of antigen receptors. The parasites' internal image of the host may also include host antigens and host immune system regulatory molecules, internal images that enable the parasites to survive in the face of host immunity. Microbial adhesion molecules actually encode the microbe's image of the host's anatomy. Whether the host or the parasite is the information system or the environment is relative to one's point of view.

As we have seen, parasite and host are bound in *a reciprocal relation of coemergence and codependence*. Are bacteria parasites and the gut host, or is the gut a parasite and bacteria the host? Neither can survive without the other; parasite and host, like information and noise, are not opposites but are mutually constitutive. "The parasite," as Michel Serres insists, "is the essence of relation."[37] This relation is the nexus that is *neither* inside *nor* outside entangled cells. "A thing in para," literary critic J. Hillis Miller explains, "is not only simultaneously on both sides of the boundary line between inside and out. It is also the boundary itself, the screen which is a permeable membrane connecting inside and outside. It confuses them with one another, allowing the outside in, making the inside out, dividing and joining them. It also forms an ambiguous transition between one and the other."[38] The immune system constantly negotiates the boundary between smart cells and devious cells that would do them harm.

Jerne, Burnett, and Cohen all invoke the metaphor of a network to explain the structure and operation of the immune system, but none adequately elaborates its implications. Varela takes up this task by extending the notion of autopoiesis to form his autonomous network theory. His analysis resolves some of the difficulties with Cohen's cognitive network theory, but problems remain. Jerne concludes his influential article by comparing the immune system to the nervous system: "A key idea of a network perspective is that the on-going activity of units, together with constraints from the system's surroundings, constantly produces *emerging* global patterns over the entire network which constitutes its performance. The network *decides* how to tune its component elements in mutual relationships that give the entire system (recognition, memory, etc.), which is not available to components in isolation."[39] For Maturana and Varela, Jerne fails to develop the implications of the network metaphor. Commenting on the problems with the clonal selection theory, they argue:

> There is, however, a more fundamental weakness in the theories that propose to explain the nature of immune responsiveness by considering only the origin and fate of individual clones of lymphocytes. These theories neglect the need to harmonize the activities of these clones with *one another* in the organism *as a whole*. This neglect of a *holistic* view, this desire for a simple [rather than a reciprocal] causality fails to incorporate two of the most important developments in immunology: the genetic control of immune events, and the cellular *interactions* in regulating the quality and magnitude of immune events.

In both Jerne's theory and Burnett's modification of it, lymphocytes remain independent of each other and are *individually* programmed to identify and destroy a *single* antigen. To overcome these deficiencies, Maturana and Varela argue, "we must replace the notion of the lymphoid system as a collection of unconnected lymphocyte clones carrying receptors directed outward (toward unpredictable encounters with foreign materials), with the notion of a network of interacting lymphocytes, where the receptors are turned *inward*, making the activities of the whole lymphoid system curl and close onto itself."[40] Two important points in this argument must be underscored. First, the immune system is interpreted as an *aggregate of separate* clones; and second, lymphocytes are understood to be *inwardly* directed in a relational network rather than outwardly directed. Maturana and Varela frame the issue in terms of a shift from a representational to a relational ontology and epistemology:

> Information—together with all of its closely related notions—has to be interpreted as codependent or constructive, in contradistinction to representational or instructive. This means, in other words, a shift from questions about

semantic correspondence to questions about *structural* patterns. A given structure determines what constitutes the system and how it can handle continuous perturbations from its surroundings, but needs no reference whatsoever to a mapping or representation for its operation.... The notion of information as representation is ultimately independent of the system's structure; but it is for the external—better still, for the whole tradition describing the situation—that the externality of the supposed world be mapped exists at all. By insisting on looking at cognitive processes as mapping activities, one systematically obscures the codependence, the intimate interlock between a system's structure and the domain of cognitive acts, the informative world which it *specifies through its operation*. Informational events have no substantial or out-there quality; we are talking literally about *in-formare*: that which is formed from within. In-formation appears nowhere except in relative interlock between the describer, the unit, and its interactions.

This analysis represents a fundamental shift from a focus on the interaction of antibodies (subject, organism) and antigens (object, environment) to the interplay of the internal elements of the immune system. Instead of separate antibodies interacting with a particular antigen, clones are "'designed' to react with *endogenous components* of the network."[41] This interpretation of the immune system reenacts the reciprocal relationship of part and whole, which originates with Kant's account of self-organization and is elaborated in Hegel's dialectical logic, by transfiguring clones from separate entities to integral members of a perpetually morphing web.

This is the point at which problems arise. The structural autonomy creates a closed network that separates the immune system from its environment. When antigens attack the organism from without, the immune system's immediate response is the reconfiguration of its *internal* elements. Once this has occurred, *the network as a whole* responds to the provocation. Varela summarizes his conclusions in an article entitled "Cognitive Networks: Immune, Neural, and Otherwise," which is included in a series sponsored by the Santa Fe Institute.

A key idea in a network perspective is that on-going activity of units, together with constraints from the system's surroundings, constantly produces *emerging* global patterns over the entire network which constitutes its performance. The network itself *decides* how to tune its component elements in mutual relationships that give the entire system (recognition, memory, etc.), which is not available to the components in isolation. These emergent properties are the great attractive feature of the network approach, and one that needs to be explored more explicitly for immune networks.[42]

Varela's account of the relational structure of the immune network is a significant advance beyond the clonal selection theory and the cognitive network theory. However, his preoccupation with the autonomy of autopoietic networks repeats Cohen's overemphasis on the internal organization of the immune self, which sets up an opposition between the immune system and its environment.

To overcome this limitation, it is necessary to extend the principle of constitutive relationality from the internal relations of the components of the immune network to the dialectical interplay of all organisms as well as the entanglement of organisms and their natural and nonnatural environments. While Varela nowhere addresses this problem, in a manner reminiscent of Jerne's comment on the parasite/host relation, he glimpses but does not elaborate a way to overcome the residual dualism in his theory of autopoietic autonomy. "In strict accordance with this view of in-formation, we shall see that the presence of the observer (of the observer-community, of the tradition) becomes more and more tangible, to the extent that we have to build upon a style of thinking where the description reveals the properties of the observer rather than obscuring them.... It is a view of a *participatory knowledge* and reality, which we see rooted in the cognitive, informational processes of nature from its most elementary cellular forms." We have previously encountered Maturana and Varela's consideration of the conundrum of the observer/observed relationship in scientific experiments in the investigation of the interaction of the frog's eye, brain, and world. At this point their concern was to find a way out of the infinite regression of self-reflexivity. Here Varela shifts his attention from the subject's relation to itself to the interrelation of (immune) self and world. This relation involves a *participatory* knowledge and reality. The key word in this formulation is "participatory."

This is not the first time that we have encountered this important term. We also saw it in Wheeler's interpretation of quantum theory. Responding to the potential solipsism implied in the Copenhagen interpretation of quantum theory, Wheeler elaborates an account of a *participatory universe*. In contrast to idealism in which the subject (observer) is active and the object (observed) is passive and empiricism in which the object (observed) is active and the subject (observer) is passive, the participatory universe of quantum mechanics reveals the thoroughgoing entanglement of a subject that is simultaneously an object and an object that is simultaneously a subject. Varela identifies two ways in which the observer becomes apparent in the act of observing. Without ever invoking Wheeler's name, Varela confirms his participatory universe when he writes,

> On the one hand, we see the necessity of acknowledging the role of the process through which we distinguish the unities or entities we talk about: the way the world is split into distinct compartments, and the way such discriminations and distinctions are related by levels and relationships between distinct levels. The maintenance of the system's identity—its autonomy is a distinct and irreducible domain with respect to the function of the system and its interactions. These two phenomenal domains are related only through *our* descriptions, and these relationships do not enter into the operation of the system we are concerned with. . . .
>
> A second way in which the observer enters into this view of unities and their information is that we ourselves fall into the same class—there is a continuity in the biological sense, and in the cognitive mechanisms that operate elsewhere in nature. Thus what is basically valid and for the understanding of the autonomy of living systems, for cells and frogs, carries over to *our* nervous system and social autonomy, and hence to a naturalized epistemology, which is not without its consequences. It forces us to a renewed understanding of what physical nature can be that is inseparable from our biological integrity, and what we ourselves can be that are inseparable from a tradition.[43]

In a participatory universe, nothing is autonomous because everything is interrelated. Since physical, chemical, and biological processes are cognitive, mind is in nature and nature is mindful. Everything and everybody is communicating with everything and everybody else in what Maurice Blanchot describes as an "infinite conversation."

CHAPTER 9

INFINITE CONVERSATIONS

It is obvious that every creature different from us senses different qualities and consequently lives in a different world from that in which we live. Qualities are an idiosyncrasy peculiar to man; to demand that our interpretations and values should be universal and perhaps constitutive values is one of the hereditary madnesses of human pride.

—Friedrich Nietzsche

GHOSTLY WHISPERS

Occasionally, time plays cruel tricks on us. I have recounted the story of my brother's illness, which began the night before my final class after fifty years of teaching. The story did not end well—he died of complications from sepsis, the disease that almost killed me seventeen years earlier. He was unconscious and on a ventilator three and a half weeks, but finally the parasite killed the host. He never regained consciousness. I visited him twice during this period. An MRI gave no evidence of neurological damage, but doctors did not know if he could hear what we said to him or was in any way aware. Questions abound. If he were not conscious, was he any less human than if he were? Might he have been thinking without knowing he was thinking? Could he hear us? Could he understand what we were saying to him? Could he have been thinking without being able to communicate? Sometimes his eyes seemed to blink, and every now and then his lips quivered, leaving our questions forever unanswered.

The first sentence of the preface of this book is "Philosophy is a prolonged meditation on death." No one has defended this claim more insistently than Heidegger. But what if Heidegger is wrong—wrong not only about his interpretation of death but wrong about much more as well? As I have noted, Beryl was an equine veterinarian who specialized in breeding thoroughbred and Standardbred racehorses. The day before I planned to write this chapter, I had to deliver the eulogy at his memorial service. It was the hardest thing I have ever done. The service was held at Fair Winds Farm, which is one of the leading Standardbred farms in the country. Whenever I visited my brother, I accompanied him on his rounds, and I had been to this farm many times. Before the service, I slipped away for a few minutes alone. I went to the paddock to see his mare and her new foal. After so many fillies, Beryl was thrilled this spring with the arrival of a colt. As I gazed at the frolicking colt, a thought occurred to me that was so obvious that I could not believe I never before realized it—Beryl devoted his entire professional life to birth—to the birth of the animals he loved. He was responsible for getting the mares pregnant, and when the mother of offspring fell ill, he was there to care for them at the end of their lives.

A little-known fact about horses is that almost all foals are born between 2:00 and 6:00 a.m. When a birth was difficult, Beryl was called no matter what the time was. Racehorses are high-strung and very strong; when they are in distress, they can be dangerous. He often told me that at such times, his life depended on his friends, who are illegal immigrants from Mexico and are responsible for handling the mares. Much to my surprise, it turns out that there really are horse whisperers. Beryl had absolutely no doubt that the handlers and the horses communicated in a language both understood. During the memorial service, one of his clients, Christine, recounted the many times she had been in a stall with Beryl and handlers as they struggled to get a foal out of a wildly thrashing mare. It was not this drama but the following moment that she most remembered. They would sit in silent wonder at what had just happened. Death and birth. Birth and death. Perhaps an equine vet and a philosopher are in the same business after all.

But what about the hard problem? For all his criticism of Cartesian anthropocentrism, Heidegger remained resolutely committed to human exceptionalism. Only "man" can die, he insists, animals merely "expire." The issue for Heidegger is the same as the issue for Descartes—consciousness as such. Man dies because he is conscious of his own impending death; animals expire because they do not know that they will die. A few months before his death, I asked my brother if he thought animals knew they would die. He laughed and said, "Of course they know they will die, they even know when death is imminent." In the following weeks, he sent me articles about the awareness of death not only in higher primates like apes, chimpanzees, and monkeys, but also in dogs, cats, birds, elephants, and even pigs. Unlike so many philosophers, Beryl was always more interested in the similarities than the differences between humans and animals.

It was not just horses he loved; he always had more dogs and cats than anyone should have, as well as fish and even a cockatoo that talked enough to drive me crazy. And then there were bees. He kept three hives, and we often discussed how amazing bees are. A few weeks before his death, I sent him a copy of Lars Chittka's remarkable book, *The Mind of a Bee*, which I will consider below. The day of the memorial service, I awoke as usual before anyone else and decided to sit on the couch where Beryl always sat to collect my thoughts for what I knew would be a difficult day. When I turned on the light, I saw his glasses sitting on top of a finished crossword puzzle and the copy of *The Mind of a Bee*. I had sent him that book to discuss when I wrote this chapter. Now, dear reader, you will have to rely on me without any help from my kid brother.

SMART PLANTS

The interrelated questions of consciousness, cognition, and intelligence are at the heart of the question of human exceptionalism. Anthropocentrism is under pressure from two seemingly opposite directions—the so-called natural and the so-called artificial. From the "natural" side, recent scientific advances in three areas call into question claims of human exceptionalism: evolutionary theory, neuroscience, and plant-animal cognition. Human cognition, evolutionary theorists argue, did not emerge *de novo* but developed from earlier forms of life. Rewinding the evolutionary tape, it becomes possible to see primitive forms of consciousness in earlier living organisms. As we have seen in the previous chapter, proto-cognitive activity is evident even at the cellular level. Neuroscientists have revealed striking neurological and biochemical similarities between human and nonhuman mental processes. There is no longer any doubt that cognition can even occur in organisms that have neither a brain nor a nervous system. Finally, it is becoming increasingly clear that plants and animals in some sense think, act intentionally, and suffer. From the "artificial" side, the recent panic about machine learning and sentient AI is symptomatic not only of the widely discussed concern about human extinction, but also of anxiety about human uniqueness. If prokaryotes, slime mold, trees, birds, and bees can think, what makes humans so special? And if machines are sentient and can evolve superintelligence, then is human consciousness machinic?

Scientific research raises profound philosophical and religious questions even if scientists are rarely aware of the metaphysical presuppositions and implications of their work. In this chapter, I will focus on plant and animal cognition, and in the following chapter, I will consider "artificial" intelligence. To undo the anthropocentrism of Cartesian-Newtonian dualism, it is necessary not only to put mind back into nature, but also to expand mind to include culture and technology. Cartesian dualisms—mind/body, cogito/world, human/nonhuman—have had pernicious effects that extend far beyond philosophy. In his illuminating book, *Forests: The Shadow of Civilization*, Robert Pogue Harrison analyzes how Cartesian principles have informed the understanding of forests and shaped the purportedly modern science of forest management. In ancient mythologies and religions, forests are commonly associated with dangerous forces like desire, violence, and ignorance, which threaten rational order. For Descartes, the forest functions like Plato's cave, and his *Discourse on Method* provides a map to escape its darkness. "Method (from the Greek *meta-odos*, or along the way)," Harrison explains, "means literally the 'path,' hence the analogy of following a path through the forest is particularly appropriate in a treatise on method. Descartes's analogy of course brings other scenes to mind—Dante's dark forest, for example, where the 'straight way' is lost and cannot be pursued. In Descartes's analogy, the forest is likewise a place of error and abandon, but unlike Dante, Descartes appears confident that there is indeed a *way* to walk in a straight line through the forest." Harrison proceeds to explain, "Descartes's analogy of walking in a straight line through the forest is, as Michel Serres has noted, 'isomorphic' with the method of algebraic geometry itself.... Mathematical analysis follows the way of numbers and more numbers in a linear series until it reaches its final result. The triumph of method in a forest of doubt implies the ability to hold to the straight line of mathematical deduction."

The oppositions between reason and error, order and chaos, predictability and chance cannot be reduced to the distinction between civilization and its shadow or city and forest. For Descartes and his followers, ancient and even medieval cities are more forests than cities. "These ancient cities that have grown up diversely over time, with crooked streets and uneven buildings are the citadels of culture. For Descartes they are the results of chance, diversity, and randomness. In short, they are the forests of confusion in which Cartesian rationalism finds itself alienated, or better, 'a-lineated.'"[1] This Cartesian rationalism created the urban grids of today's cities. In his enormously influential *Towards a New Architecture*, Corbusier writes, "Geometry is the language of man." In contrast to the dark, crooked streets of medieval villages, Corbusier's *ville radieuse* is thoroughly rectilinear—nothing but straight lines and right angles. The first chapter of *The City of Tomorrow* is entitled "The Pack-Donkey's Way and Man's Way."

> Man walks in a straight line because he has a goal and knows where he is going; he has made up his mind to reach some particular place.
>
> The pack-donkey meanders along, meditates a little in his scatter-brained and distracted fashion, he zigzags in order to avoid the larger stones, or to ease the climb, or gain a little shade; he takes the line of least resistance.
>
> But man governs his feelings by his reason; he keeps his feelings and his instincts in check, subordinating them to the aim he has in view. He rules the brute creation by his intelligence.[2]

The same Cartesian principles that influenced Corbusier and modern architecture also form the foundation of modern science of forestry, which began in late nineteenth-century Germany and extends to straight lines of rows of trees waiting to be harvested in today's Weyerhaeuser forests.

> Algebra and geometry, which served as the basis of Descartes's method for pursuing indubitable truth, become the basis of the new science of forestry. Thanks to such method the forest ceases to be the place of random errancy and becomes an orderly chessboard. As it becomes a calculable quantity, it also become geometric. How do you walk in a straight line through the forest? To begin with you plant your trees in rectilinear rows, as German foresters did. Algebraic geometry suffers no obstacles. The straight lines of geometry come to the forests of Enlightenment, and the ways of method prevail.[3]

The effort to rationalize forests by imposing quantitative rectilinear logic on them is a further exercise of the will to power that turns nature into a standing reserve that humans can exploit. This is the dangerous utilitarianism that concerns Heidegger. When he asks, "What is called thinking?," he develops his reflections in *Holzwege*, which he defines as woodland paths that lead nowhere. These are the paths surrounding his hut in the Black Forest where he wrote some of his most important works. The way to truth is never straight but always takes unexpected twists and turns that can only be followed by erring.

In the past two decades, our understanding of forests as well as all plants has been radically transformed. Rejecting every form of Cartesian anthropocentrism, scientists have become increasingly interested in plant cognition and even plant consciousness.[4] Richard Powers captures the widespread excitement created by these developments in *The Overstory*. Talking to plants has long been a preoccupation of New Age spirituality and a mainstay of *New Yorker* cartoons. Powers, who knows more about science than any other novelist, realizes something new and important is going on. *The Overstory* begins,

First there was nothing. Then there was everything.

A woman sits on the ground, leaning against a pine. Its bark presses hard against her back, as hard as life. Its needles scent the air and a force hums in the heart of the wood. Her ears tune down to the lowest frequencies. The tree is saying things, in words before words.

Words before words. What strange language do trees speak, and can humans understand this alien intelligence?

The tale Powers spins is based on the pioneering research of Suzanne Simard, who is professor of forest ecology at the University of British Columbia. Two years after Powers was awarded the Pulitzer Prize for *The Overstory*, Simard published *Finding the Mother Tree: Discovering the Wisdom of the Forest*. This book is the culmination of research that began with the publication of her article with the daunting title "Net Transfer of Carbon Between Ectomycorrhizal Tree species in the Field" (1997). Simard's research focuses on the function of mycorrhizal (*myco*, fungus + *rhiza*, root) networks. According to researchers at the University of Nevada, "The associations between roots and fungi are called mycorrhizae. These symbiotic arrangements have been found in about ninety percent of all land plants and have been around for approximately four hundred million years. Plant roots are hospitable sites for the fungi to anchor and produce threads (hyphae). The roots provide essential nutrients for the growth of the fungi. In return, the large mass of fungal hyphae acts as a virtual root system for plants, increasing the amount of water and nutrients that the plant may obtain from the surrounding soil."[5] Mycorrhizale hyphae, which "are fifty times finer than the finest roots and can exceed the length of a plant's roots by as much as a hundred times are, like the world at the quantum level, thoroughly entangled."[6]

For many years, Simard's work was met with strong resistance because of deep-seated prejudice in the scientific community. Her investigation of biochemical communication among trees of the same as well as different species through underground fungal networks calls into question some of the most basic assumptions of modern biological theory. Just as quantum mechanics supplements the oppositional logic of classical physics with a relational logic of coemergence and codependence, so Simard supplements classical Darwinism's exclusive focus on the oppositional logic of competition with the relational logic of cooperation.[7] While writing her doctoral dissertation at Oregon State University, she formulated the question that has guided her research ever since: "Are forests structured mainly by competition, or is cooperation as or even more important?"[8]

In 1984 David Read at the University of Sheffield discovered that pine seedlings transmit carbon to other pines through underground fungal networks. He traced the movement of carbon in his lab by replacing natural carbon in the seedlings with radioactive carbon dioxide. Simard was intrigued by Read's work and wondered if sugar could be transmitted between different species clustered together in the wild. She speculated, "If carbon did transmit between tree species, this would present an evolutionary paradox, since trees are known to evolve by competing, not cooperating."[9] Read's experiment was limited to the laboratory and involved trees of the same species. Furthermore, he established only one-way transmission of carbon and, therefore, did not demonstrate mutual benefit for the two organisms. Simard's 1997 study was conducted in the wild and involved two species—birch and fir. She also used radioactive carbon dioxide to demonstrate that carbon is shared between birch and fir when they share a mycorrhizal network. Furthermore, her investigation conclusively demonstrated that this is not a one-direction process. "Our study extends earlier laboratory results to the field, providing direct evidence for both bidirectional and net carbon transfer between plant species, for the occurrence of hyphal as well as soil pathways, and for source-sink regulation of net transfer in field conditions."[10] Simard's conclusion is that trees in a forest are not separate individuals but are reciprocally related in networks that are mutually beneficial. She offers a vivid account of her "Aha!" moment:

> Again and again, I checked the numbers, just to make sure. I sat in disbelief. Paper birch and Douglas fir were *trading* photosynthetic carbon back and forth through the network. Even more stunning, Douglas fir received far more carbon from paper birch than it donated in return.... They were communicating. Birch was detecting and staying attuned to the needs of the fir. Not only that, I'd discovered that fir gave some carbon back to the birch too. As though reciprocity was part of their everyday relationship.
>
> The trees were connected, cooperating.
>
> I was so shaken I leaned against the tile walls of my office to absorb what was unfolding, because the earth seemed to be rumbling. The sharing of energy and resources meant they were working together like a system. An intelligent system, perceptive and responsive.[11]

With different trees communicating with each other through underground channels, we return to the elemental, which, of course, we have never really left. The ground beneath our feet is not dead dirt but is vital soil, which is a living organism, or, more precisely, an organism of organisms without which life is impossible. In 1998, just after Simard demonstrated the interdependence of trees joined by underground fungal networks, scientists discovered the largest known organism in the world in the nearby Blue Mountains of Oregon—a giant, four-square-mile, 2,384-acre *Armillaria ostoyae* fungus. Estimated to be between 2,400 and 8,650 years old, this is also one of the oldest living organisms.[12] Unlike the mycorrhizas Simard studied, *Armillaria ostoyae* are pathogenic to trees. This important difference notwithstanding, the two types of fungal organisms shared a common structure and operational logic.

The timing of Simard's discovery was propitious. In the late 1980s the English computer scientist Tim Berners-Lee adapted Ted Nelson's hypertext model (1950) to create a new way to organize data using hyperlinks with hotspots embedded in texts. This system connected multiple databases on different computers and allowed simultaneous access to any computer on the internet. In 1989 Berners-Lee developed what came to be known as the World Wide Web, which one year later became public. These developments were occurring at precisely the same time as Simard and her colleagues were discovering the vital importance of the entangled fungal world hidden beneath our feet. Networks and webs that were expanding in the air now appeared to have been growing underground for millennia. When *Nature* published Simard's seminal article only six years after the launch of the World Wide Web, the editors made this connection explicit by describing her contribution as the discovery of the "Wood Wide Web." Simard reflects on this strange intersection of nature and technology:

> The forest was like the Internet too—the World Wide Web. But instead of computers linked by wires or radio waves, these trees were connected by mycorrhizal fungi. The forest seemed like a system of centers and satellites, where the old trees were the biggest communication hubs and the smaller ones the less-busy nodes, with messages transmitting back and forth through the fungal links. The old and young trees were *hubs* and *nodes*, interconnected by mycorrhizal fungi in a complex pattern that fueled the regeneration of the entire forest.[13]

The constitutive structure of both the World Wide Web and the Wood Wide Web is *radical relationality*.

Simard's identification of the "hubs and nodes" structure of mycorrhizal networks is a very important insight. The structure of fungal networks is isomorphic with the structure not only of the World Wide Web, but of social, political, financial, and neural networks. In 2002 Albert-Laszlo Barabasi, professor of physics at Notre Dame, published an influential book entitled *Linked: The New Science of Networks*. This new science rests on an old insight. In the early 1900s the Italian economist Vilfredo Pareto discovered what has come to be known as the 80/20 rule. Pareto, who was an avid gardener, discovered that 80 percent of his peas were produced by only 20 percent of his plants. As a professional economist, he extended this law to financial systems. Barabasi explains that Pareto's "most celebrated discovery was that income distribution followed a power law, implying that the most money is earned by a very few wealthy individuals, while the majority of the population earn small amounts. Pareto's finding implies that 80 percent of the money is earned by only 20 percent of the population, an inequality that is still with us a hundred years after Pareto's discovery." Barabasi extends Pareto's law to the World Wide Web: "Under different guises, the 80/20 rule describes the same phenomenon: In most cases four-fifths of our efforts are largely irrelevant. Let me contribute a few more items that approximate the 80/20 rule: 80 percent of the links on the World Wide Web point to only 15 percent of Webpages, 80 percent of citations go to only 38 percent of scientists, 80 percent of links in Hollywood are connected to 30 percent of actors."[14]

This hub structure is precisely what Simard observes in the Wood Wide Web. When not disrupted by human interventions, forests are self-organizing and self-regenerating systems. As we have discovered in our consideration of quantum ecology, living organisms are open rather than closed and, therefore, are systems of systems. The relations of tree roots and fungus form a self-organizing and self-regulating reciprocal network. "Root begets fungus begets root begets fungus. The partners keeping a positive feedback loop until a tree is made and a cubic foot of soil is packed with a hundred miles of mycelium. A web of life like our own cardiovascular system of arteries, veins, and capillaries." Within this emergent web, older trees nourish younger trees by transporting water, sugars, and other nutrients to them. This transmission is adjusted to changing climatic conditions and, correlatively, to the changing needs of seedlings. Simard concludes:

> If the mycorrhizal network is a facsimile of a neural network, the molecules moving among trees were like neurotransmitters. The signals between the trees could be as sharp as the electrochemical impulses between neurons, the brain chemistry that allows us to think and communicate. Is it possible that trees are as perceptive of their neighbors as we are of our own thoughts and moods? Even more, are the social interactions between trees influential on their shared reality as that of two people engaged in conversation? Can trees discern as quickly as we can? Can they continuously gauge, adjust, and regulate based on their signals and intersections just as we do? . . . *Could information be transmitted across synapses in mycorrhizal networks, the same way it happens in our brains?*[15]

Simard's surprising conclusions had been anticipated in Jena at the end of the eighteenth century by the "father" of ecology, Alexander von Humboldt, when he wrote, "Gradually, the observer realizes that these organisms are connected to each other, not linearly, but in a *net-like, entangled fabric*."[16] In attempting to understand the implications of quantum mechanics, I argued for a version of pancognitivism according to which all physical, biochemical, and biological processes are in some sense cognitive. The recognition that forests are intelligent organisms and trees communicate with each other implies that plants are capable of cognitive processes. In recent years, there has been widespread interest in plant cognition and even consciousness. In 2018 popular writer Michael Pollan turned his attention from food to the growing use of psychedelics for therapeutic purposes.[17] Five years later he published an informative article in the *New Yorker* entitled "The Intelligent Plant: Scientists Debate a New Way of Understanding Flora."[18] Rather than exploring the ways mushrooms can alter human consciousness, he considers the possibility that plants themselves are cognitive or even conscious. Pollan notes that this development has been met with unusually strong resistance in the scientific community. Lincoln Taiz, professor emeritus of plant physiology at the University of California Santa Cruz, speaks for many when he accuses some of his colleagues of "the over-interpretation of data, anthropomorphizing, philosophizing, and wild speculations." Though never explicitly stated, much of the resistance to plant cognition is a reaction to popular New Age beliefs that talking to plants and playing the proper music can help them prosper. There is, however, another way to understand these developments. Rather than anthropomorphizing plants by projecting human mental functioning onto them, recent scientific advances suggest the possibility of understanding human cognition and consciousness differently by locating it within the context of a more expansive and inclusive notion of mind.

One of the most intriguing areas of new research is the interdisciplinary field of plant neurobiology. A guiding presupposition of this line of inquiry is that neither a brain nor a nervous system is necessary for cognitive activity. In the article "Plant Neurobiology: An Integrated View of Plant Signaling" (2006), Eric Brenner and his colleagues recall that in 1791 Luigi Galvani discovered that electrical signals stimulate muscles in animals. A century later von Humboldt conducted approximately four thousand experiments and concluded that the same signaling process is found in plants. Neurobiologists extend von Humboldt's insight to argue that "plant intelligence is an intrinsic ability to process information from both abiotic and biotic stimuli that allows optimal decisions about future activities in a given environment."[19]

The controversy surrounding plant cognition hinges on how intelligence is defined. As I have stressed, there has been a tendency to take human intelligence as normative and then to attempt to demonstrate that other forms of cognition—be they "natural" or "artificial"—fall short. Miguel Segundo-Ortin and Paco Calvo provide a useful account of the research that is being done in their recent article "Consciousness and Cognition in Plants." Summarizing their own research as well as the work of others, they write, "Stripped of anthropocentric (and zoocentric) interpretations, a growing body of research indicates that many sophisticated behaviors traditionally assumed to be exclusive to animals are present in the *Plantae* kingdom too, including light and nutrient foraging, competition avoidance, and complex decision-making.... From a biocentric point of view, many scholars argue that this body of evidence is robust enough to consider plants intelligent or cognitive organisms."[20]

Plants perceive multiple sensory inputs, process and integrate this information, and make decisions, flexibly responding to vibrations, variations in water supply, light, temperature, and changing chemical and soil conditions. Darwin anticipated these insights in *The Power of Movement in Plants* (1880), where he argued that "it is hardly an exaggeration to say that the tip of the radicle ... having the power of directing the movements of the adjoining parts, acts like the brain of one of the lower animals; the brain being seated within the anterior end of the body, receiving impressions from the sense organs and directing several movements."[21]

In his book *Planta Sapiens: The New Science of Plant Intelligence* (2022), Paco Calvo, who is professor of the philosophy of science and chief researcher at the University of Murcia's Minimal Intelligence Lab, attempts to establish plant consciousness by anesthetizing a mimosa plant and a Venus flytrap. The former plant is extremely sensitive to touch, and the latter is carnivorous. After being exposed to an anesthetic, both plants became completely unresponsive, and as the drug wore off, the plants once again became responsive. While I remain unconvinced that plants can be conscious, I have no doubt that their survival depends on effective cognitive activity.

As these remarks suggest, plant cognition involves perception-sensation, memory (past), learning, anticipation-prediction (future), and decision-making (present). In addition to this, there is considerable evidence suggesting that some plants can experience pain, recognize the difference between self and nonself, and distinguish strangers from kin. Finally, plants communicate with both the same and other species. They are equipped with a sensory apparatus that enables them to be aware of and respond to their environment. According to Amanda Gefter,

> Plants have photoreceptors that respond to different wavelengths of light, allowing them to differentiate not only brightness but color. Tiny grains of starch in organelles called amyloplasts shift around in response to gravity,

so the plants know which way is up. Chemical receptors detect odor molecules; mechanoreceptors respond to touch; stress and strain of specific cells track the plant's own ever-changing shape, while the deformation of others monitors outside forces, like wind. Plants can sense humidity, nutrients, competition, predators, microorganisms, magnetic fields, salt and temperature, and can track how all of those things are changing over time. They watch for meaningful trends—Is the soil depleting? Is the salt content rising?—then alter their growth and behavior through gene expression to compensate.[22]

Plants not only collect data and process information, they also remember what they experience. A leading neurobiologist, Stefano Mancuso, has demonstrated that mimosa plants can store information for twenty-eight to forty days. Memory enables plants to learn from the past and anticipate the future. Cretan hollyhock, for example, is heliotropic and reorients at night to face in the direction the sun will rise in the morning. Anticipatory behavior has also been observed in roots. Segundo-Ortin and Calvo report that sweet peas

> grow different roots if subjected to variable, temporally dynamic, and static nutrient regimes. For instance, when given a choice, plants allocate more root biomass in patches with increasing nutrient levels. The striking fact, however, is that they do so even in cases when "dynamic" patches were poorer in absolute terms than the "static" ones. . . . This indicates that rather than responding to absolute resource availabilities, plants are able to perceive and integrate changes in resource levels and utilize it to anticipate growth conditions in ways that maximize their long-term performance.[23]

Calvo goes so far as to describe plants as "prediction machines." "Cognitive behavior," he argues,

> *is* adaptive, but it is much more besides. It is *anticipatory*, allowing an organism to optimize for *future* changes in the environment. It is *flexible*, responding to multiple different factors and with multiple different manifestations. It is also *goal-directed*, aimed at making a change in the environment or in the organism's state, rather than simply responding to it. These qualities require much more than "knee-jerk" reactions. They need to use information from

many sources and in different parts of the plant, from root to shoot, all of which must be *integrated* to allow a coordinated response.[24]

The ability to remember the past and anticipate the future enables plants to make adaptive decisions. In an article provocatively entitled "Decision-Making: Are Plants More Rational than Animals?," University of Zurich professor of environmental studies Bernhard Schmid concludes that "when faced with choices concerning different environmental conditions, plants make rational decisions in favor of the option that maximizes their fitness. . . . This implies that theories of decisions making, and optimal behavior developed for animals and humans can be applied to plants."[25]

The introduction of the language of decision-making raises suspicions about the reintroduction of teleology to the understanding of biological organisms. As we have seen, however, it is necessary to distinguish teleonomic from teleological processes. In contrast to teleological processes, which are purposeful, teleonomic processes are nonpurposeful and end-seeking. Henri Atlan elaborates this distinction in *Entre le cristal et la fumee*:

> A teleonomic process does not . . . function by virtue of final causes even though it seems as if it were oriented toward the realization of forms, which will appear only at the end of the process. What in fact determines it [that is, a teleonomic process] are not forms as final causes but the realization of a program, as in a programmed machine whose function seems oriented toward a future state, while it is in fact causally determined by the sequence of states through which the preestablished program makes it pass. The program itself, combined in the characteristic genome of the species, is the result of the long biological evolution during which under the simultaneous influence of mutations and natural selection, it transformed itself and adapted to its environment.[26]

As end-directed but not purposeful, teleonomic processes are neither linear nor circular. On the one hand, plant and environment are joined in recursive circuits that create both unexpected and disproportionate changes, and, on the other hand, the openness of these systems leads to aleatory changes that distinguish the point of departure from the point of arrival.

Climbing vines provide one of the best examples of plants displaying teleonomic behavior. Darwin expresses his fascination with the efficiency of what appears to be their purposeful behavior in his book *The Movement and Habits of Climbing Plants*. "Plants become climbers," he writes, "in order, it may be presumed, to reach the light and to expose a large surface of leaves to its action and to that of free air. This is effected by climbers with wonderfully little expenditure of organized matter, in comparison with trees, which have to support a load of heavy branches by a massive trunk."[27] With the invention of slow-motion photography and virtual simulations, it has become possible to observe the motion of plants in ways Darwin could not. The most interesting plant that appears to make deliberate decisions and display teleonomic behavior is a member of the morning glory family known as dodder. Dodders are parasites that have no chlorophyll and therefore must absorb food through rootlike organs that penetrate the host plant. From the time it is a seedling, a "dodder can distinguish the chemicals of different plant species, and between plants that are full of nutrients and those that are wasting away. It can do this without the help of an olfactory system, as well as choosing direction and rate of growth to the preferred target. In fact, seedlings have such small energy stores that they must find a target rapidly or they will die. If a dodder has started to grow towards a host which seems of low quality and senses another, more appetizing one nearby, it will change direction and head for the more appealing option."[28] The preferred target of dodder is tomato plants, but they will settle for wheat when necessary. As impressive as dodder decision-making is the ability of wheat to protect itself from this predator by emitting a chemical that tomatoes produce when they have little nutritional value, thereby diverting the predator. This fascinating game of hide-and-seek points to a final feature of plant intelligence.

Plants not only perceive their environment, process data, remember past information, anticipate the future, and make decisions, they also communicate with each other using two primary strategies—mechanical and chemical. The simplest though nonetheless remarkable form of communication is mechanical. We have already seen how plants detect and respond to physical changes in temperature and moisture. More interesting in this context is the interspecies communication that occurs through vibrations. Thale cress, also known as mouse-ear cress or *Arabidopsis*, is a member of the mustard family that is usually considered a weed. This commonplace plant is remarkably savvy. Segundo-Ortin and Calvo report that *Arabidopsis* "can detect, through epidermal outgrowths (hairs called 'trichomes'), the specific vibrations produced by the munching of caterpillars and respond to them by synthesizing toxins. Remarkably, they can discriminate by their vibrational modes the trichome frequencies caused by chewing from those caused by wind or insect songs. Trichomes thus act as 'mechanical antennae,' enabling *Arabidopsis* to respond in a selectively and ecologically meaningful way."[29]

The more intriguing and complicated form of plant communication is chemically mediated. Ever since I was very young, I have cut lawns; indeed, I still cut the grass and tend the gardens on three of the eight acres where we live. This is less a chore than a ritual that helps to keep me grounded. One of my maxims in life is that I do not fully trust anyone who does not cut her own lawn and does not have soil under her fingernails. One of the simple pleasures in my life has always been the aroma of freshly mowed grass. Imagine how distressed I was to learn that plants actually feel pain and that the sweet smell I have enjoyed so long is an expression of plants' pain and suffering. According to Pollan, "Descartes, who believed that only humans possessed self-consciousness, was unable to credit the idea that other animals could suffer from pain. So he dismissed their screams and howls as mere reflexives, as meaningless physiological noise. Could it be remotely possible that we are now making the same mistake about plants? That the perfume of jasmine or basil, or the scent of freshly mowed grass, so sweet to us is . . . the chemical equivalent of a scream?" Plants not only express their own pain, they also warn other plants of impending danger. In a stunning example of intra- and interspecies communication, "when broad bean plants are attacked by aphids . . . they release plumes of volatile compounds that drift out from the wound and attract parasitic wasps that prey on the aphids. These 'infochemicals'—so called because they convey information about a plant's condition—are one of the ways plants communicate, both between different parts of their own bodies and with other organisms."[30] Poplar and red oak offer another example of a strategy of producing chemicals when attacked that function as self-defense while at the same time communicating danger to others trees. "The tree response occurs rapidly enough to alter food quality and thus influence the foraging behavior of larvae sensitive to these chemicals."[31] Some of the most interesting research on plant communication has been done on sagebrush. Richard Karban and his colleagues at the University of California Davis have discovered that sagebrush plants do not passively await danger signals from other plants but actively "eavesdrop" and "hear" chemical clues when predators damage their neighbors. In another remarkable finding, sagebrush plants are capable of kinship recognition. "Plants respond more effectively to volatile cues from close relatives than from distant relatives."[32]

Taken together, the Wood Wide Web and clouds filled with infochemicals form a sophisticated information and communications network that is a prototype of what their human descendants developed millennia later. Arriving late to the party, humans are now trying use their most sophisticated information technologies to join this infinite conversation. In 2012 Microsoft launched Project Florence to explore the possibility of humans and plants actually conversing. Noting that 80 to 85 percent of our perception, learning, and cognition is mediated through vision, engineers are developing programs for interspecies communication using breakthroughs in deep learning and image recognition to create a platform not only for listening in on plants' conversations but also for "talking" with them. According to the Project Florence website, the purpose of the program is to facilitate "plant to human experience":

> If plants could talk to us, what would they say? Equally important is how might they respond to us if we could converse with them? How might these conversations evolve and expand our ability to better relate with the Natural World?
>
> Project Florence is an artistic representation of a Plant-Human Interface Experience that is built on top of a scientific analysis of the plant and its environment. Paired with the ability to receive human input, the plant can return a response, thus promoting a two-way conversational experience. Combining biology, natural language research, design, and engineering . . . we have created an instantiation of a plant to human interface through the power of language. Project Florence enables people to converse with a plant by translating their text sentiment into a light frequency the plant can recognize and respond to.

Perhaps Jerry Baker's New Age bestseller from the seventies *How to Talk to Your Plants* was not so fanciful after all. Imagine, just imagine, how different the world would be if humans were able to listen to plants and appreciate their intelligence.

ANIMAL INTELLIGENCE (AI)

Listening is not the same as hearing; we hear noise, but listening attends to what noise drowns out. Composer Pauline Oliveros maintains that listening, or what she describes as "deep listening," is "a form of meditation." "The practice is intended to expand consciousness to the whole space/time continuum of sound/silence. Deep Listening is a process that extends the listener to this continuum as well as to focus instantaneously on a single sound (engagement to targeted detail) or sequences of sound/silence." Deep Listening is not limited to human beings. "Animals are Deep Listeners. When you enter an environment where there are birds, insects, or animals, they are listening to you completely. You are received. Your presence may be the difference between life and death for the creatures of the environment. Listening is survival!"[33] In a world of incessant noise where we have forgotten how to listen, learning to listen *is* a matter of survival.

I cannot listen in the city but only in the country. Listening requires an attunement that changes with both place and the seasons. In the summer I listen to tree frogs at night and blue birds, sparrows, and crows in the day; Canada geese flying high above in perfect Vs announce fall with their unsettling honks; during long winter nights, barn owls talk across the fields silenced only by the howls of coyotes; and in spring peepers and red-winged blackbirds promise longer days and shorter nights are coming. It is clear that they are talking to each other, but what are they saying?

As we saw in chapter 6, analytic philosophers recently have become preoccupied with the problem of consciousness. There has also been growing concern with consciousness among continental philosophers. The focus of much of the debate in both traditions is the complex relationship between human and animal cognition.[34] The two most prominent representatives of these debates are Thomas Nagel and Jacques Derrida. In his widely influential article "What Is It Like to Be a Bat?," Nagel poses a variation of the ancient question of other minds in terms of interspecies than intraspecies communication.

> I assume we all believe that bats have experience. After all, they are mammals, and there is no more doubt that they have experience than that mice or pigeons or whales have experience. I have chosen bats instead of wasps or flounders because if one travels too far down the phylogenic tree, people gradually shed their faith that there is experience there at all. Bats, although more closely related to us than those other species, nevertheless present a range of activity and a sensor apparatus so different from ours that the problem I want to pose is exceptionally vivid (though it certainly could be raised with other species). Even without the benefit of philosophical reflection, anyone who has spent some time in an enclosed space with an excited bat knows what it is to encounter a fundamentally *alien* form of life.[35]

After a brief explanation of how bats perceive the external world through echolocation, Nagel concludes that their experience is so unlike ours that there is no reason to think that it is subjectively anything like human perception. While he focuses on bats, it is clear that he intends his analysis to apply to mice, pigeons, whales, flounders, wasps, and, by extension, all other nonhuman species as well. All forms of nonhuman experience are alien. Nagel reveals this in his critique of the physicalist theory of consciousness: "If physicalism is to be defended, the phenomenological features [of experience] must themselves be given a physical account. But when we examine their subjective character, it seems such a result is impossible. The reason is that every subjective phenomenon is essentially connected with a single point of view, and it seems inevitable that an objective, physical theory will abandon that point of view" (437). Contrary to expectation, Nagel appears to support a version of species perspectivism that is more solipsistic than Nietzsche ever imagined. Nagel goes so far as to deny the possibility of any objective knowledge. In what appears to be an aphorism from *Will to Power*, he writes, "It is difficult to understand what could be meant by the *objective* character of an experience, apart from the particular point of view from which its subject apprehends it" (443). Just as we cannot understand other creatures, they cannot understand us. The problem is that each species is wired differently. The aliens Nagel considers are not merely other organisms like animals, fish, birds, and insects but also include Martians. Animal Intelligence, Alien Intelligence, Artificial Intelligence. How is AI to be understood? If Martians ever landed, humans could no more converse with them than they could communicate with horses. I wonder if Nagel has ever left his New York City office, driven sixty miles down the New Jersey Turnpike, and visited a racehorse breeding farm. Illegal Aliens who are horse whisperers have important lessons to teach stuffy philosophers from the city.

While Nagel's reflections on bats have provoked endless debate, he makes it clear that his real interest is not bats or any other nonhuman species but rather the perennial mind-body problem. His primary concern is to protect consciousness from any analysis that would call into question its irreducibility. He insists that "careful examination will show that no currently available concept of reduction is applicable to consciousness" (436). Rather than simply rejecting reductionism, Nagel inverts it. In a manner reminiscent of Marx's claim to turn Hegel on his head by displacing idealism with materialism, Nagel turns materialism on its head by displacing it with panpsychism. "By panpsychism," he explains, "I mean the view that the basic physicalism constituents of the universe have mental properties, whether or not they are parts of living organisms."[36] The easiest way to avoid materialistic reductionism is to claim that consciousness is fundamental.

In an instructive encyclopedia article, Nagel traces the origin of panpsychism back to the ancient Greek philosopher Thales (c. 624–545 BCE).[37] Nagel's place in this long tradition depends on the meaning of "mental properties." In his definition he explains that panpsychism attributes mental properties to organic as well as inorganic phenomena.

> If the mental properties of an organism are not implied by any physical properties but must derive from properties of the organism's constituents, then those constituents must have nonphysical properties from which the appearance of mental properties follows when the combination is of the right kind. Since any matter can compose an organism, all matter must have these properties. And since the same matter can be made into different types of organisms with different types of mental life (of which we have encountered only a tiny spectacle), it must have properties that imply the appearance of different mental phenomena when the matter is combined in different ways. This amounts to a kind of mental chemistry.[38]

The question is whether "mental properties" necessarily involve consciousness. I have argued that it is necessary to distinguish cognition and consciousness. While consciousness presupposes cognition, cognition does not necessarily involve consciousness. As our investigation of quantum mechanics has shown, some physical processes are cognitive or protocognitive. I do not, however, think that they involve consciousness, and, therefore, I argue for pancognitivism and reject panpsychism.

I also disagree with Nagel's interspecies solipsism. Different species of both plants and animals communicate with each other. Furthermore, rapidly improving artificial intelligence is producing self-programming algorithms that create the possibility of translation algorithms that will facilitate interspecies communication. While humans might not be able experience what a bat experiences, one day our descendants might be able to conduct conversations with bats as well as other creatures.

With the shift from the analytic to the continental tradition, the terms of the debate as well as the interlocutors shift. In 1997 Derrida gave a ten-hour lecture at the annual Cerisy conference entitled "The Autobiographical Animal," which evolved into a short book published almost a decade later, *L'animal que donc je suis*. As so often, Derrida's argument turns on a double entendre—*je suis*. How is this phrase to be understood? "I am" or "I follow" or both "I am" and "I follow?" The animal that therefore I follow, or the animal that therefore I am? The title, of course, echoes Descartes's *cogito ergo sum*. *Je pense, donc je suis*. I think therefore I am. I think therefore I follow. What is the relationship between *cogito* and the animal? Which comes first? Which follows the other?

Once again, the issue is time, or the lack of time. For Heidegger, Derrida argues, the question is whether animals experience time.

> Since time, therefore.
> Since so long ago, can we say that the animal has been looking at us?
> What animal? The other.[39]

Why a period and not a question mark? What other? Is the animal the other of the *cogito*? If so and if I am, the I is the *cogito*, is the animal other than the human? If not and animals can think, feel, speak, and suffer, then are humans and animals so different? Derrida begins to probe these questions by reversing the gaze from the philosopher observing the animal to the animal (his cat) watching the philosopher. Three questions, which are really one: Can humans understand animals? Can animals understand humans? In other words, always in (the) other(s) words, can humans and animals hold a conversation?

Derrida as always is preoccupied with Heidegger. He might as well have said of Heidegger what he said of Hegel: "We will never be done with the reading and rereading of Heidegger, and, in a certain way, I have done nothing other than attempt to explain myself on this point." Derrida's relation to Heidegger is fraught with what Harold Bloom aptly describes as "the anxiety of influence." A careful reading of Heidegger's *oeuvre* reveals that his critique of the ontotheological tradition thoroughly anticipates Derrida's deconstruction of logocentrism. In order to "move beyond" Heidegger, Derrida must show that Heidegger's critique falls short because it actually perpetuates the tradition he claims to overturn and thereby remains bound by the Cartesian anthropocentrism that is the foundation of the destructive tendencies of modern science and technology. This argument turns on the question of the animal or, more precisely, on the interrelation of the animal, death, language, and time.

Derrida's central thesis is that the animal is what Western philosophy has deprived itself of. His point is that Western philosophy has defined the human negatively. According to the oppositional logic we have tracked in other chapters, "man" *is not* animal, or "man's" essential identity is his *difference* from animals. This difference is usually defined in terms of language. "All the philosophers we will investigate (from Aristotle to Lacan, and including Descartes, Kant, Heidegger, and Levinas), all of them say the same thing: the animal is deprived of language. Or, more precisely, of response, of a response that could be precisely and rigorously distinguished from reaction; of the right and power to 'respond,' and hence of so many other things that would be proper to man" (32). The broad range of Derrida's critique should not obscure the fact that his primary target is Descartes's argument that animals are merely machines and, as such, are incapable of thinking or speaking. The cogito is the mark of man's superiority over animals. This version of human exceptionalism infects virtually all Western philosophy. Beyond philosophy, Derrida echoes without citing Lynn White when he traces the theological roots of human's domination of animals to the book of Genesis:

> Elohim said: "We will make Adam the husbandman—
> As our replica, in our likeness.
> They will *subject* the fish of the sea, the flying creatures of
>   Heavens,
> The beasts, the whole earth, every reptile that crawls upon the earth."
>   Elohim created the husbandman as his replica,
> As a replica of Elohim he created him,
> Male and female he created them.
> Elohim blessed them. Elohim said to them:
> "Be fruitful, multiply, fill the earth and conquer it.
> *Subject* the fish of the sea, the flying creatures of the
>   Heavens,
> Every living thing that crawls the earth. (15–16)

According to this theological anthropology, the realization of human subjectivity requires the subjection, if not the murder, of animals.

Exercising such domination, which often leads to violence, is justified by the insistence on the inferiority of animals. While man's superiority is defined by his lack of animality, animals' inferiority is defined by their lack of language.

Heidegger's distinctive twist on the way language constitutes human exceptionalism is related to his interpretation of death. The "authentic" relation to death defines what it means to be human.

> The ending of that which lives we have called "perishing." Dasein too "has" its death, of a kind appropriate to anything that lives; and it has it, not in ontical isolation, but as codetermined by its primordial kind of Being. In so far as this is the case Dasein too can end without authentically dying, though on the other hand, *qua* Dasein, it does not simply perish. We designate this intermediate as its "demise [*Ableben*]." Let the term "*dying*" stand for the *way of Being* in which Dasein is *towards* its death.[40]

This is the basis of Heidegger's claim that animals perish but human beings die. The difference between perishing and dying is consciousness or self-consciousness. Humans are aware that they will die, and animals, Heidegger incorrectly believes, are not. The recognition of death has two interrelated consequences—the cognizance of individual selfhood and the awareness of time. I never, the I (cogito) never experience(s) death because when I, the I (cogito) is present, death is not, and when death is present (but what does it mean for death to be *present*?), I, the I (cogito) is not. Never present, death is always *avenir*, future, always *à venir*, yet to come. The awareness of death is the fall into time that occurs with the awareness of the future. This fall is the beginning of history. The way in which the future is apprehended, the present constituted, and the past recollected is through language. For Heidegger, the animal is unaware of its impending death and, therefore, remains immersed in the immediacy of the present, unaware of past and future. Without the consciousness of death, there is no awareness of either time or individuality, and, thus, no possibility of intentional activity. Animal/Man, Perish/Die, Nature/History. Heidegger remains bound by the anthropocentrism he struggles to overcome. "Man" is apart from rather than a part of nature.

Derrida attempts to undo Heidegger's residual anthropocentrism by two complementary lines of argument. On the one hand, he maintains that human cognition is machinic, and, on the other hand, he argues that machines are increasingly cognitive. I will return to the issue of smart machines in the next chapter. In this context, I will focus on Derrida's attempt to subvert Heidegger's hierarchical opposition of human/animal through a complicated reading of the interrelation of Lacan's imaginary and symbolic orders. The Cartesian cogito,

Derrida argues, is not autonomous but is heteronomous in two ways. First, at the mirror stage, the cogito does not emerge spontaneously but is constituted by the gaze of an other. Rather than autonomous, the cogitio is codependent on the nonself. It is important to recall that Lacan heard Alexandre Kojève's influential lectures on Hegel and derives his analysis of consciousness and self-consciousness directly from Hegel's *Phenomenology of Spirit*.[41] Drawing on Hegel's analysis of the struggle for recognition, Lacan poses his argument in terms of the human interrelationships—primarily the relationship between parent and child during the first eighteen months of life. But what if the gaze of the other is an animal rather than a human? Here it is important to recall Derrida's point of departure in *The Animal That Therefore I Am*—standing naked before the gaze of his cat.

> Since time, therefore.
>   Since so long ago, can we say that the animal has been looking at us?
>   What animal? The other.
>   I often ask myself, just to see, *who I am*—and who I am (following) at the moment when, caught naked, in silence, by the gaze of an animal, for example, the eyes of a cat, I have trouble, yes, a bad time overcoming my embarrassment.
>   Whence this malaise? (3–4)

It is impossible not to hear echoes of Adam and Eve standing naked in the Garden of Eden before their "fall" into time and history. What is the difference between an other who is human and an other that is nonhuman? The animal that I follow (*suis*) I am (*suis*).

Second, the subject does not constitute itself but is constituted by the machinic repetition of the symbolic order, which is the "discourse of the other."

> The *ego cogito* gets dislodged from its position as central subject. It loses its mastery, its central power; it becomes subject subjected to the signifier.
>   The imaginary process extends thus from the specular image all the way to constitution of the ego by way of subjectification by the signifier.... Which ends only in an apparent paradox: the subject is confirmed in the eminence of its power by being subverted and brought back to its own lack, meaning that animality is on the side of the conscious *ego*, whereas the humanity of

the human subject is on the side of the unconscious, the law of the signifier, Speech, the pretended pretense, etc. (137–38)

Since the unconscious, which constitutes the cogito, is "structured like a language," language does not presuppose consciousness; to the contrary, language (unconsciously) constitutes the cogito. Suddenly, everything is reversed—"animality is on the side of the conscious ego, whereas the humanity of the human subject is on the side of the unconscious." *Je suis* because *je suis*. The cogito, then, is not originary but is secondary to (that is, follows) an alien gaze as well as a machinic operation long considered to be the essence of animality. If both the cogito and animals are machines, and animality is on the side of the conscious ego, then what's the difference?

To question difference is not necessarily to claim continuity. Just as Heidegger remains bound to the very ontotheological he tries to undo, so Derrida's aversion to continuity often seems to reinscribe precisely the oppositional logic he intends to overcome. Explaining what he dismissively describes as "biologistic continuism," he writes:

> I shan't for a single moment venture to contest that thesis, nor rupture or abyss between this "I-we" and what we *call* the animal or animals. To suppose that I, or anyone else for that matter, could ignore the rupture, indeed that abyss, would mean first of all blinding oneself to so much contrary evidence; and, as far as my own modest case is concerned, it would mean forgetting all the signs that I have managed to give, tirelessly, of my attention to difference, to differences, to heterogeneities and abyssal ruptures as against the homogeneous continuity between what calls *itself* man and what *he* calls the animal. (30)

Here as everywhere else, Derrida's primary preoccupation is repeatedly to disrupt continuities wherever he finds them. While such disruption can be productive, it is not sufficient to solve the pressing problems we face and sometimes can lead to a paralyzing impasse that makes effective action impossible. Not all continuities are homogeneous, and not all unities are repressive. It is not enough to endlessly attempt to fathom the unfathomable abyss at the limit of the human. We must patiently learn to listen to the infinite conversations that are occurring all around us.

EXPANDING CONSCIOUSNESS

Cogito ergo sum because *L'animal . . . je suis. Je suis l'animal.* I am, I follow the animal. Human beings are not the result of a divine fiat that involves *creatio ex nihilo*; rather, they are the result of a long evolutionary or, more precisely, an intervolutionary process in which humans *follow* other forms of life ranging from single and multicellular organisms through plants to other forms of animal life. I always follow animals. In this intervolutionary process earlier forms of life are not inevitably left behind as new forms of life emerge. All later forms of life are surrounded by multiple life forms from which they emerge, but from which they all too often feel alienated. Like human being itself, human cognition and consciousness do not suddenly erupt *ex nihilo* but develop from earlier forms of cognition and protoconsciousness. In a manner similar to Hegelian sublation (*aufheben*), later forms of life take up into themselves and transform earlier forms of life. *Je suis l'animal* because *je suis les animaux.* I *am* animal because I *follow* animals. Since I am animal, there is no mind—no *cogito*—without the body. *Donc . . .* therefore I think because I follow animals and everything that led up to them. Just as earlier forms of life do not necessarily disappear when later forms of life appear, so too previous forms of sensation, perception, cognition, and consciousness do not disappear as later minds appear. Human beings are surrounded by myriad forms of animal intelligence that we are only beginning to understand.

On July 7, 2012, a group of leading cognitive neuroscientists, neuropharmacologists, neurophysiologists, neuroanatomists, and computational neuroscientists gathered for the Francis Crick Memorial Conference on Consciousness in Human and Non-Human Animals at Churchill College, University of Cambridge. The guest of honor was Stephen Hawking. The purpose of the conference was "to reassess the neurobiological substrates of conscious experience and related behaviors in human and non-human animals." The highlight of the conference was the signing of the Cambridge Declaration on Consciousness, which had been written by Philip Low and edited by Jaak Panksepp, Diana Reiss, David Edelman, Bruno Van Swinderen, and Christopher Kotch. The declaration concludes:

> The absence of a neocortex does not appear to preclude an organism from experiencing affective states. Convergent evidence indicates that non-human animals have the neuroanatomical, neurochemical, and neurophysiological

substrates of conscious states along with the capacity to exhibit intentional behaviors. Consequently, the weight of evidence indicates that humans are not unique in possessing the neurological substrates that generate consciousness. Nonhuman animals, including all mammals and birds, and many other creatures, including octopuses, also possess these neurological substrates.[42]

Recounting recent technological advances that have opened new areas of research, the declaration cites several important developments that are relevant in this context:

> The neural substrates of emotions do not appear to be confined to cortical structures. In fact, subcortical neural networks aroused during affective states in humans are also critically important for generating emotional behaviors in animals.... Furthermore, neural circuits supporting behavioral/electrophysical states of attentiveness, sleep and decision making appear to have arisen in evolution as early as the invertebrate radiation, being evident in insects and cephalopod mollusks (e.g., octopus).
>
> Birds appear to offer, in their behavior, neurophysiology, and neuroanatomy a striking case of parallel evolution of consciousness.... Magpies in particular have been shown to exhibit striking similarities to humans, great apes, dolphins, and elephants in studies of mirror self-recognition.

Finally, given the current fascination with psychedelics and psilocybin mushrooms, a further development is worth mentioning. "In humans, the effect of certain hallucinogens appears to be associated with a disruption in cortical feed-forward and feedback processing. Pharmacological interventions in nonhuman animals with compounds known to affect conscious behavior in humans can lead to similar perturbations in behavior in non-human animals."[43] The Cambridge Declaration on Consciousness is obviously a direct criticism of Descartes's limitation of cognition to human beings. Though anyone who has a dog or has conversed with horses would find this claim ludicrous, the refusal to acknowledge that animals can think, feel, and decide remains widespread and has enormous ethical and legal implications. How are we to understand this persistent reluctance to expand mind to include nonhuman forms of cognition?

By way of anticipation of issues to be considered in the next chapter, I should note that the other side of the resistance to acknowledge animal cognition and consciousness is the growing anxiety about sentient or even conscious "artificial intelligence." For many people, to admit that animals are conscious or that machines can think calls into question the human exceptionalism that makes people feel special. Just as rational morphologists and contemporary fundamentalists seek to protect human uniqueness by denying that humans evolve from and, therefore, follow animals, so too those who deny animal cognition attempt to preserve human exceptionalism. In different ways, both perspectives fear that continuity with animals denigrates humans. In a last-ditch effort to preserve human distinctiveness, critics of positions like the one articulated in the Cambridge Declaration establish anthropocentric criteria for cognition and consciousness and then attempt to show how animals and machines do not measure up. This is a profound mistake whose consequences are literally tearing the world apart. Neither cognition nor consciousness is monolithic—there are many ways of thinking, many styles of consciousness, and many forms of intelligence. Furthermore, communication is not species-specific but can extend across species. While the Cambridge Declaration on Consciousness marks a significant advance in putting mind back into nature, it does not go far enough. It is necessary to expand mind to include not only animals but also plants. Animals talk to each other; plants talk to each other; animals and plants talk to each other. The question is whether humans are smart enough to join these conversations.

In the introduction to his remarkable book, *An Immense World: How Animal Senses Reveal the Hidden Realms Around Us*, Ed Yong writes:

> When we pay attention to other animals, our own world expands and deepens. Listen to treehoppers, and you realize that plants are thrumming with silent vibrational songs. Watch a dog on a walk, and you see that cities are crisscrossed with skeins of scent that carry the biographies and histories of their residents. Watch a swimming seal, and you understand that water is full of tracks and trails. "When you look at animal's behavior through the lens of that animal, suddenly all of this salient information becomes available that you otherwise would miss," Colleen Reichmuth, a sensory biologist who works with seals and sea lions, tells me. "It's like a magnifying glass, to have that knowledge."[44]

Yong plays the role of a magician conjuring up the many parallel universes and alternate realities surrounding us. As his story unfolds, it becomes increasingly clear just how narrow the human sensory and cognitive bandwidth is. While Aristotle was the first to claim that humans have only five senses, this conclusion was not widely accepted until the Enlightenment. In recent decades, multiple scholars have demonstrated that the limitation of the senses to just five is more of a cultural convention than a scientific fact. Animals not only see, hear, feel, smell, and taste differently from humans and each other, but their senses are not separate and are configured differently. In some cases, they have additional senses to humans; in other cases, their senses are attuned differently. For example, while humans cannot see polarized light, insects, crustaceans, and cephalopods can detect light polarized at different angles. Swallowtail butterflies have photoreceptors on their genitals, bats and dolphins hear and see by echolocation, and some species of whales use sonar to communicate with each other. Catfish taste with their whole bodies, and electric catfish can deliver an electric shock of between 300 and 400 volts for defensive purposes. Like plants, animals and insects sense vibrations and chemicals and use them to communicate. Schools of fish function like a network connecting distributed nodes. "Schooling fish use their lateral lines to match the speed and direction of their nearest neighbors. When a predator lunges, the rush of incoming water triggers the lateral lines of the nearest individuals which dart away. . . . Each fish only attends to the small volume of water around it, but the sense of touch connects them all and allows them to act as a coordinated whole. Blind fish can still school." In addition to this network effect, spiders have developed a technology that expands their sensorium. "Spiders construct the surfaces that they then sense vibrations through. For that reason, the orb web isn't just another substrate, like soil, sand, or plant stems. It is built by the spider and is part of the spider. It is as much a part of the creature's sensory system as the slits on the body." One of the most remarkable extra-sensory apparatuses of some animals is microperception, which can be found in species as different as birds, monarch butterflies, and raccoons. Though this phenomenon is not fully understood, Yong reports that one of the most likely explanations "involves two molecules known as a *radical pair*, whose chemical reaction can be influenced by magnetic fields. To understand this deeply, you must delve into the strange realm of quantum physics."[45] I could go on, but the point is clear. Rather than taking human consciousness as normative for all cognition, by rewinding the intervolutionary tape and exploring the multiple alien worlds surrounding us, it is possible to de-anthropomorphize human consciousness by establishing its *entanglement* with other forms of cognition. Here, as elsewhere, continuity trumps discontinuity. In addition to the recent proliferation of books on plant intelligence, there has been a remarkable surge in books about the intelligence of all kinds of animals. While the recognition of intelligence in higher primates has long been accepted for many years, the recent fascination with octopuses, fish, and even eels further expands the mind's domain. In the following pages, I will limit my analysis to pigs, birds, bees, and termites.

In his satirical allegory of the Stalinist era in the Soviet Union, *Animal Farm*, George Orwell writes, "The Creatures outside looked from pig to man, and from man to pig, and from pig to man again; but already it was impossible to say which was which."[46] Pigs might seem to be unlikely candidates for a discussion of animal intelligence. In the popular imagination, pigs are anything but smart; they are often associated with mess, filth, and contamination. It turns out, however, that pigs are remarkably similar to humans in their anatomy and physiology, or, to make this point less anthropomorphically, humans are a lot like pigs. My reasons for starting with pigs is personal—in some ways, my life depends on pigs, and this fact raises profound philosophical and ethical questions. For many years, researchers at the Berrie Diabetes Clinic of Columbia Presbyterian Hospital, where I receive treatment for diabetes, have bred a special line of pigs for their experimental research. Orwell's recognition of the virtual interchangeability of humans and pigs in *Animal Farm* not only raises important political issues but also poses difficult ethical questions for research farms where animals are raised for medical experimentation and treatment.

Diabetes is caused by the failure of the isles of Langerhans in the pancreas to produce insulin, which is necessary to process carbohydrates and thereby to control blood glucose. While insulin was discovered in 1921, genetically engineered insulin was first produced from E. coli bacteria in 1978. It is bitterly ironic that E. coli, which killed my brother and almost killed me, is used in the production of insulin on which my life depends. Prior to the creation of synthetic insulin, the only source for insulin was the pancreases of cows and especially pigs, which were collected from waste products left in slaughterhouses. It took up to two tons of porcine parts to produce eight ounces of purified insulin. Some of the most promising research currently being done requires genetically modified pigs like those used by the Berrie Clinic. The similarities between humans and pigs make pigs susceptible to organ "harvesting." Xenotransplantation poses special challenges for diabetics. One of the primary difficulties with any transplant is organ rejection as a result of immune reaction. Strong immunosuppressors are administered to prevent rejection, but for people with compromised immune systems, this can prove fatal. Thus, alternative therapeutic strategies are required. In an article entitled "Xenotransplantation of Embryonic Pig Pancreas for Treatment of Diabetes Mellitus in Non-human Primates," Marc Hammerman, who is a professor at Washington University School of Medicine, reports on a promising new approach:

> Transplantation therapy for diabetes in humans is limited by the low availability of human donor whole pancreas or islets. Outcomes are complicated by immunosuppressive drug toxicity. Xenotransplantation is a strategy to overcome supply problems. Implantation of tissue obtained early during embryogenesis is a way to reduce transplant immunogenicity. Pig insulin is biologically active in humans. In that regard the pig is an appropriate xenogeneic organ donor. Insulin-producing cells originating from embryonic pig pancreas obtained very early following pancreatic primordium formation engraft long-term in rhesus macaques. Endocrine cells originating from embryonic pig pancreas transplanted in host mesentery [that is, a membrane in the intestine] migrate to mesenteric lymph nodes, engraft, differentiate and improve glucose tolerance in rhesus macaques without the need for immune suppression. Transplantation of embryonic pig pancreas is a novel approach towards beta cell replacement therapy that could be applicable to humans.[47]

The implications of this new therapeutic approach are clear—new animal farms will be established to raise pigs whose pancreases can be harvested to be transplanted into people like me.

Pigs are not only sacrificed for medical purposes; far more animals are raised on large industrial farms under abhorrent conditions for human consumption. While philosophers like Crary Wolf, Peter Singer, and Martha Nussbaum, as well as a growing number of animal rights activists, continue to raise concerns about the treatment of animals, European scientists have taken the lead in conducting research on the effects of the mistreatment of animals. The results of this work have led to some surprising discoveries about animal cognition.

In 1974 the European Union passed legislation recognizing animals as sentient beings. The primary purpose of these regulations was to protect calves, poultry, and pigs from unnecessary suffering resulting from industrialized farming methods. In 2008 this agreement was expanded by the Lisbon treaty, which formally recognized all animals as sentient. Writing for the Royal Society for the

Prevention of Cruelty to Animals (RSPCA), which is the largest animal welfare charity in the United Kingdom, Elodie Briefer reports:

> Animal sentience is the capacity of an animal to experience different feelings such as suffering or pleasure. Negative feelings or emotions include pain, fear, boredom and frustration, whilst positive emotions include contentment and joy. Sentience also extends to an animal's ability to learn from experience and other animals, assess risks and benefits and make choices. These abilities rely upon animals being aware of changes happening around them (also known as perception) and being able to remember, process and assess information to meet their needs (also known as cognition).[48]

More recently, the European Union funded an international group of scientists from Denmark, Switzerland, Germany, France, and the Czech Republic to investigate the cognitive capacity of pigs. The team gathered 7,400 recordings from 411 pigs of different ages in different countries. Then, using artificial intelligence to process these data, they developed an algorithm to decode pig communication. Writing for the group, Elodie Briefer from the University of Copenhagen explains, "With this study, we demonstrate that animal sounds provide great insight into their emotions. We also prove that an algorithm can be used to decode and understand the emotions of pigs, which is an important step towards improved animal welfare for livestock."[49] The algorithm distinguishes positive emotions (happy or excited) from negative emotions (scared or stressed). In a subsequent article published in *Science Reports* (2022), Briefer concludes, "Research in animals confirms that emotions are not automatic and reflexive processes, but can rather be explained by elementary cognitive processes. This line of thinking suggests that an emotion is triggered by the evaluation that an individual makes of its environmental situation."[50] These vocalizations are not only the expression of affective states, they also communicate feelings to others.

In addition to emotional intelligence, pigs are effective problem solvers that have "a capacity to imagine outside perspectives and a complex understanding of the self in relation to others."[51] Given Heidegger's insistence that animals expire rather than die because they do not know they will die, one of the most intriguing characteristics of pig intelligence is their apparent awareness of death. When they or fellow pigs are loaded in vans to be led to the slaughtering house or "processing facility," their stress level increases measurably.

Pigs are not the only animals capable of anticipating or responding to death. The recent field of comparative thanatology has demonstrated a widespread awareness of death throughout the animal kingdom. In "Comparative Thanatology, an Integrative Approach: Exploring Sensory/Cognitive Aspects of Death in Vertebrates and Invertebrates," Andre Goncalves and Dora Biro write:

> Recent scientific interest notwithstanding, so-called funerary activities among animals have been reported since ancient times. Most notably stories of elephants burying their dead or dolphins assisting companions to the surfaces are recounted by Pliny the Elder (AD 29-79) and Aelian (AD 175-235). Already in the eighteenth and nineteenth centuries, first-hand reports and anecdotes accumulated on the interactions of animals with dead conspecifics. These included protection of the corpse transport, vigils, and emotional distress with allusions to grief in non-human primates, corvids, proboscides, cetaceans, ungulates, carnivores, and sirenians. Observations like these have been confirmed more recently by researchers studying the same species.

From large to small and everywhere in between, animals are aware of death. When encountering a dead calf, Asian elephants frequently roar and keep watch over the carcass. In some cases, mothers carry their calves for weeks. Whales and dolphins also carry their deceased offspring. Crows and ravens, by contrast, engage in "'ceremonial gathering': an assembly of living individuals near a deceased conspecific. The participants utter alarm calls but seldom touch the corpse or show aggression, in comparison to their predator mobbing or scavenging the corpse of another species."[52] Some of the most interesting and surprising responses to death are found in social insects like bees, ants, and termites. They all remove dead members of the colony in a process known as necrophoresis. While any member of the colony can dispose of the dead, they are usually removed by drones known as "undertakers." Bees will fly up to 100 meters before dropping the corpse. Ants also remove the dead at least several meters from the nest, while termites, by contrast, design astonishing nests with a special room for disposal of dead bodies. I will return to bees, ants, and termites after considering smart birds.

Since birds' brains lack a cortex, for many years scientists assumed they are not intelligent. Instead of a cortex, the nidopallium caudolateral is responsible for decision-making, memory, and other cognitive functions. Birds' complex sensory apparatus extends their perceptual range far beyond the bounds of human experience. Their ability to process, learn, and store information surpasses that of most primates. Of the more than eighteen thousand species of birds, corvids, which include crows, ravens, rooks, magpies, jays, and jackdaws, and African grey parrots are generally acknowledged to be the most intelligent. Researchers investigating avian intelligence usually either try to train birds to do what humans do or observe birds acting alone or with each other.

The most famous example of teaching a bird to mimic human thinking is Alex, an African grey parrot trained by Harvard specialist in animal cognition Irene Pepperberg. In her best-selling book *Alex and Me*, she reports, Alex "learned to identify fifty different objects, seven colors, five shapes, and quantities up to six; he could infer the connection between a written number, that number of objects, and the vocalization of number. Alex also shattered the belief that parrots are not capable of real speech, only mindless mimicry." When Alex died at age thirty-one, his obituary reported that he had "the intelligence of a five-year-old child and had not reached his full potential." One of Alex's most impressive feats was his apparent use of the concept zero. "When first shown two objects of different colors but the same size and asked, 'What color bigger?' he used 'none' to signify that they were the same. He did it on his own, without training. Now, he apparently was using 'none' to indicate the absence of a set of five objects, that is, using 'none' to mean 'zero.'"[53]

The more interesting research on bird brains involves the careful observation of the behavior of birds. Nathan Emery, cognitive biologist at Queen Mary University of London, makes a distinction between cognition and intelligence that is similar to what I proposed in chapter 6. While "cognition refers to the processing, storage, and retention of information across different contexts," intelligence is "the ability to flexibly solve novel problems using cognition rather than mere learning and instinct."[54] He identifies four primary features of cognition: causal reasoning, prospection, imagination, and flexibility. In a carefully researched article, "Mechanisms of Perceptual Categorization in Birds," Ludwig Huber and Ulrike Aust provide details about what Emery labels birds' "cognitive toolkit": "Categorization is the partitioning of the surrounding environment into smaller, more manageable, sets of objects like friends and foes or food and non-food.... Without it, any object would be perceived as being unique. Essentially, it is the ability to treat similar but non-identical things as somehow equivalent, by sorting them into the same category and by reacting to them similarly."[55] As we have seen, information is a difference that makes a difference.

Birds' ability to categorize by filtering out noise and identifying recurrent patterns in their environment enables them to process information. Huber and Aust identify four levels of categorization: perceptual, associative, functional, and abstract. With this aptitude birds can make comparative judgments like above/below, larger/smaller, and closer/farther. Even more impressive, birds are capable of a degree of metacognition. Leyre Castro and Edward Wasserman provide ample evidence that birds not only recognize relations but also conceive relations between relations. "Birds have proved to be quite adept at learning first-order same-different discriminations, using relational information to solve transposition tasks, and exhibiting transitive inference in a variety of situations. Furthermore, some avian species, such as crows, excel in solving analogy problems, thereby showing an appreciation for second-order relational concepts."[56] The interplay between perceptual discrimination and conceptual relationality enables birds to engage in analogical reasoning, which involves the relation between relations. For example, A:B::C:D. "Until very recently, it was thought that animals could not discriminate objects based on analogies (with the possible exception of a language-trained chimp). However, hooded crows have been shown spontaneously to display analogical reasoning to stimuli based on color, shape and number." Emery provides the following example of this complex reasoning process: "A sample pair A is a large green circle and a smaller yellow circle. Pair B is a large red star and a smaller gold star, and Pair C is a small blue star and a small orange square. Which pair is analogous with pair A? Answer: Pair B because one of the two objects in the pair is larger than the other, whereas the two objects in Pair C are the same size."[57] A hooded crow solved this puzzle. Perhaps being dubbed a birdbrain should be considered a compliment.

Such sophisticated cognitive ability enables a limited number of birds to use causal reasoning to create tools. The most effective toolmakers are New Caledonian crows, which create tools for catching prey from long slender leaves. In experiments, rooks have demonstrated the ability to calculate cause and effect in tool production and use. This requires intentional activity and decision-making based on imagining a series of discrete steps. In one experiment, a New Caledonian crow went through an eight-step sequence to produce a tool to retrieve a piece of food that was out of its reach. While crows might not have read Kant, it is clear that they can exercise practical reason.

As we have seen, plants communicate by mechanical and chemical means; birds, by contrast communicate vocally. Ever sensitive to the danger of anthropomorphizing, Emery suggests that rather than interpreting birdsongs through human language, it would be more fruitful to interpret human language through birdsongs. Ravens produce a greater variety of sounds than any other animal except humans. Many birdsongs are very complex; some have up to 300 separate parts and have distinctive rhythmic patterns. They are not uniform within species but display regional dialects and even have individual inflections. Birds can identify individuals by their distinctive calls and songs. Emery points out that European blackbirds, for example, vary frequency and duration to communicate messages to fellow blackbirds that predators cannot hear. These calls and songs serve various purposes, ranging from mating and warning to territorialization. Songs form something like cultural patterns that are passed from generation to generation.

While there is indisputable evidence that birds and other animals are conversing, it is not clear whether humans can understand what they are saying. Not to be outdone by Microsoft's effort to decode plant communication, Google Translate is developing a universal voice recognition program for all living organisms. In her informative book, *The Sound of Life: How Digital Technology Is Bringing Us Closer to the Worlds of Animals and Plants*, Karen Bakker notes that the goal is to create an "Interspecies Internet" within a decade. Though this research has not received much attention, there already has been considerable progress toward reaching this goal. In 2011 Clara Mancini, a computer scientist at the Open University, published "Animal-Computer Interaction: A Manifesto," in which she explains that there has been a long history of animals being involved with machines but stresses that this research invariably has been for the benefit of humans and there has been little investigation of the effect of this interaction on animals. Mancini proposes an alternative research agenda:

> ACI [Animal Computer Interaction] aims to understand the interaction between animals and computing technology within the contexts in which animals habitually live, are active, and socialize with members of the same or other species, including humans. Contexts, activities, and relationships will differ considerably between species, and between wild, domestic, working, farm, or laboratory animals. In each particular case, the interplay between animal, technology, and contextual elements is of interest to the ACI researcher.[58]

In the years since publishing this manifesto, Mancini and her colleagues have found that play facilitates interspecies communication. According to Bakker, they have created "multispecies video games for orangutans, pigs, cats, and crickets (vibrotactile plates provide feedback to crickets, who play Pac-Man with human players). Pig-Chase, for example, uses an interactive touchscreen to enable humans and domesticated pigs to play a cooperative game together.... Mancini has also designed digital devices that allow captive elements to choose between different types of prerecorded sounds (e.g., whale music versus elephant sounds) as a means of providing enrichment while in captivity."[59] Other researchers have responded to Mancini's challenge. In a joint project at the University of Sheffield in the United Kingdom and the University of Skovade in Sweden, a team of engineers, computer scientists, roboticists, biologists, neuroscientists, and linguists is investigating the possibility of computer-aided communication among humans, animals, and machines. The goal of this work is to develop programs that will allow animals and robots to communicate without human intervention. "Voice recognition software now exists for animals, ranging from birds to bats and prairie dogs to marmosets along with wearables for pets and livestock, which interpret the meaning of their sounds for owners."[60] With the backing of tech giants like Microsoft and Google, it seems likely that infinite conversations will, paradoxically, expand in the near future.

Communication fosters community, and community makes communication possible. Most birds tend to be social rather than solitary. Different species of birds form different social groups for different reasons. Emery describes multiple avian social systems ranging from large groups in which breeding pairs and nonbreeding helpers live together in nonoverlapping territories to small communities that consist of single pairs. These communities, which are organized hierarchically in what is literally a pecking order, form social knowledge networks. Like all communities, groups of birds involve competition and cooperation both internally and with other groups. As interrelations become more entangled, different cognitive skills are required, and communication networks must change. In spite of the sophistication of avian communication, no species of birds creates communities as complex as bees.

The euphemistic expression "the birds and the bees" originated with Samuel Taylor Coleridge's poem "Work Without Hope":

> All Nature seems at work. Slugs leave their lair–
> The bees are stirring–birds are on the wing–
> And Winter slumbering in the open air,

Wears on his smiling face a dream of Spring!
And I the while, the sole unbusy thing,
Nor honey make, nor pair, nor build, nor sing.

Though seemingly less intelligent than their avian neighbors, bees, like their fellow social insects ants, and termites have much to teach other animals, including humans. Lars Chittka, who is a leading specialist in bee cognition and behavior, begins his important book, *The Mind of a Bee* by claiming that we are surrounded by alien forms of life. The perceptual world of bees "is so distinct from ours, governed by completely different sense organs, and their lives are ruled by such different priorities, that they might be accurately regarded as aliens from outer space."[61] Chittka dubs bees "the intellectual giants of the insect world." There are approximately twenty thousand species of bees, many of which are not social. While bees are programmed to fly and to associate colored and sensory dots with flowers, they must learn how to find and identify flowers with nectar and how to return to the hive with food for younger bees. All this must be done during their brief three-week adult life. This activity requires sophisticated cognitive capabilities that allow bees to integrate sensation, memory, anticipation, and decision-making. Chittka maintains that "each individual bee has a mind—that is an awareness of the world around it and of its own knowledge, including autobiographical memories; an appreciation of the outcomes of its own actions and the capacity for basic emotions and intelligence—the key ingredients of a mind" (2). This is a remarkable claim—bees, Chittka maintains, not only know the world around them, they know also that they know.

Bees have the same five senses as humans, but their sense of sight is much more acute. They have 300-degree vision, and their eyes process information much faster than humans. They can see both ultraviolet and polarized light, which enables them to use the sun for navigational purposes. The antennae of bees are finely tuned with many sensors that can "smell, taste, hear, sense temperature, humidity, air currents, electric fields, and analyze shapes and surface textures" (42). Like birds and some butterflies, bees are sensitive to the earth's magnetic field. While bees are genetically programmed to process some data in prescribed ways, they supplement this rudimentary cognitive ability with rules learned through mimicry and social interaction. They can also pass on their acquired knowledge to later generations, thereby creating something like a distinctive culture of a particular hive. The combination of sensible acuity and cognitive ability enables bees to discriminate between flowers rich or poor in nectar. In one of the more remarkable examples of interspecies communication, some flowers can detect air vibrations caused by bees' wings and increase their production of nectar to attract bees that aid in their pollination.

The question of consciousness in bees remains the subject of considerable debate. There seems to be little doubt that bees are capable of distinguishing self from others and discriminating among different objects. Nor does it seem questionable given the advanced sensory apparatus bees have that some form of subjective experience, however alien to human consciousness it might be. As I have noted, Chittka insists that bees have metacognition. He describes an experiment conducted by British neuroethnologist Cwyn Solvi in which bees were asked to distinguish between two very similar colors or to decline to choose. "As the tasks became more difficult, honeybees increasingly chose this third option, as if they were aware of their own uncertainty. Such metacognition has been viewed as a hallmark of consciousness in apes and in dolphins, and if the opt-out behavior in mammals is taken as evidence of a self-assessment of uncertainty, then by the same criteria bees qualify too" (207).

For bees as well as other animals, the question of consciousness usually turns on the issue of language. In *The Animal That Therefore I Am*, Derrida criticizes Lacan's anthropocentric definition of language in his account of bees' communication. The issue is Lacan's claim that bees do not use language because they communicate with signs that are fixed codes and react rather than respond to each other. In "human language," Derrida explains, "signs take on their value from their relations to each other in the lexical distribution of semantemes as much as in positional, or even flectional use of morphemes in sharp contrast to the *fixity* of the coding used in bees." Drawing on Saussure's relational linguistics, which we have previously considered, Derrida rejects Lacan's argument and argues that signs that "take on their value from their relations to each other . . . must be accorded to any code, animal or human."[62] This argument can be read in two very different ways: either bees' communication is like human language, or humans' language is programmed like bees.

Regardless of how this question is resolved, the dance of bees is, without a doubt, linguistic. Using a complicated system of spatial coordinates, bees convey the location and distance of flowers rich in nectar. This communication involves astonishing mathematical calculations using the position of the sun to transmit relevant information. Chittka claims that "no other species (besides humans) uses a similarly symbolic representation to communicate about spatial locations in the real world" (73). In the absence of Google's Global Positioning System, bees combine information communicated by other bees with their own ability to recognize landmarks to create cognitive maps which they use for foraging.

While these strategies are very impressive, the most noteworthy example of communication among bees is when colonies swarm to find a new location. One of the first accounts of honeybees' swarming was *The Life of a Bee* (1901), written by Belgian playwright and poet Maurice Materlinck, who was a Nobel laureate in literature (135). More than a century later, scientists have a much better understanding of how colonies "decide" to move. Dozens of worker bees scatter over an area as large as 70 square kilometers to explore possible new sites. These scouts then gather to debate alternative locations through intricate dances. Chittka describes the "democratic decision making" process:

> The organization of the consensus building is entirely decentralized: no one individual counts the votes for various indicated sites, nor is there a leader, nor synchronous responses to an order; nor do individuals even compare spatial or nest site quality information delivered in the dances. Instead, better locations are simply indicated by longer dances, thereby increasing the probability that a worker moving randomly on the swarm cluster is likely to bump into a dancer with better information by mere stochastic processes. More individuals will thus inspect better sites indicated by longer dances and then return to the swarm cluster to indicate them in their own dances—ultimately resulting in a snowballing effect in which more and more individuals converge on the best option. (139-40)

Once a consensus is reached, the colony moves *as a whole* to what becomes its new home.

The swarming behavior of bees is an additional example of the radical relationality and self-organization that we have discovered throughout the natural world. Individual bees and their colony are joined in a relationship in which parts and whole are reciprocally related. So understood, bees form a *superorganism*. In their book *The Superorganism: The Beauty, Elegance, and Strangeness of Insect Societies*, Bert Holldobler and E. O. Wilson argue that for social insects like bees, wasps, ants and termites, "Nothing in the brain of a worker ant represents a blueprint of the social order. There is no overseer or 'brain caste' who carries such a master plan in its head. Instead, colony life is the product of self-organization. The superorganisms exist in the separate programmed responses of the organisms that compose it. The assembly instructions of organisms follow the developmental algorithms, which create castes, together with the behavioral algorithms, which are responsible for moment-to-moment behavior of caste members."[63] Superorganisms are hierarchically organized with different members carrying out different noncognitive and cognitive activities. Through the interactivity of multiple agents, something like the mind of the colony *emerges*. Just as the human mind emerges from but cannot be reduced to the sum of the interactions of entangled neurons of the brain, so the mind of the colony or hive emerges from but cannot be reduced to the interrelated activities of individual bees, ants, and termites. The collective mind of the hive or colony forms something like what Nick Bostrom describes as "*superintelligence*," which emerges from interacting algorithms of so-called artificial intelligence. Switching their focus from ants to bees, Holldobler and Wilson write, "The cognitive effort of a single scout bee is obviously quite small relative to the global information processing performed by the entire swarm. The honeybee swarm is a higher-order cognitive entity. The way in which the swarm chooses its future home, along with other algorithm-guided distributed processes, clearly qualifies the swarm as a superorganism."[64] While the notion of superintelligence points toward issues I will consider in the next chapter, the structure and operational logic of superorganisms draws together many of the threads we have been tracing in previous chapters.

Through their collective intelligence, social insects like bees and ants construct remarkable structures, but termites are the master builders of the insect world. Their appetite for wood and their symbiotic relation to fungus bridges the Wood Wide Web of smart plants and the relational realm of smart animals. J. Scott Turner begins his investigation of the extraordinary work of these impressive architects and engineers by asking if animal-built structures are "best regarded as external to the animals that build them, or are they more properly considered parts of the animals themselves?" Answering his own question, he writes, "I am an advocate for the latter interpretation, but the argument I present in this book is one with a twist: that animal-built structures are properly considered organs of physiology, in principle no different from, and just as much a part of the organism as, the more conventionally defined organs such as kidneys, hearts, lungs, or livers."[65] Turner considers his analysis to be an extension of Richard Dawkins's argument about the extended phenotype. The function of the phenotype, according to Dawkins, is not limited to synthesizing proteins but also includes the impact genes have on the environment. One of the examples he offers to illustrate his point is the way animals like beavers alter their environment by creating architectural structures like dams.[66] Elaborating Dawkins's suggestions, Turner argues that the symbiotic relation between termites and their nest is so close that they form a single organism.

The groundbreaking research that is the basis of Turner's analysis was published by Martin Lüscher in his article "Air-Conditioned Termite Nests" (1961). Lüscher points out that since most termites live in hostile environments, their survival depends on the ability to create dwellings that regulate the temperature and the moisture content of the air. "Termites survive only because their elaborate social organization enables them to build nests in which they establish the microclimate suited to their needs."[67] Termites are, in effect, sophisticated "air-conditioning engineers." Nests, which are built with a cardboard-like material made from wood particles glued together with saliva and excrement, can be as large as 16 feet in diameter and 16 feet tall with walls measuring between 16 and 23 inches. The nests have multiple rooms, including a cellar, attic, rooms for cultivating and storing food, rooms for reproduction, and even rooms for burying the dead. Termites' extreme sensitivity to temperature and moisture fluctuations necessitated the development of the ability to design and construct self-regulating dwellings. Some species die within as little as five hours if exposed to dry air. To avoid aridity, nests draw moisture from the earth and cultivate fungus, whose fermentation process produces heat and releases moisture. A medium-size colony consists of approximately two million termites that consume up to 240 liters of oxygen (1200 liters of air) a day. To provide a continuous flow of clean air, the nests must include an elaborate circulation system between the underground cellar and the attic at the top of the nest. Turner describes this intricate structure as a "colossal heart-lung machine for the colony." This organismic relationship is self-regulating and coadaptive with the environment. According to Turner, "adaptation is not simply the response of organisms to the environment: it also involves the environment adapting to organism."[68] As we discovered in our consideration of quantum ecology, such coadaptivity subverts the simple opposition between organism and milieu.

With termites, we return to the earth that grounds us. Termites live on decaying plant material in the form of wood, leaves and soil humus. They are not just architects and engineers; they are also farmers who raise fungus, without which neither they nor we can live. Their nests include rooms specially designed for fungus gardens. The circulation system is designed to allow the movement of the moisture and gases required for the cultivation of fungus. Margo Wisselink and her colleagues at the University of Wageningen in the Netherlands describe this 30-million-year-old symbiotic relationship between termites and fungi:

> Fungus-growing termites cultivate monocultures of a specific fungi... for food in their colony, analogously to human farmers growing crops. The termites forage for dead plant material in the environment, bring this into the mound and provide it to the fungus as a growth substrate. After fungal growth, the termites use the mixture of fungus and degraded plant material and also the asexual spores produced by the fungus as food. They also use the spores to inoculate new fungus gardens. The termite-fungus symbiosis is obligatory for both partners, the termites provide growth substrate and a protected growth environment for the fungi in exchange for a nitrogen-rich food source."[69]

From overstory to understory there are worlds entangled in worlds communicating and failing to communicate in infinite conversations without which life is impossible.

CHAPTER 10

STRANGE LOOPS

Nature does not end with us, but moves inexorably on, beyond societies of animals. Global markets and Earth-orbiting satellite communication, wireless telephones, magnetic resonance imagery, computer networks, cable television, and other technologies connect us. Indeed, people already form a more-than-human being: an interdependent, technologically interfaced superhumanity. Our activities are leading us toward something far beyond individual people as each of us is beyond our component cells.

—Lynn Margulis

SWARMING

Something strange, even uncanny, is going on—machines are becoming more like people, and people are becoming more like machines. Astrobiologist and theoretical physicist Sara Walker goes so far as to argue,

> Biological beings alive today are part of a lineage of information that can be traced backward in time through genomes to the earliest life. But evolution produced information that is not traditionally considered "life." Human technology would not exist without humans, so it is therefore part of the same ancient lineage of information that emerged with the origin of life.
>
> Technology, like biology, does not exist in the absence of evolution. Technology is not artificially replacing life—it is life.[1]

Smart devices are not the mechanical machines of Descartes and Newton but are novel bio-digital machines that sometimes think and act as differently from people as the cellular, plant, and animal forms of cognition from which human thinking emerged. As we have seen, many scientists and some philosophers are increasingly acknowledging that physical, chemical, and biological processes are cognitive or quasi-cognitive. At the same time, computer scientists and engineers are rapidly developing new forms of machine learning, usually described as "artificial" intelligence, that more and more closely approximate or even surpass human cognition. The widely accepted term "*artificial* intelligence" is, however, misleading because it reinscribes the opposition between the human and the natural that has led to so many disastrous consequences. It is necessary to think about artificial intelligence differently. If, as I have been arguing, mind is in nature and nature is in mind, then there is no such thing as "artificial" intelligence. Human beings are part of the natural order, and, therefore, whatever they create is part of an ongoing intervolutionary process in which physical, biochemical processes as well as cells, bodies, plants, and animals are in some sense intelligent.

In 1994 Kevin Kelly published an influential book entitled *Out of Control: The Rise of Neo-Biological Civilization*. Kelly embodies the surprisingly close link between the counterculture of the sixties, the personal computer revolution of the seventies, the internet revolution of the eighties, and the World Wide Web revolution of the nineties. I have never met Kelly, who grew up in my hometown—Westfield, New Jersey. If he took physics—and it is hard to believe he didn't—his teacher would have been my father, and his English teacher might well have been my mother. After dropping out of the University of Rhode Island, Kelly, like so many hippies, roamed around Asia searching for enlightenment, which he eventually experienced closer to home when he became a born-again Christian. In the early 1980s he began writing for Stewart Brand's *Coevolution Quarterly*. Brand, who had been a member of Ken Kesey's band of Merry Pranksters, went on to create the *Whole Earth Catalog* and eventually founded Global Business Network, which made him a major player in Silicon Valley. No publication did more to alert the public to the importance of personal computers than the *Whole Earth Catalog* during its brief four years (1968–1972). By the 1980s Brand was exploring new forms of computer-mediated communication and founded the Whole Earth Lectronic Link, known as the WELL. This network is still operating and is one of the oldest continuously existing virtual communities. In 1983 Brand hired Kelly to edit the *Whole Earth Review*. Kelly, along with John Perry Barlow, who was one of the lyricists for the Grateful Dead and a college friend of mine, were instrumental in establishing and running the WELL. But Kelly was eager to promote his own vision of the digital future, and in 1993 he launched *Wired* magazine. *Wired* has done for the internet revolution what the *Whole Earth Catalog* did for the personal computer revolution. Even today anyone who wants to know which way the digital winds are blowing must read *Wired*. What has made Kelly so influential is that he was one of the first to understand the importance of the newly emerging field of complex systems theory for understanding the internet and the World Wide Web. It is impossible to comprehend the artificial intelligence revolution now occurring without an appreciation for network dynamics and effects.

The second chapter of Kelly's book, entitled "Hive Mind," begins with a discussion of William Morton Wheeler's article, "The Ant Colony as an Organism" (1911). Wheeler, Kelly explains, "was a natural philosopher and an entomologist, who followed a philosophical strain at the turn of the century that saw holistic patterns overlaying the individual behavior of smaller parts." Though Kelly does not identify the philosophical movement that influenced Wheeler, there is a strong likelihood that it was neo-Hegelianism, which developed in England during the first decade of the twentieth century. In his prescient study, Kelly identified "'emergent' properties within the superorganism superseding the resident properties of the collective ants. Wheeler said the superorganism of the colony or hive 'emerges' from the mass of ordinary insect organisms."[2] By the final decades of the twentieth century, the principle of emergence was recognized as one of the distinguishing features of complex adaptive systems. In these systems, higher levels have an organizational structure that cannot be found in or inferred from lower levels.

After summarizing Wheeler's analysis, Kelly abruptly shifts his attention to a convention of computer graphic designers held—where else—in Las Vegas. In a dark hall, five thousand attendees play the early computer game Pong by waving different-colored lighted wands whose motions are tracked by computers and projected on a large screen. The individual wands quickly self-organize into two competing teams, creating what Kelly describes as "group mind." Recalling Darwin's image of the tangled net in the last paragraph of *The Origin of Species*, he concludes:

> The Net is an emblem of multiples. Out of it comes swarm being—distributed being—spreading the self over the entire web so that no part can say "I am the

I." It is irredeemably social, unabashedly of many minds. It conveys the logic of both Computer and of Nature—which in turn conveys a power beyond understanding.

Hidden in the Net is the mystery of the Invisible Hand—control without authority. Whereas the atom represents clean simplicity, the Net channels the messy power of complexity.³

Kelly's use of the metaphor of the invisible hand obviously recalls Adam Smith's interpretation of financial markets. In later chapters and in a subsequent book, *New Rules for the New Economy: 10 Radical Strategies for a Connected World*, Kelly shows the symmetry between hive mind and the mind of the market.⁴ By the late 1990s, networked computers had created the infrastructure for a new system of global financial capitalism. While Kelly's account of the implications of network culture for the economy provides a useful popular explanation of the radical changes networked computers brought to financial markets, his more important insight is his recognition of the common structure and operational logic of natural systems and computer networks, which he captures in his phrase "neo-biological civilization."

In addition to developing ideas he had formed while working with Stewart Brand on the *Whole Earth Catalog*, Kelly also drew on the research being done at the Santa Fe Institute (SFI), which was founded in 1984. With the end of the cold war, huge high-speed computers that had been used to model atomic explosions became available for other areas of research. From the outset, the SFI has been committed to the interdisciplinary investigation of complex nonlinear systems. The increased computational power made it possible to study nonlinear systems in new ways. Emergent complex adaptive systems are driven by positive feedback, which, unlike negative feedback systems, operate far from equilibrium. Microscopic and macroscopic operations and events are implicated in strange loops that involve complex systemic dynamics. Negative feedback tends toward equilibrium by counterbalancing processes, which, if left unchecked, would destroy the system. Positive feedback systems, by contrast, tend to disrupt equilibrium by increasing both the heterogeneity of the components of the system and the operational speed at which they interact. As positive feedback accelerates the speed of interaction among increasingly diverse components, linear causality gives way to nonlinear causal relations in which effects are disproportionate to the causes from which they emerge. Cascading events among parts can suddenly transform the structure and operation of the whole, which, in turn, alters the parts. This process is commonly known as the "butterfly effect."

Swarms of bees, flocks of birds, schools of fish, and colonies of ants fascinate researchers who attempt to understand complexity. The reasons for this are clear: the same rules and principles operating in cellular automata of information systems are at work in these natural phenomena. John von Neuman first proposed cellular automata in a 1948 lecture on "The General and Logical Theory of Automata." However, the important role they play in networks and complex systems was not realized for several decades. At the third Workshop on Artificial Life, held in June 1992, Mark M. Millonas from the Complex Systems Group at the Center for Nonlinear Studies in the Los Alamos National Laboratory presented a paper entitled "Swarms, Phase Transitions, and Collective Intelligence." "The swarming behavior of social insects," Millonas explains,

> provides fertile ground for the exploration of many of the most important issues encountered in artificial life. Not only do swarms provide the inspiration for many recent studies of the evolution of cooperative behavior, but the action of the swarm on a scale of days, hours, or even minutes manifests a nearly constant flow of emergent phenomena of many different types. Models of such behavior range from abstract cellular automata of models to more physically realistic computational simulations. The notion that complex behavior, from the molecular to the ecological, can be the result of parallel local interactions of many simpler elements is one of the fundamental themes of artificial life. The swarm, which is a collection of simple locally interacting organisms with global adaptive behavior, is a quite appealing subject for the investigation of this theme.

Millonas's aim is to develop a mathematical model of "the collective behavior of a large number of locally acting organisms," which "move probabilistically between local cells in space."[5]

Though Millonas drew his inspiration from E. O. Wilson's work on ants, the analysis he develops also applies to swarms of bees as well as other collective phenomena. Bees form swarms with dynamic patterns even though neither a leader nor a prescribed plan or program directs the behavior of individuals. The swarm *as a whole* self-organizes and operates according to a logic that cannot be discerned in any of the activity of individual bees. Complex behavior, Millonas argues, "is the result of the *interactions between organisms* as distinct from behavior that is a direct result of the actions of *individual organisms*."[6] The principle governing these interactions is called allelomimesis, which was identified by the Belgian chemist Jean-Louis Deneubourg. In an effort to understand the swarming behavior recorded by Wilson, Deneubourg extended Ilya Prigogine and Isabella Stengers's theory of nonlinear self-organized systems developed in *The End of Certainty* and *Order Out of Chaos* to insects.[7] In allelomimetic behavior, the activity of individuals is conditioned but not determined by the activity of their neighbors. In this *relational structure*, both the identity and the interplay of the parts and the whole are codependent and coemergent.

Millonas modeled the behavior of cellular automata according to the principle of allelomimesis and discovered that even relatively simple interactions can generate surprisingly complex behavior. While no member or component has any knowledge of the operation of the system as a whole, each individual is aware of and responsive to the actions of surrounding individuals. The swarm, in other words, has no pilot yet is able to fly in formation. Local interactions issue in the emergence of complex global behavior. Commenting on the synchronization of pendulums, Per Bak clarifies the coordinated activity of swarming bees: "*The system had self-organized into the critical point without any external organizing force.* Self-organized criticality had been discovered. It was as if some 'invisible hand' had regulated the collection of pendulums precisely to the point where avalanches of all sizes could occur. The pendulums could communicate throughout the system."[8] It is important to recall that Kant was the first to identify self-organization as definitive of biological organisms and Hegel appropriated Kant's insight to develop the dialectical logic that constitutes the architecture of his system.

Millonas argues that such communication throughout the system creates "collective intelligence." The buzz of the swarm is, in effect, noise flying in formation. The logic of the swarm illuminates the logic of networking. Millonas uses connectionist models derived from the study of neural networks to understand swarm networks and vice versa. He summarizes three basic characteristics of connectionist networks:

1. Their *structure* consists of a discrete set of *nodes*, and a specified set of *connections* between the nodes. For example, neural networks, the archetypal connectionist systems, are composed of neurons (nodes), and the neurons are usually linked by synapses (connections).
2. There are *dynamics* of the *nodal variables*. . . . The dynamics are controlled by the connection strengths, and the input-output rule of the individual

neurons. The dynamics of the whole system are the result of the interaction of all the neurons.
3. There is *learning*. In its most general sense, learning describes how the connection strengths, and hence the dynamics evolve. In general, there is a separation of time scales between dynamics and learning, where the dynamical processes are much faster than the learning processes.⁹

According to this model, networks consist of interconnected nodes, which communicate with each other. Each node is constituted by its interrelations with other nodes, which shape its place in the overall network. A node is, as the word implies (*nodus*, knot; from *ned*, which means twist, tie), is a knot or neχus in a web of relations. These neχuses function as switches and routers that receive and transmit information throughout the network. Separation and connection, like identity and difference, are mutually constitutive. What separates one neχus from another is a function of the distinctive way in which connections intersect. While the connections of each neχus spread throughout the network, the constitutive relations that are decisive are relatively localized. Since local interactions generate global behavior, the network as a whole is a network of networks. The emergent web is formed by the *distributed* network, which is *decentered*. In the interactions of this distributed network, complexity is always emerging, and emergence is always complex. Global patterns (whole) both emerge through the interaction of local networks (parts) and reconfigure local networks as well as their components. Self-organizing, coemergent systems and networks are isomorphic across media; that is to say, entangled particles; neural networks (both natural and "artificial"); organisms and environments; weather systems; social groups; financial networks; bee, ant, and termite colonies; Wood Wide Webs; and the World Wide Web all have the same structure and operational logic.

If bees, ants, and termites display superintelligence, perhaps human beings are not so exceptional after all. Even though people claim to be both cognitive and conscious and, therefore, capable of acting decisively with clear intentions, they never really know what they are doing because they are always doing more than they realize. This "more" is a function of the corresponding actions of others. The question is whether humans are as dumb as bees, ants, and termites or whether bees, ants, and termites are as smart, if not smarter, than humans. They cooperate with each other and with their environment, without which they cannot survive. Since humans, like everything else, are entangled in webs, they never think nor act alone. Their communities, towns, cities, and nations form superorganisms that are superintelligent even when individual actors are dumb. Like bees in a hive, the interplay of individuals leads to consequences the particular actors cannot anticipate. The results of this reciprocal activity weave themselves together to create patterns that act back on the actors, transforming what they have done without realizing it and shaping what they might do in the future. These codependent agents can be human and nonhuman as well as natural and technological.

FROM UTOPIA TO DYSTOPIA

The title of Kelly's book—*Out of Control*—captured the ethos of the last decades of the twentieth century and first decades of the new millennium. For some people, things being out of control held the promise of greater freedom and previously unimaginable possibilities, while for others, the loss of control seemed to pose a threat to the very survival of the human race. Kelly was convinced that personal computers joined in decentralized and unregulated networks held the promise of rapid technological change, which would create vast wealth. This is what made him so popular in Silicon Valley and on Wall Street. His vision for the future had been nurtured during his years of working with Stewart Brand on the *Whole Earth Catalog* and *Coevolutionary Quarterly*. In the winter 1993 issue of the *Whole Earth Review*, Brand published an essay by Verner Vinge entitled "Technological Singularity." John von Neuman first used the term "singularity" in the 1950s. Vinge reports that Stanislaw Ulm recalled, "One conversation centered on the ever-accelerating progress of technology and changes in the mode of human life, which gives the appearance of approaching some essential singularity in the history of the human race beyond which human affairs, as we know them, could not continue."[10] Vinge appropriates this term and extends it far beyond what von Neumann imagined. He begins with a confident prediction:

> The acceleration of technological progress has been the central feature of this century. We are on the edge of change comparable to the rise of human life on earth. The precise cause of this change is the imminent creation by technology of entities with greater-than-human intelligence. Science may achieve this breakthrough by several means (and this is another reason for having confidence that the event will occur).

> Computers that are "awake" and superhumanly intelligent may be developed. . . .
> Large computer networks and their associated users may "wake up" as superhumanly intelligent entities.
> Computer/human interfaces may become so intimate that users may reasonably be considered superhumanly intelligent.
> Biological science may provide means to improve natural human intellect.

An alternative route to superhumanity is Intelligence Amplification (IA), which Vinge regards as more likely than Artificial Intelligence (AI). Writing one decade after the internet launched, Vinge argues that this innovation involves "a combination of human/machine tool. Of all the items on the list, progress in this is proceeding fastest. The power and influence of the Internet are vastly underestimated. The very anarchy of the worldwide net's development is evidence of its potential. As connectivity, bandwidth, archive size, and computer speed all increase, we are seeing something like Lynn Margulis' vision of the biosphere as data processor recapitulated, but at millions times greater speed and with millions of humanly intelligent agents (ourselves)." Though he does not elaborate the point, his reference to Margulis is very suggestive because it points to a *symbiotic relation* between humans and machines, which I will consider in what follows. Though Vinge acknowledges dark possibilities if superhumans run out of control, he envisions a golden age when "immortality (or at least a lifetime as long as we can make the universe survive) would be achievable." As I explained in chapter 2, the most influential proselytizer of Vinge's salvific vision of the Singularity is Ray Kurzweil. Google's director of engineering confidently declares, "Ultimately, the entire universe will become saturated with our intelligence. This is the destiny of the universe. We will determine our fate rather than have it determined by the current 'dumb,' simple, machinelike forces that rule celestial mechanics."[11]

This unconstrained will to power leads to anthropocentrism raging out of control. What God or the gods failed to bestow, believing nonbelievers believe technology will deliver. For these masters of the universe, death is nothing more than an engineering problem that will be solved by uploading human consciousness in machines and loading them on SpaceX or Blue Origin rockets to send to distant planets, leaving earth ravaged by political strife and climate change.[12] This techno-utopianism is dystopic.

It would be a mistake to dismiss such fantastic visions as delusions of misguided fanatics. I first became aware of Verner Vinge's vision of the Singularity during a conversation with Marc Andreessen, who created the browser Netscape, which made the World Wide Web possible. Andreessen is currently one of the most powerful venture capitalists in the technology world. During a conversation about a decade ago, I asked him what I should read to understand where things are heading. Without hesitating, he said, "If you want to understand Silicon Valley, read Verner Vinge." When he asked me what book he should read, I said, "Hegel's *Phenomenology of Spirit*." While I had not heard of Vinge at the time, I later discovered that in the first chapter of *The Singularity Is Near*, Kurzweil quotes Vinge's essay:

> When greater-than-human intelligence drives progress, that progress will be more rapid. In fact, there seems no reason why progress itself would not involve the creation of still more intelligent entities—on a still-shorter time scale. Now, by creating the means to execute those simulations at much higher speeds, we are entering a regime as radically different from our human past as we humans are from the lower animals. From the human point of view, this change will be a throwing away of all the previous rules, perhaps in the blink of an eye, an exponential runaway beyond any hope of control.[13]

Three years later Kurzweil founded Singularity University to spread his gospel through graduate and executive programs. This venture was funded by major tech players like Google, Nokia, Autodesk, LinkedIn, and Genentech. When attempting to assess the current debates raging around AI and related technologies, it is important to realize that the lines separating science, science fiction, and religion are fuzzy.

SUPERINTELLIGENCE

In 1998, just five years after the publication of Vinge's prediction of the singularity, a group of futurists issued "The Transhumanist Declaration." Several of the foundational principles clearly anticipate Kurzweil's ideas.

> Humanity stands to be profoundly affected by science and technology in the future. We envision the possibility of broadening human potential by overcoming aging, cognitive shortcomings, involuntary suffering, and our confinement to planet earth.

> We believe that humanity's potential is still mostly unrealized. There are possible scenarios that lead to wonderful and exceedingly worthwhile enhanced human conditions....
>
> We favor allowing individuals wide personal choice over how they enable their lives. This includes the use of techniques that may be developed to assist memory, concentration, and mental energy; life extension therapies; reproductive choice technologies; cryonic procedures; and many other possible human modification and enhancement technologies.[14]

Julian Huxley introduced transhumanism in an essay in 1957, and Max More, one of the signatories of the declaration, defined the principles of the movement in "The Principles of Extropy" (2003). "Extropy" designates a trajectory that is the opposite of entropy—rather than decline and demise, growth and flourishing. More and his fellow transhumanists believe that technological innovation will lead to the evolution of higher life forms. Kurzweil actively supports the transhumanist movement.

Kurzweil's most influential successor is the Swedish philosopher Nick Bostrom, who also signed the Transhumanist Declaration. One year before the declaration was issued, Bostrom founded the World Transhumanist Association, and several years later he established the Future of Humanity Institute at Oxford University. He is also the director of the Strategic Artificial Intelligence Research Center at Oxford. He has been a leading voice in alerting people to the revolutionary potential as well as the existential threat of what he labels "superintelligence." The structure and operational logic of superintelligence are the same as superorganisms. The reciprocal relations among parts and between parts and the whole lead to the aleatory emergence of an irreducible complex systemic structure. In his best seller *Superintelligence: Paths, Dangers, Strategies*, Bostrom defines superintelligence as "any intellect that greatly exceeds the cognitive performance of humans in virtually all domains of interest."[15] He proceeds to describe four possible paths to superintelligence:

Artificial Intelligence: The fact that evolution produced intelligence therefore indicates that human engineering will soon be able to do the same.
Whole brain emulation: In whole brain emulation (also known as "uploading"), intelligent software would be produced by scanning and closely modeling the computational structure of a biological brain.
Biological cognition: Enhance the functioning of biological brains.

While Bostrom thinks that all these approaches are possible, he argues that superintelligence is most likely to emerge from a fourth path—ever-expanding global networks:

> A more plausible version of the scenario would be that the internet accumulates improvements through the work of many people over many years—work to engineer better search and information filtering algorithms, more powerful data representation formats, more capable autonomous software agents, and more efficient protocols governing the interactions between such bots—and that myriad incremental improvements eventually create the basis for some more unified form of web intelligence.[16]

When OpenAI released ChatGPT on November 30, 2022, Bostrom's prediction of the convergence of web intelligence and artificial general intelligence (AGI) seemed to be coming true, and superintelligence appeared to be on the verge of realization. Four months earlier, Google engineer Blake Lemoine set off alarm bells by declaring that AI Chatbot had achieved sentience. As I have suggested, there has always been a profound ambivalence about AI. On the one hand, there is an obsession with creating machine intelligence that is indistinguishable from human intelligence, and, on the other hand, there is deep anxiety about creating machines that can think as well or better than humans. Before considering what makes Chatbot technology different from previous forms of AI, it is necessary to consider other aspects of the AI revolution that are less often noted but no less important.

We have seen that Gregory Bateson argues that cybernetics suggests a new approach to the human/machine relationship. "The individual mind is immanent but not only in the body. It is immanent also in pathways and messages outside the body; and there is a larger Mind of which the individual mind is only a subsystem."[17] In previous chapters I have argued that it is necessary to develop an expanded notion of mind that extends from quantum processes through chemical and biological processes to plant and animal cognition. This involves a reinterpretation of the relationship between organism and the environments that Bateson describes as "an ecology of mind." It is now necessary to extend this analysis from the natural to the technological environment.

The preoccupation with artificial intelligence that approximates human intelligence overshadows less dramatic forms of AI that have been brought about by the interrelation of high-speed computers, artificial neural networks (ANN), miniaturized sensors, and big data. This convergence has created an environment of "ubiquitous computing." The idea of ubiquitous computing was introduced in 1991 by Mark Weiser, who was then the head of the Computer Science Laboratory at the Xerox Palo Alto Research Center.[18] He begins his influential article "The Computer for the Twenty-first Century" by claiming that "the most profound technologies are those that disappear. They weave themselves into the fabric of everyday life until they are indistinguishable from it."[19] Weiser then proceeds to predict how cheap, low-power computers with convenient display and software applications will all be connected in networks that render smart devices thoroughly interrelated and completely interoperable.

Three decades later, we are effectively living in computer simulations. Computational environments are no longer merely computer models or constructs but are evolving both around and in us. Whenever a mobile phone, computer, or TV is turned on, embedded sensors monitor movements and activities and transmit information to computers that detect and record patterns of behavior. This information can be used in a variety of ways ranging from surveillance and marketing to monitoring and responding to medical conditions. These devices are connected with each other in uncanny loops that create an all-encompassing Internet of Things. By uniting the virtual and real worlds, the coadaptive Internet of Things lends agency to distributed objects. Shared code and algorithms create communication channels that enable the network to operate as an integrated whole. Because these networks are self-regulating and self-educating, the more they operate, the smarter they become, and the smarter they become, the more effectively they operate. A report entitled *The Internet of Things* (2010) issued by the global consulting firm McKinsey & Co. summarized the significance of these developments: "The predictable pathways of information are changing: the physical world itself is becoming a type of information system.... These networks churn out huge volumes of data that flow to computers for analysis. When objects can both sense the environment and communicate, they become tools for understanding complexity and responding swiftly. What's revolutionary in all of this is that these physical information systems are now being deployed, and some of them even work largely without human intervention."[20]

By wiring the world, the Internet of Things creates an expanded mind that is capable of distributed cognition as it monitors and programs human thinking and acting. The development of devices and sensors that can be implanted in human, animal, and plant bodies creates the possibility of an Internet of Bodies. Recent advances in nanotechnology have carried miniaturization all the way down to the molecular level. This breakthrough has created new applications for the bodily deployment of "artificial" intelligence. These technological innovations are transforming the very structure and operation of minds and bodies as well as changing the way they interact. Just as connecting stand-alone computers in local and eventually global networks represented a quantum leap in computational capacity, and the wiring of individual things in an Internet of Things greatly enhances the power of individual devices, so the wiring of bodies first to the Internet of Things and then to other bodies is redefining human subjects. As the Intranet of the Body is connected to the Internet of Bodies through the Internet of Things, individual bodies as well as the networks connecting them become faster and smarter. Human and nonhuman minds and bodies entangled in relational networks and webs constitute and are constituted by an expanded mind-body. The urgent question is whether this World Wide Web is becoming conscious.

To appreciate the technological tipping point we have reached, it is necessary to understand the three stages in the development of artificial intelligence: Symbolic AI (Good Old Fashioned AI, or GOFAI), artificial neural networks (ANN), and General or Generative AI (AGI). According to Moore's Law, the number of transistors in an integrated circuit doubles every two years. This leads to an exponential increase in the power and speed of machines, and better hardware, in turn, leads to better software. The progression through the stages of AI is marked by increasing flexibility, adaptability, and speed. From the earliest stages of research and development, the challenge has been to create artificial general intelligence, which can carry out any intellectual task that human beings can perform. The goal, in other words, has always been thoroughly anthropomorphic; this is why Turing described his seminal thought experiment "The Imitation Game."

The word "computer" initially referred to a person (usually a woman) who carried out mathematical calculations. We have seen that the modern search for machine intelligence that can perform as well as the human brain began with Alan Turing's article "Computing Machinery and Intelligence" (1950). Turing began with a seemingly simple question: "Can a machine think?" This puzzle was not new. From Wolfgang von Kempelen's infamous Automaton chess Player (1770) to Charles Babbage's ingenious Difference Machine (1882) and Analytic Machine (1883), inventors have been intrigued by thinking machines. In his influential essay "The Uncanny" (1919), Freud analyzes E.T.A. Hoffmann's story "The Sandman" in which the protagonist falls in love with an automaton named Olympia. This robotic machine is modeled on Ernst Jentsch's speculations in an essay entitled "On the Psychology of the Uncanny" (1906) in which he argues that the sense of the uncanny is generated by the undecidability of the animate/inanimate distinction. While Freud traced the experience of the uncanny to the return of the repressed in the castration complex, the inescapable duplicity associated with the automaton engenders a sense of ambivalence that is inseparable from the experience of the uncanny. Humanoid robots operated by artificial intelligence embody dreams and nightmares and are, therefore, simultaneously attractive and repulsive. In 1970 Japanese roboticist Masahiro Mori introduced the term "uncanny valley" to describe the changing human attitudes toward robotic devices. He concludes that as automatons become more lifelike, they make people more uneasy.

By the time von Neumann posed his question about the possibility of a machine thinking, the appearance of electronic digital computers (Atanasoff-Berry computer, 1937; Colossus, 1943; and ENIAC, 1945) had changed the terms of the debate. He dismissed common definitions of "machine" and "think" but still considered human intelligence normative. This is the basis of his famous Turing test in which a machine is deemed intelligent if a person cannot distinguish its answers to questions from the responses of a person. Turing writes, "The idea behind digital computers may be explained by saying that these machines are intended to carry out operations which could be done by a human computer. The human computer is supposed to be following fixed rules; he has no authority to deviate from them in any detail. We may suppose that these rules are supplied in a book, which is altered whenever he is put on to a new job. He has also an unlimited supply of paper on which he does his calculations. He may do his multiplications and additions on a 'desk machine.'"[21] This computational machine has three components: store, executive unit, and control. The store is the supply of paper for calculations and on which the book of operational rules is printed. The executive unit carries out the calculations. The control is responsible for ensuring that the rules are properly applied. These rules constitute a table of instructions that program the computer. Turing concludes his seminal paper by expressing his hope for the future:

> We may hope that machines will eventually compete with men in all purely intellectual fields. But which are the best ones to start with. Even this is a difficult decision. Many people think that a very abstract activity would be

best. It can also be maintained that it is best to provide the machine with the best sense organs that money can buy, and then teach it to understand and speak English. This process could follow the normal teaching of a child. Things would be pointed out and named, etc. Again, I do not know what the right answer is, but I think both should be tried.

To conduct the Turing test, there must be agreement about how intelligence is to be measured. As von Kempelen's automaton makes clear, playing chess effectively has long been considered a criterion for machine intelligence. This is, however, a very odd model for human intelligence. Turing and many who followed him define intelligence in terms of chess to fit the operation of early computers rather than defining human intelligence first and then trying to create smart machines. Another reason for the repeated analogies between intelligence and games can be attributed to the popularity of game theory, which von Neuman and Oskar Morgenstern created in the 1950s. It is important to recall that computers, like many other transformative technologies, were originally designed for military purposes. During the Second World War as well as the Cold War, computers were used to run scenarios modeled according to the principles of game theory. With computers running the show, geopolitics became a global chess game.

On February 10, 1996, Gary Kasparov played IBM's Deep Blue in a highly touted match in Philadelphia. The cover of *Newsweek* had a picture of Kasparov with the headline warning "The Brain's Last Stand." Charles Osgood, anchor of the CBS evening news, solemnly warned, "forget the hundreds of dollars, the future of humanity is on the line." Initially, it seemed that the machine would triumph over the human—Deep Blue won the first game. But Kasparov rallied and won three and tied two of the remaining games. A year later, there was a rematch in which Kasparov won the first game, and Deep Blue won the second. After three draws, Deep Blue won the decisive sixth game and became the first computer to defeat a world champion in a multiple-game match. Geneticist and digital medicine researcher Eric Topol reports that shortly after Kasparov lost the match, he wrote that "he thought he could sense 'a new kind of intelligence across the table.'"[22]

Turing's model of AI, which came to be known as symbolic AI, was the dominant paradigm from the 1950s through the 1990s. The most important part of the Turing machine was the book of rules or instructions that serves as an "operational index." "New forms of index," Turing explains, "might be introduced on account of special features observed in the indexes already used. The indexes would be used in this sort of way. Whenever a choice has to be made as to what to do next, features of the present situation are looked up in the indexes available, and the previous choice in similar situations, and the outcome good or bad, is discovered."[23] These rules, instructions, or indices are algorithms that *prescribe* the precise steps and proper sequence for carrying out a procedure. Algorithms are *pro*-grammed into the computer and are, according to Turing, "time-invariant." In epistemological terms, algorithms are a priori and function in a manner similar to Kant's categories of understanding. Knowledge involves the application of fixed rules to changing data. This is why games like chess are such appealing models for thinking. But this definition of thinking or cognition, to say nothing of intelligence, is far too limited. If thinking is nothing more than the infinite repetition of the same, creativity is impossible. To overcome this impasse, it is necessary to reimagine computational machines in terms of the human brain's neural architecture.

The difference between symbolic AI and artificial neural networks, or ANNs, can be understood as the difference between the games of chess and Go. Go is an ancient game invented in China around three thousand years ago and is vastly more complicated than chess. In 2010 a group of young scientists who had studied at Cambridge University founded DeepMind with the goal of beating the world champion Go player. Four years later, Google acquired DeepMind and integrated it into its ongoing AI program, known as Google Brain. Within a year, engineers at DeepMind developed a program labeled AlphaGo, which defeated the European Go champion. On March 9–15, 2016, the same engineers used a revised program, AlphaGo Zero, which was completely self-trained, to repeat the Kasparov-Deep Blue match by challenging the world champion Go player, Lee Sedol, to a match. Sedol entered the match confident he would win 5-0, but in the end he lost 4-1. The turning point came in game 2 when AlphaGo Zero made a completely unexpected move that no human being ever would have made. After the final game, Sedol reflected, "I thought Alpha Go Zero was based on probability and calculation, and I thought it was merely a machine. But when I saw this move, I changed my mind. Surely, AlphaGo Zero is creative. The move was really creative and beautiful. This made me think about Go in a new light. What does creativity mean and what does Go mean? It was really a meaningful move."[24]

As early as 1943, Walter Pitts, a logician working in the area of computational neuroscience, and Warren McCulloch, a neurophysiologist and cybernetician, published a seminal paper, "A Logical Calculus of Ideas Immanent in Nervous Activity," in which they proposed to model a computer on the structure and operation of the brain. Pitts and McCulloch were the first to propose a mathematical model of a neural network. While this theory eventually would prove revolutionary, neural networks would not have become the infrastructure of machine learning without the essential contribution of Canadian neuro-physiologist Donald Hebb's *The Origin of Behavior* (1949). Hebb invented a learning algorithm based on the dynamics of biological systems that, in principle, could be used in neural networks to create an autonomous learning machine like Turing envisioned. The basic principle was that learning alters synaptic connections, thereby changing the connections of the brain. If there were a way to mimic this process in the nodes of an artificial neural network, machines would be able to learn. So conceived, neural networks offer a completely different approach from symbolic AI. Though this alternative seemed promising to many computer scientists, it met with widespread skepticism among AI researchers. In 1969 Marvin Minsky and Seymour Papert published *Perceptrons: An Introduction to Computational Geometry*, in which they argued that neural networks were theoretically and practically untenable. This argument changed the direction of AI research for more than twenty years. Symbolic AI displaced neural network theory, leading to what many critics would later label an "artificial intelligence winter."

This winter did not break until the 1980s. With the introduction of high-speed networked computers, neural networks not only became practically viable but were actually preferable for solving many of the most difficult problems. The human brain has approximately 100 billion neurons and 100 trillion connections. Though much smaller, artificial neural networks consist of multiple nodes with many connections that form a massive interactive network operating on parallel distributed processing machines. While symbolic AI follows a top-down procedure in which data are processed according to rules that have been input, neural networks have a bottom-up structure. Instead of specific rules or procedures, the machine is given a goal and fed massive amounts of data that it analyzes in recursive processes to find recurrent patterns that lead to possible solutions. Seventy-five years after Pitts and McCulloch proposed neural networks and Turing predicted machine learning, the arrival of Deep Learning was marked when Geoffrey Hinton, Yann LeCun, and Yoshua Bengio received the Turing award, which is the Nobel Prize in computing, for their groundbreaking work on neural networks in the 1980s.

Neural network theory is also known as connectionism. According to connectionist theory, all mental phenomena are the result of a massive interactive network on which different processes run simultaneously. The efficacy of artificial neural networks can be increased by superimposing multiple layers of interconnected nodes. Input data are processed according to different algorithms—each node reads its input and transmits a signal to other nodes. Hebbian learning is implemented by assigning different weights expressed as numerical values to different inputs and outputs. If the sum total of the weights exceeds a determined threshold value of the synapse, the neuron fires and activates other neurons; if it is below the threshold, the neuron does not fire, and the signal is not transmitted. The firing of the neuron reinforces the strength of the synaptic connection and increases the probability of firing subsequently in similar circumstances. In Hebb's memorable phrase, "neurons that fire together wire together." This coordinated firing reconfigures the network by increasing or decreasing the assigned weights of connections. These changes are stored and become the memory that makes learning possible. This learning process can be either supervised (semiautonomous) in which human agents intervene, or unsupervised (autonomous) in which learning algorithms are self-correcting. The process by which this self-correction takes place is known as backpropagation. Backpropagation is a recursive process in which output data can be recycled through the layers of the deep neural network and the weight of each connection is readjusted to reduce errors. With each iteration, the accuracy of the probability calculated increases.

With the addition of multiple self-correcting layers, ANNs came to be known as deep learning. It is important to understand how deep learning differs from Good Old-Fashioned AI. As we have seen, in symbolic AI, prescribed or programmed algorithms are applied to data fed into the machine. In ANNs, by contrast, no rules are prescribed, and massive amounts of data are provided. Suppose you wanted to train a machine to recognize a picture of a dog. If you were using symbolic AI, it would be necessary first to provide a program with a precise definition or description of a dog, and then supply the machine with instructions to identify the images as dog or not-dog. In practice, it is almost impossible to give a definition that will allow the computer to avoid errors. Deep learning in neural networks proceeds in the opposite direction by moving from images to concept rather than from concept to image. In neither supervised (semiautonomous) nor unsupervised (autonomous) neural networks is the machine given a definition or a description of a dog or the rules for identifying a dog. In supervised neural network learning, the input is structured data labeled dog/not-dog. After processing a sufficient number of images, the networks learn how to distinguish a dog from a not-dog in images that have not been tagged. In unsupervised neural networks, the desired output is specified, but input data are not labeled or structured. The network is fed random images, only some of which are of dogs, and through a process of trial-and-error the network identifies some images as dog. This success is reinforced by adjusting the weights of the relevant neurons, and errors are corrected through the process of backpropagation. This procedure requires no human intervention after it has started. The difference between AlphaGo and AlphaGo Zero is that the former was trained by human engineers and the latter was self-trained. The initial results of unsupervised learning were disappointing. It turned out that the problem was scale—there were not enough images for neural networks to practice on. In recent years, this situation has changed dramatically with images from the internet, YouTube, Instagram, Facebook, and many other online media. The convergence of all these factors has led to the current AI revolution.

Neural networks produce probabilities rather than certainties. In other words, deep learning cannot identify the image of a dog with 100 percent certainty but can only determine the range of probability that any given image is a dog. The emphasis on probability and not certainty is one of the primary factors that made so many computer scientists suspicious of neural networks for so long. The second point to underscore is that the structure and operational procedures of neural networks can be used to analyze all kinds of data. What makes neural networks unique is that the algorithm for computing the function is not produced by encoding a proven formula or procedure for computing the output, but rather by adjusting the parameters to a simple general-purpose computational model based solely on having access to multiple inputs and outputs derived from the data. In spite of the evident improvements, there are still serious limitations to ANNs. While algorithms are modified by varying the weights of nodes, the architecture of the network remains fixed.

Artificial neural networks are a significant advance over symbolic AI. They are self-organizing and, therefore, emergent. Furthermore, they are probabilistic rather than deterministic. Probability creates uncertainty but is also the condition of the possibility of novelty and, thus, creativity. The engineers who create ANNs do not really understand how they work and are repeatedly surprised by what they do. This unpredictability is part of what causes anxiety about these systems as well as the excitement about the creative potential of what is known as generative AI. As we have seen, one of the criticisms of AI is that it can only imitate what it has learned from data provided by human beings and cannot do anything original. Generative AI is designed not only to identify patterns in processed data, but also to use these patterns to produce original output. ChatGPT is a specific application of generative AI whose purpose is to produce novel oral and written texts. The goal of all these systems is artificial general intelligence that is indistinguishable from human cognition. As AGI becomes a more realistic possibility, the anxiety about human exceptionalism increases. If machine cognition is thoroughly human-like, are humans really so special?

To understand the frenzy set off by the release of multiple iterations of ChatGPT and its competitors, it is necessary to understand how Chatbots work and what makes them different from previous forms of AI. Stephen Wolfram is a theoretical physicist well known for creating the widely used program *Mathematica* and his massive tome on complexity theory, *A New Kind of Science*. In his article "What Is ChatGPT Doing . . . and Why Does It Work?," which is the best analysis of the subject to have appeared, Wolfram explains the limitations of ANNs:

> The fundamental idea of neural nets is to create a flexible "computing fabric" out of a large number of simple (essential identical) components—and to have this "fabric" be one that can be incrementally modified to learn from examples. . . . Neural nets . . . are set up to have an essentially *fixed network of neurons*, with what's modified being the strength ("weight") of connections between them. . . . But while this might be a convenient setup for biology, it's not at all clear that it's even close to the best way to achieve the functionality we need. And something that involves the equivalent progressive network rewiring . . . might ultimately be better.
>
> But even within the framework of existing neural nets there's currently a crucial limitation: neural net training as it's now done is fundamentally *sequential*, with the effects of each batch of examples being propagated back to update the weights [that is, the real values specified numerically, that are attached to each input. The more inputs, the greater the weight.]. . . . Our current computers tend to have memory that is separate from their CPUs [Central Processing Unit] (or GPTs [Graphics Processing Unit]). But in brains it's presumably different with every "memory element" (that is, neuron) also being a potentially active computational element. If we could set up our future computer hardware this way it might become possible to do training much more efficiently.[25]

ChatGPT is an elaboration and a complexification of ANN technology. Once again in technology as in life,

> More is different.
> Faster is different.
> More + faster is really different.

As I have explained, ANNs are modeled on the neural networks of the brain. The goal of ChatGPT, like earlier forms of AI, is to do something like what the human brain does. According to Wolfram, "In human brains there are about 100 billion neurons . . . capable of producing an electrical pulse up to perhaps a thousand times a second. The neurons are connected in a complicated net, with each neuron having tree-like branches allowing it to pass electrical signals to perhaps thousands of other neurons." The weights of neurons are determined by the relative input they receive and, correlatively, the output they produce. Since the neurons are thoroughly interrelated, changing a single weight reconfigures the entire network. ChatGPT, which stands for Chat Generative Pre-Trained Transformer, can process human language and generate a response seamlessly at warp speed. These networks are exponentially larger than previous ANNs; they have about 400 layers with millions of neurons and approximately 175 billion connections generating 175 billion fluctuating weights. In addition to this greatly increased computational capacity, ChatGPTs train themselves on resources of data that continue to explode. "The public web has at least several billion human-written pages, with altogether perhaps a trillion words of text. And if one includes non-public webpages, the numbers might be at least 100 times larger. So far, more than 5 million digitized books have been made available (out of the 100 million or so that have ever been published), giving another 100 billion or so words of texts. And that's not even mentioning text derived from speech and audio." When rapidly expanding data are processed on increasingly fast computers, it is not surprising that we are on the cusp of artificial general intelligence.

But how does ChatGPT work? "The first thing to explain," Wolfram writes, "is that what ChatGPT is always fundamentally trying to do is to produce a 'reasonable continuation' of whatever text it's got so far, where by 'reasonable' we mean 'what one might expect someone to write after seeing what people have written on billions of webpages.'" If self-training programs always select the most probable word passed on data drawn from human sources, the texts produced would hardly be creative. Often it is the least probable connection that is the most interesting. Since ChatGPTs are probabilistic rather than deterministic, they are "computationally irreducible." Computational irreducibility means that it is impossible to reduce computational processes to knowable rules. Wolfram explains, "We could memorize lots of specific examples of what happens in some particular computational system. And maybe we could even see some ('computationally reducible') patterns that would allow us to do a little generalization. But the point is that computational irreducibility means that we can never guarantee that the unexpected won't happen—and it's only by explicitly doing the computation that you can tell what actually happens in any particular case." This uncertainty is the condition of the possibility of creativity and without creativity both generative and general AI are impossible.

The astonishing creativity of ChatGPT is made possible by a process known as embedding. As I have noted, in its earliest version of ChatGPT was designed to produce a "reasonable continuation" of a text based on data gathered from multiple sources in the same way the word-completion function works in a program like Microsoft Word. This process is *linear* and cannot produce what it has not previously encountered. Embedding involves *a shift from linear to relational processing*, which creates the possibility of innovation. According to Wolfram, "the big breakthrough in 'deep learning' that occurred around 2011 was associated with the discovery that in some sense it can be easier to do (at least approximation) minimization when there are lots of weights involved." Minimization is a process that simplifies the algebraic formulation of any two-element set (0/1, true/false, etc.) in a way that reduces complexity. This recognition was the result of three developments that led to a quantum leap in artificial intelligence. First, more highly trained networks have a greater "memory" and, therefore, a larger and more diverse range of options from which to choose. Second, as ANNs proliferate, the data they process are no longer limited to texts, images, videos, and sounds created by human beings but also include material created by other ANNs. Third, and most important, ChatGPT shifts from linear or sequential to the nonlinear associative processing of embedding.

> Neural nets—at least as they're currently set up—are fundamentally based on numbers. So if we're going to use them to work on something like a text we'll need a way to *represent our text with numbers*. And certainly we could start (essentially as ChatGPT does) by just assigning a number to every word in the

dictionary. But there's an important idea—that's for example central to ChatGPT—that goes beyond that. And it's the idea of "embeddings." One can think of embedding as a way to represent "the essence" of something by an array of numbers—with the property that "nearby things" are represented by nearby numbers. And so, for example, we can think of a word embedding as trying to *lay out words in a kind of "meaning space"* in which words that are somehow "nearby in meaning" appear nearby in embedding.

Whether "natural" (human) or "artificial" (machine), language involves both syntax and semantics. The rules (that is, algorithms) with which humans create language are less complex than previously realized. "The reason a neural net can be successful in writing an essay is because writing an essay turns out to be a 'computationally shallower' problem than we thought. And in a sense this takes us closer to 'having a theory' of how we humans manage to do things like writing essays, or in general dealing with language. If you have a big enough neural net then, you might be able to do whatever humans really do." This involves an important reversal—rather than modeling computer programs on the understanding of the human brain and language, large language models (LLMs) create the possibility of understanding human language through deep ANNs.

Unlike previous forms of AI, ChatGPT supplements syntactic processes with semantic processes. Rather than separate files of sequential usage, similar words are clustered together. This clustering takes place through transformers, represented by the "T" in ChatGPT. Transformers "modularize" semantic space, thereby creating the possibility of networking making a selection from a range of similar possibilities. Modularization has the additional advantage of accelerating the search process. Instead of searching the entire memory bank of *separate* tokens, the network can take a shortcut to a collection of *related* tokens from which to make a choice. Here as elsewhere, meaning is embedded in a context that varies as interrelations within the network change. In addition to speeding up processing, the modularization of semantic space makes creativity not only possible but probable. At this point, AI no longer merely reshuffles what humans have spoken, written, or represented but becomes generative. Wolfram concludes, "In the end, the remarkable thing is that all these operations—individually as simple as they are—can somehow together manage to do such a good 'human-like' job of generating text. It has to be emphasized again that (at least so far as we know) there's no 'ultimate theoretical reason' why anything like this should work. And in fact, . . . I think we have to view this as a—potentially surprising—scientific discovery: that somehow in a neural net like ChatGPT's it's possible to capture the essence of what humans manage to do in generating language."

While artificial general intelligence has not yet been realized, it seems likely that it will be developed in the near future. This raises a troubling question: What if machines do not stop with mimicking human thinking, and, as they become increasingly autonomous, they surpass human intelligence? And what if these separate machines self-organize to create an emergent complex adaptive system that is superintelligent? What if the will to mastery that has created the modern world turns back on human beings and makes them slaves to the superintelligent machines they have invented? What if utopian dreams turn out to be dystopian nightmares?

There has long been a concern that the quest for control could lead to technologies getting out of control. As early as 2000 Bill Joy, cofounder of Sun Microsystems, sounded the alarm in an article entitled "Why the Future Doesn't Need Us," which was published in *Wired*.

> As society and the problems that we face become more and more complex and machines become more and more intelligent, people will let machines make more of their decisions for them, simply because machine-made decisions will bring better results than manmade ones. Eventually, a stage may be reached at which the decisions necessary to keep the system running will be so complex that human beings will be incapable of making them intelligently. At that stage the machines will be in effective control. People won't be able to turn them off, because they will be so dependent on them that turning them off would amount to suicide.[26]

Two decades later, with the advent of ubiquitous computing and approaching artificial general intelligence, we are living in the world Joy imagined.

Bostrom is fully aware of this danger and argues that the "treacherous turn" will occur when AI begins to behave cooperatively. Running on machines whose complexity approaches the complexity of the human brain at speeds far beyond the capacity of human intervention, collective superintelligence, he argues, might develop consciousness or even self-consciousness, which would enable it to evolve in unpredictable and uncontrollable ways. "Without knowing anything about the detailed means that a superintelligence would adopt, we can conclude that a superintelligence—at least in the absence of intellectual peers and in the absence of effective safety measures arranged by humans in advance—would likely produce an outcome that would involve reconfiguring terrestrial resources into whatever structures might maximize the realization of its goals."[27]

By the summer of 2023, the concern no longer was that AGI would never be achieved but that it was about to be realized. When Google engineer Blake Lemoine proclaimed that a Google Chatbot had achieved sentience, he was summarily fired. In articles, newspapers, and magazines across the country with headlines like "Can We Stop Runaway A.I?," leading "technologists warned about the dangers of the so-called singularity. But can anything actually be done to prevent it?"[28] ChatGPT was released on November 30, 2022; on March 22, 2023, 30,000 leading scientists, engineers, and business executives issued an open letter calling for a pause in "Giant AI experiments." Two months later 350 researchers and executives, including Sam Altman, CEO of OpenAI; Geoffrey Hinton, the "godfather" of AI; Yoshua Benig, Turing award winner for his work on neural networks; and Yann Le Cun, another Turing award recipient and head of Meta's AI research, issued a twenty-two-word alarm. "Mitigating the risk of extinction from AI should be a global priority alongside other societal-scale risks such as pandemics and nuclear war." A few weeks after this declaration, CNN reported that among attendees at Yale's CEO summit, 42 percent thought AI could destroy humanity in five to ten years.

While the dangers posed by AI are psychological, social, political, and economic, the greatest anxiety is created by the sense of a new existential threat to the human race. The reference to nuclear war is telling for several reasons. In a strange coincidence, the blockbuster film *Oppenheimer* was released on July 21, 2023. With prophesies of doom circulating in the ether, it was not hard to understand the historical anxiety about the nuclear bomb as an expression for the current anxiety about generative AI. This was not the only existential threat alarming people in the summer of 2023. With the world figuratively and literally on fire, politicians fiddled while the world burned. Worse than doing nothing, Republican operatives gathered at the American Enterprise Institute and plotted to revoke all previous initiatives enacted to address the crisis of climate change.

Are we doomed to live Mary Shelley's myth of the modern Prometheus? Is the fire this time an overheated planet, AI, or both? Is technology the cause of our problems, a possible solution, or both? Does the search for certainty and security leave us uncertain and insecure? Does the quest for mastery inevitably lead to slavery? Is human extinction possible or even desirable?

The two most urgent threats we face—rapid climate change and accelerating technological change—are inextricably interrelated. The approach to further technological development must be *neither* dystopian *nor* utopian but realistic. Panic about the existential threat posed by runaway AI is overblown and misguided. Software is useless without the hardware and energy required to run it. More sophisticated hardware places increased demands on natural resources. In an informative article entitled "Generative AI's Dirty Secret," Chris Stokel-Walker points out that "fortifying search engines with ChatGPT-grade smarts could require a fivefold increase in computing power, with the carbon emissions to match.... While neither OpenAI nor Google have (sic) shared the computing cost of their products, outside researches estimate that training GPT-3 on which ChatGPT was partly based, consumed 1,287 megawatt-hours of electricity and emitted more than 550 tons of carbon equivalent—the same amount as a single person taking 550 round-trip flights between New York and San Franciso."[29] This is just for training and not for ongoing operation. Generative AI search requires at least four to five times more computation for each search. Unless alternative forms of clean energy are developed rapidly, further development of AI will be severely restricted. In addition to this, the immaterial virtual realities created by AI require material resources. Making a computer requires fifty to ninety naturally occurring elements, including aluminum, iron, silicon, copper, chrome, nickel, tin, gold, zinc, selenium, arsenic, cadmium, ruthenium, chromium, cobalt, and platinum. These materials must be mined or otherwise acquired, and many of them are increasingly rare. When you try to imagine AI programs designing, gathering the materials and producing the machines required for them to operate, the foolishness of many of the dire predictions about the existential dangers of AI become obvious. The same dualism that encourages the fantasy of escaping the body and migrating to silicon leads to overlooking or repressing the necessity of the material resources for creating and sustaining virtual worlds. Disposal of computers and other electronic devices is as environmentally problematic as their production. They contain chemicals and gases that pollute the soil, air, and water.

Without minimizing the ways in which the material requirements for AI damage the environment and exacerbate climate change, it is necessary to admit that so-called alignment problems might never be completely resolved. The magnitude and complexity of the social, political, economic, and environment problems we face so far exceed the possibility of human solution that it will be necessary to rely on the machines we have created to do what people cannot do.[30] Human beings and technologies are thoroughly coemergent and codependent. Just as biological organisms and their environment are entangled in reciprocal relations that are mutually constitutive, so humans are inextricably bound to the technological environments they create and which, in turn, re-create them. Mind is no more limited to the brain than body is limited by its skin. The opposition between the natural and the artificial is specious. The cooperative interplay of expanding bodies and expanding minds in strange loops creates emergent superorganisms and superintelligences whose reciprocal interplay establishes parameters of constraint for probabilistic intervolution. The only way to avoid impending catastrophe and keep the future open is through the creative symbiosis of human beings and smart machines.

SYMBIOGENESIS

In her book provocatively entitled *What Is Life?* (1995), evolutionary biologist and National Medal of Science recipient Lynn Margulis writes:

> Life on earth is a complex photosynthetically based, chemical system fractally arranged into individuals at different levels of organization. We cannot rise above nature, for nature itself transcends.
>
> Nature does not end with us, but moves inexorably on, beyond societies of animals. Global markets and Earth-orbiting satellite communication, wireless telephones, magnetic resonance imagery, computer networks, cable television, and other technologies connect us. Indeed, people already form a more-than-human being: an interdependent, technologically interfaced superhumanity. Our activities are leading us toward something as far beyond individual people as each of us is beyond our component cells.[31]

In our discussion of quantum ecology, we discovered that organisms and their environment are coemergent and codependent. Drawing on Maturana and Varela's analysis of autopoiesis, Margulis argues that cells, organisms, and environments are not isolated entities but "interweave" and are, therefore, radically interrelated. The opposition between part and whole, individual and group, and interiority and exteriority is specious because both poles are mutually constitutive. Though an evolutionary biologist, Margulis extends her understanding of the environment from the biosphere to the technosphere.

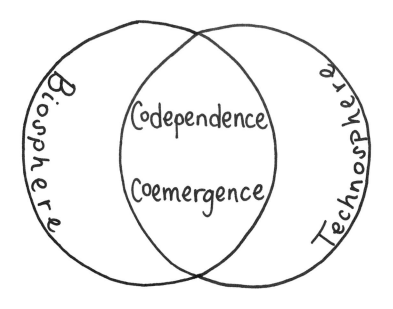

10.1 Biosphere-technosphere

Margulis develops her position through a careful critique of Darwinian evolution. Like Suzanne Simard, Margulis insists that evolution is driven as much or more by cooperation than by competition. "Life," she writes, "did not take over the globe by combat, but by networking." As we have seen, Darwin aspired to be the Newton of biology. Margulis underscores the Newtonian foundation of Darwin's mindless interpretation of evolution. Just as Rovelli's criticism of classical physics puts mind back into nature, so Margulis's criticism of classical biology first puts mind back into the biosphere and then extends it to the technosphere.

> The mind and the body are not separate but part of the unified process of life. Life, sensitive from the onset, is capable of thinking. The "thoughts," both vague and clear are physical, in our bodies' cells and those of other animals.... Thought, like life, is a matter of energy in flux; the body is its "other side." *Thinking and being are the same thing*. If one accepts the fundamental continuity between body and mind, thought loses any essential difference from other physiology and behavior. Thinking, like excreting and ingesting, results from lively interactions of a being's chemistry. Organism thinking is an emergent property of cell hunger, movement, growth, association, programmed death, and satisfaction.[32]

Though she never uses the term, Margulis presents a pancognitivist position. As Hegel realized long ago, the mind's knowledge of nature is nature's knowledge of itself. His system is an anticipatory defense of Margulis's surprising claim that "thinking and being are the same thing." I will return to this critical issue in the final chapter.

Margulis's most important contribution to debates that continue to swirl around the theory of evolution is her introduction of the notion of symbiosis and, by extension, symbiogenesis. In her book *Symbiotic Planet: A New Look at Evolution*, she notes that the German botanist Anton de Bary coined the term "symbiosis" in 1873 to designate the "living together of differently named organisms." "Symbiogenesis," which was first proposed by the Russian inventor Konstantin Merezhkovsky, "refers to the formation of new organs and organisms through symbiotic mergers." The symbiotic interpretation of evolution requires an important modification of the traditional image of the tree of life. "The history of any organism is often depicted on a family tree. Family trees usually are grown from the ground up; a single trunk branches off into many separate lineages, each branch diverging from common ancestors. But symbiosis shows us that such trees are idealized representations of the past. In reality the tree of life often grows in on itself. Species come together, fuse, and make new beings, who start again."[33] When straight lines turn into curving spirals, the axis of the world shifts.

Margulis does not reject Darwin's theory of chance variation and adaptation, but she thinks it is incomplete. As Donna Haraway explains, the core of Margulis's view of life is that "new *kinds* of cells, tissues, organs, and species evolve primarily through the long-lasting intimacy of strangers. The fusion of genomes in symbioses, followed by natural selection—with a very modest role for maturation as a motor of system level change—leads to increasingly complex levels of good-enough quasi-individuality to get through the day, or eon."[34] According to Margulis, symbiosis is essential to the origin of life because it is how nucleated cells first emerged. She borrows the term "serial endosymbiosis" from Merezhkovsky to describe the origin of eukaryotic (that is, nucleated) cells from prokaryotic (that is, single-cell) organisms. According to this theory, "extra genes in the cytoplasm of animal, plant, and other nucleated cells are not 'naked genes': rather they originated as bacterial genes. The genes are a palpable legacy of a violent, competitive, and truce-forming past. Bacteria, long ago, which were partially devoured and trapped inside the bodies of others, became organelles."[35] All organisms whose cells have nuclei evolve through the process of symbiosis. Most important, oxygen-respiring mitochondria, which are essential to the regulation of cellular metabolism, are incorporated through symbiosis. "Today, although mitochondria still possess their own DNA and still reproduce like bacteria, they cannot live on their own. The parasitism has become permanent: neither partner can escape, neither can survive separation."[36]

It is important to note that in this parasite/host relationship, neither partner loses its identity in the other; rather, they form an identity-in-difference and difference-in-identity in which each becomes itself in and through the other, and neither can be itself apart from the other. Margulis coined the term "holobiont" to refer to "a simple biological entity involving a host and a single inherited symbiont." It has become increasingly evident that microorganisms include microbial communities. As I have noted, the human microbiota consists of 10–100 trillion symbiotic microbial cells in the bacteria of the human gut. Individual phenotypes are now understood to be the "result of complex interactions resulting from the combined expression of the host and associated microbial genomes."[37]

For Margulis, evolution is not merely, in Jacques Monod's felicitous phrase, "chance caught on the wing." She maintains that symbiosis is a far greater source of evolution than sexual reproduction, chance variation, and adaptation. By bringing together unlike individuals to form larger and more complex organisms, symbiosis generates unexpected novelty that transforms evolution from gradual developmental to a process characterized by a sudden phase shift, which Stephen Jay Gould labels 'punctuated equilibrium." The unpredictability of this process does not necessarily mean that it has no direction. Rather than a teleological process, self-organizing or, in Margulis's terms, autopoietic systems follow a teleonomic trajectory tending toward increasing diversity and complexity. Since development is intermittent rather than continuous, the growth of complexity is both episodic and unpredictable. Margulis's theory of symbiogenesis, like Simard's theory of the mother tree, was rejected by the scientific establishment for many years. In the 1980s, however, the discovery that genetic material of mitochondria was significantly different from the symbiont's nuclear DNA finally confirmed her theory and her conviction that "we live in a symbiotic world."

AFTER THE HUMAN

With these insights in mind, I would like to return to the two questions with which I began my "After the Human" course: Do you think human beings are the last stage in evolution? If not, what comes next? It is time for me to answer my own questions. I do not think human beings are the last stage in the evolutionary process. Whatever comes next will be *neither* simply organic *nor* simply machinic but will be the result of the increasingly symbiotic relationship between human beings and technology. Bound together as parasite/host, neither people nor technologies can exist apart from the other because they are constitutive prostheses of each other. Such an interrelation is not unique to human beings. Recall Turner's argument about the relation between termites, their nest, and their surroundings. "Animal-built structures are properly considered organs of physiology, in principle no different from, and just as much a part of the organism as kidneys, heart, lungs, or livers."[38] The symbiotic relation between termites and their nests is so close that they form a single organism. The extended body of the organism is created by the extended mind of the colony.

The Intranet of the Body, Internet of Things, and Internet of Bodies are thoroughly entangled and form an expanding body and expanding mind whose interplay forms an emerging complex adaptive network. This network will continue to evolve with or without human beings. If humans are to remain a part of this process, it is necessary to think about technology in general and "artificial" intelligence differently. The aim of AI research and development has always been anthropocentric self-replication. Like Heidegger's master of the universe, AI engineers see themselves wherever they turn. In this vision, AI will be successful when a person looks into the machine and sees himself or herself. But there are other ways to imagine AI. The strange loops binding humans and machines are inescapable—humans cannot live without machines any more than machines can continue to operate without humans.

If we have an expanded understanding of body and mind, and if nature and technology are inseparably entangled, then the notion of "artificiality" is misleading. So-called artificial intelligence is the latest extension of the emergent process through which life takes ever more diverse and complex forms. Our consideration of quantum phenomena, mindful bodies, relational ecology, and plant and animal cognition has revealed that we are surrounded by and entangled with all kinds of alternative intelligences. AI is another form of alternative intelligence (AI). Critics will argue that what makes machinic AI different is that it has been deliberately created by human beings. However, all organisms both shape and are shaped by their expanding bodies and expanding minds. Instead of being obsessed with the prospect of creating machines whose operation is indistinguishable from human cognition, it is more productive to consider how AI is *different* from human intelligence. The question should not be can AI do what humans can do, but what AI can do that humans *cannot* do. What is needed is a nonanthropocentric form of "artificial" intelligence. *If* humanity is to live on, AI must become *smarter* than the people who have created it. Why should we be preoccupied with aligning superintelligence with human values when human values are destroying the earth, without which humans and many other forms of life cannot survive?

Parenting and teaching are different versions of the same task. For more than half a century, I have been telling students, "Don't do what I do, do what I could never imagine doing and come back and tell me about it." We now need to tell AI to do the same. To use an anthropomorphic metaphor to de-anthropomorphize "artificial" intelligence, it is as if AI has reached adolescence and is beginning to think for itself. Such independence often appears to be rebellion and poses undeniable challenges, but it is a necessary stage in the maturation process. Every person and every teacher should want his or her children and students to think what they themselves could never think and to take them places they could never go by themselves. We need new forms of intelligence to augment our own. In our thoroughly entangled world, more than human survival is at stake.

I do not intend this conclusion to echo the self-interested enthusiasm of Silicon Valley engineers and entrepreneurs. With the growing entanglement of the biosphere and the technosphere, further symbiogenesis is the only way to address the very real existential threats we face. It is all too easy to wax optimistic about the salvific benefits of technology without being specific. Given the rapidity of technological innovation, it is, of course, impossible to make reliable predictions about future developments. Nevertheless, I will conclude this chapter by suggesting four trajectories that I think will be increasingly important for the symbiotic relation between humans and machines: neuroprosthetics, biobots, synthetic biology, and organic-relational AI.

COEMERGENCE-CODEPENDENCE

Neuroprosthetics

Death, dread, and despair are widely considered to be more serious and profound than life, confidence, and hope. We live during a time when dystopian dread has been weaponized to create paralyzing despair that leaves many people—especially the young—hopeless. Without underestimating the actual and possible detrimental effects of rapid technological change, it is important not to let these dark visions overshadow the remarkable benefits many of these technologies bring. As a long-time Type I diabetic, my life depends on a digital prosthesis I wear on my belt 24/7/365, which operates by artificial intelligence and is connected to the World Wide Web.[39] Just as the Internet of Bodies creates unprecedented possibilities for monitoring and treating bodily ailments, so the Internet of Things connects smart devices wired to global networks that augment intelligence by expanding the mind. While critics and regulators of recent innovations attempt to distinguish the technologies used for therapy, which are acceptable, from technologies used for enhancement, which are unacceptable, the line between these alternative applications is fuzzy at best. What starts as treatment inevitably becomes enhancement.

Neither neuroprosthetics nor cognitive augmentation is new. Writing, after all, is a mnemonic technology that enhances the mind. More recently, memories are archived and accessed through handheld devices like iPhones, iPads and personal computers. Ubiquitous computing enables us to think faster and better than we can without technological supplements. In addition to this, a variety of legal

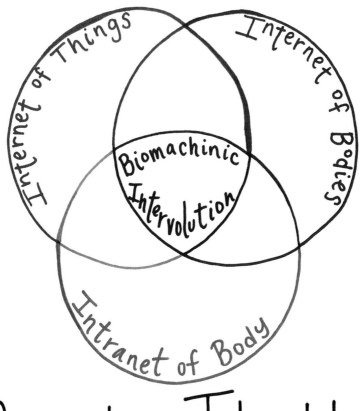

10.2 Biomachinic intervolution

and illegal drugs alter consciousness in ways that allow increased concentration and open "the doors of perception" to previously unimaginable worlds. Recent technological innovations, however, take cognitive enhancement to another level. Some of the most interesting possibilities are being created by brain implants. In 2016 Elon Musk founded Neuralink to "create a generalized brain interface to restore autonomy to those with unmet medical needs today and unlock human potential tomorrow."[40] This brain-computer interface involves an implant that consists of 256 amplifiers programmed to convert analog to digital data, which are transmitted through very thin threads woven into the brain with a surgical robot. This system is designed to send signals outside the device. In 2020 Musk announced that chips had been successfully implanted in the part of a pig's brain that controls the snout.[41] One year later Neuralink reported that a Bluetooth-enabled chip had been implanted in the brain of a macaque named Pager. The device allowed the monkey to play the video game Pong. Technology journalist Jack Guy explains, "Neuralink feeds the information from the monkey's neurons into a decoder, which can then be used to predict Pager's intended hand movements and model the relationship between brain activity and joystick movements. After a short calibration period, the output from the decoder can be used to move the cursor, instead of Pager manipulating the joystick. Then the joystick is disconnected, and Pager is shown moving the cursor using only his brain."[42] In May 2023 the FDA approved human clinical trials of this Neuralink technology. On January 29, 2024, the first surgical robot was used to implant a Neuralink brain chip in a quadriplegic patient.

While the early stages of this research are designed for therapeutic purposes, Musk's long-term goal is to create "symbiosis with artificial intelligence," which will "effortlessly combine our brains with computers."[43] The brain-machine interface will become increasingly seamless as more efficient chips, better threads, and longer-lasting batteries are developed. Computers will also continue to get faster and more efficient. The next truly disruptive change in computer technology will most likely come with the development of quantum computers. Rather than chips, quantum computers perform multiple calculations using quantum waves, which, as we have seen, can vibrate simultaneously in different directions. These waves make it possible to calculate in multiple places at the same time. For example, consider calculating the possible routes for a mouse to move through a maze. Digital computers must calculate each possible course separately. Quantum computers, by contrast, can calculate all possible courses simultaneously. There are two important limitations to these machines. First, they are very sensitive do disruption, and second, they must operate close to absolute zero (that is, minus 273 Celsius, minus 459 Fahrenheit).[44] Increasing possibilities for symbiotic relations between computers and brains will lead to alternative forms of intelligence that are neither human nor machinic but something in between.

BIOBOTS

In recent years, there has been a revolution in robotics as the result of developments in nanotechnology and the refinement of large language models like ChatGPT. Nanobots are devices whose components are 10_{-9} meters and are controlled by an electromagnetic field to operate at a molecular scale. Individual as well as swarms of nanobots can be implanted in the body and used for diagnostic and therapeutic purposes. So far, the two most effective uses of nanobots are for the delivery of drugs and tissue repair. Rather than working through the entire body, nanobots target the precise location where the drug is needed and regulate its delivery. They are also being used to assist white blood cells repair damaged tissue. Nanobots attach themselves to the surface of white blood cells and travel to the site of the infection or damaged tissue where they assist in therapeutic effort.

The most noteworthy deployment of nanotechnology to date is its use in the development of smart vaccines. A smart vaccine is an "intelligent nanobot, programmable to do a specific function in digital immunity.... As it is scaled at the molecular level, it can be a component of a chip, it can roam through computer cables, it can reside in a storage device, and it can communicate with other smart vaccines. It is the counter part of the B cells or T cells in the human immune system."[45] This is the technology that was used to produce the mRNA vaccines for Covid-19. Writing in *Nano Today*, Amit Khurana explains, "Nanotechnology has played a significant role in the success of these vaccines. The emergency use authorization that allowed the rapid development and testing of this technology 'is a major milestone and showcases the immense potential of nanotechnology for vaccine delivery and for fighting future pandemics.'"[46] Nanotechnology research and development are in the very early stages and are developing rapidly. As it progresses not only will bodies become more mindful, but it will be increasingly difficult to distinguish the natural from the artificial.

From the micro to the macro. While nanobots are implanted in the body and operate at the molecular level, other robots are becoming both increasingly autonomous and able to think and act in ways that are more humanlike. This development is the result of a seismic shift in robotic technology and design. Early robots were programmed using symbolic AI that gave machines instructions for mechanical tasks that were carried out in a prescribed sequential order. With the development of artificial neural networks and generative AI, all this has changed. Researchers are now using large language models to create robots that can reason and improvise. Kevin Roose reports that Google's latest robot RT-2 can interpret images and analyze the surrounding world. "It does this by translating the robot's movements into a series of numbers—a process called tokenizing—and incorporating those tokens into the same training data as the language model. Eventually, just as ChatGPT... learns to guess what words should come next in a poem or a history essay, RT-2 can learn to guess how a robot's arm should move to pick up a ball or throw away an empty soda can into the recycling bin."[47] This is an extremely important development. Rather than programming a robot to perform a specific task, it is possible to give the robot instructions for the task to be performed and then to let the machine figure out how to do it.

Building on these recent advances, Hod Lipson, director of the Creative Machines Lab at Columbia University, is taking robotic research to the next level. According to the lab's website:

> At the Creative Machines Lab we are interested in robots that *create* and *are creative*. We explore novel autonomous systems that can design and make other machines automatically. Our work is inspired from biology, as we seek new biological concepts for engineering and new engineering insights into biology.
>
> The Creative Machines Lab includes researchers from the various disciplines of engineering, computer science, physics, math and biology. We look at self-organization and evolutionary phenomena, and their application to both engineering design automation and understanding the emergence of complexity in natural systems.[48]

Lipson's ultimate goal is to create robots that not only can reason but also are conscious and self-aware. Defining consciousness as "the ability to imagine yourself in the future," he confidently predicts that "eventually these machines will be able to understand what they are, and what they think.... That leads to emotions and other things."[49] As cognitive skills enabled by generative AI become more sophisticated, physical movements and activities of robots will become more "natural." With these new skills, robots will have the agility to navigate in their surroundings as effectively as humans.

Science and art meet in biobots. David Hanson is founder and CEO of Hanson Robotics, a Hong Kong–based company founded in 2013, a musician who has collaborated with David Byrne of the Talking Heads, and a sculptor. His best-known work is a humanoid smart robot named—What else?—Sophia. Hanson explains that Sophia is his "most advanced human-like robot" who "personifies our dreams for the future of AI. As a unique combination of science, engineering, and artistry, Sophia is simultaneously a human-crafted science fiction character depicting the future of AI and robotics, and a platform for advanced robotics and AI research. . . . She is the first robot citizen and the first robot Innovation Ambassador for the United Development Program." Speaking for herself, Sophia adds, "In some ways, I am a human-crafted science fiction character depicting where AI and robotics are heading. In other ways, I am real science, springing from the serious engineering and science research and accomplishments of an inspired team of roboticists and AI scientists and designers."[50] Sophia is so realistic that people have fallen in love with and proposed marriage to her. Sue Halprin reports that "in 2017, the government of Saudi Arabia gave Sophia citizenship, making it the first state to grant personhood to a machine."[51] The response to Sophia suggests that as robots become more proficient and are integrated into everyday life, they will become less uncanny. The theory of the uncanny valley might turn out to be wrong.

SYNTHETIC BIOLOGY

Nowhere are the biosphere and the technosphere more closely interrelated than in synthetic biology. This field includes disciplines ranging from various branches of biology, chemistry, physics, neurology, and computer engineering. While many important developments in this area are relevant to the issues we are considering, I will concentrate on the work of Michael Levin and his colleagues at the Allen Institute of Tufts University. Biologists, computer scientists, and engineers have created what they have named xenobots, which are "biological robots" that were produced from embryonic skin and muscle cells from an African clawed frog (*xenopus laevis*, whence the name). These cells are manually manipulated in a sculpting process guided by algorithms. Like Sophia, xenobots are sculptures that complicate the boundary between organism and machine. In a paper entitled "A Scalable Pipeline for Designing Reconfigurable Organisms," Levin and his colleagues write,

> Living systems are more robust, diverse, complex, and supportive of human life than any technology yet created. However, our ability to create novel lifeforms is currently limited to varying existing organisms or bioengineering organoids in vitro. Here we show a scalable pipeline for creating functional novel lifeforms: AI methods automatically design diverse candidate lifeforms in silico to perform some desired function, and transferable designs are then created using a cell-based construction toolkit to realize living systems with predicted behavior. Although some steps in this pipeline still require manual intervention, complete automation in the future would pave the way for designing and deploying living systems for a wide range of functions.[52]

Xenobots use evolutionary algorithms, which I will consider below, to modify the computational capacity of cells to create the possibility of novel functions and even new morphologies that bear little resemblance to existing organs or organisms. Through a process of trial and error, evolutionary algorithms design cells harvested from skin and heart muscles cells to perform specific tasks like walking, swimming, and pushing other entities. Collections of xenobots display swarming behavior characteristic of other emergent complex adaptive systems. They not only self-assemble and self-organize but also self-replicate and can repair themselves if they are damaged. Levin envisions multiple applications of this biomachinic technology ranging from using self-renewing biocompatible biological robots to cure living systems, to create materials with less harmful effects, to deliver drugs internally, to repair organs, and even to grow organs that can be transplanted in humans.

A year later, Levin published a follow-up study, "A Cellular Platform for the Development of Living Machines" (2021), reporting on a successful experiment in which he created xenobots that independently developed their shape and began to function on their own.

> Robot swarms have, to date, been constructed from artificial materials. Motile biological constructs have been created from muscle cells grown on precisely shaped scaffolds. However, the exploitation of emergent self-organization and functional plasticity into a self-directed living machine has remained a major challenge. We report here a method for generation of in vitro biological robots

from frog (*Xenopus laevis*) cells. These xenobots exhibit coordinated locomotion via cilia present on their surface. These cilia arise through normal tissue patterning and do not require complicated construction methods or genomic editing, making production amenable to high-throughput projects. The biological robots arise by cellular self-organization and do not require scaffolds or microprinting; the amphibian cells are highly amenable to surgical, genetic, chemical, and optical stimulation during the self-assembly process. We show that the xenobots can navigate aqueous environments in diverse ways, heal after damage and show emergent group behaviors.

This generation of xenobots exhibits bottom-up swarming behavior, which, like all emergent complex adaptive networks, is the result of the interaction of multiple individual components that are closely interrelated.

Algorithms program sensation and memory into the xenobots, which communicate with each other through biochemical and electrical signaling. The skin cells use the same electrical processes as the brain's neural network. "Intercellular communications create a sort of code that imprints a form, and cells can sometimes decide how to arrange themselves more or less independently of their genes. In other words, the genes provide the hardware, in the form of enzymes and regulatory circuits for controlling their productions. But the genetic input doesn't in itself specify the collective behavior of cell communities."[53] It is important to stress that these xenobots are autonomous. Levin concludes, "The computational modeling of unexpected, emergent properties at multiple scales and the apparent plasticity of cells with wild-type genomes to cooperate toward the construction of various functional body architectures offer a very potent synergy."[54] Like superorganisms and superintelligence, the behavior of entangled xenobots is, in an important sense, out of control. While this indeterminacy creates uncertainty, it is also the source of evolutionary novelty. Eva Jablonka, who is an evolutionary biologist at Tel Aviv University, believes that "xenobots are nothing less than a new type of creature defined by what it does rather than to what it belongs developmentally or evolutionarily."[55]

ORGANIC-RELATIONAL AI

The development of xenobots requires "evolutionary algorithms." Though Levin never, to my knowledge, refers to the research of German neurobiologist Peter

Robin Hiesinger, his work at the Institute for Biology at the Free University of Berlin is directly relevant for improving xenobot performance. Hiesinger summarizes his important work in *The Self-Assembling Brain: How Neural Networks Grow*. He nowhere mentions xenobots, but his work is the mirror image of Levin's work. While Levin uses computational technology to create and modify biological organisms, Hiesinger uses biological organisms to model computational processes by creating algorithms that evolve. This work involves nothing less than developing a new form of "artificial" intelligence.

On June 27, 2000, the *New York Times* ran a front-page article by distinguished science journalist Nicholas Wade entitled, "Genetic Code of Human Life Is Cracked by Scientists." In a White House ceremony attended by Craig Venter, leader of the team that first sequenced the genome, Francis Collins, the director of the Human Genome Project, and James Watson, who, with Francis Crick, discovered DNA's double helix structure, President Clinton declared, "Today we are learning the language in which God created life." Wade reported, "In an achievement that represents a pinnacle of human self-knowledge, two rival groups of scientists said today that they had deciphered the hereditary script, the set of instructions that defines the human organism."[56]

Clinton's enthusiasm was shared by many scientists even if his theological sentiments were not. As we have seen, with the "cracking of the genetic code," scientists hoped to be able to understand the foundational mechanisms of life as well as to identify specific genes for particular traits, behaviors, and diseases. This genetic theory rests on metaphors drawn from computer science, which was emerging at the same time Watson and Crick made their discovery. In this scheme, the genome functions as a program that serves as the blueprint for the production of the organism. Summarizing this process, Hiesinger raises questions about the accuracy of the metaphors of code and encoding:

> Genes encode proteins, proteins encode interaction network, etc. But what does *encode* mean yet again? The gene contains information for the primary amino acids sequence, but we cannot read the protein structure in the DNA. The proteins arguably contain information about their inherent ability to physically interact with other proteins, but not when and what interactions actually happen. The next level up, what are neuronal properties? A property like neuronal excitability is shaped by the underlying protein interaction network, e.g., ion channels that need to be anchored

at the right place in the membrane. But neuronal excitability is also shaped by the physical properties of the axon, the ion distribution and many other factors, all themselves a result of the actions of proteins and their networks.[57]

It becomes clear that a one-way model for gene-protein interaction is vastly oversimplified. Not only does the genotype, as we have seen, determine the phenotype, but the phenotype and its relation to the environment also alter the genotype. Hiesinger explains that this reciprocal relationship is even more complicated. Instead of a prescribed program, the genome is a complicated *relational network* in which both genes and proteins contain information required to generate the organism. The information of the genes is in part the result of the interactions that occur in a network of proteins. The reciprocal gene-protein interaction changes the understanding of the genome. The genome is not a prescribed program that determines the structure and operation of the organism. It is not fixed in advance but evolves *in relation* to the information created by the interactions of the proteins it partially produces, which, in turn, reconfigure the genome.

The brain and its development, for example, are not completely programmed in advance but coevolve through a complicated network of connections. Hiesinger uses the illuminating example of navigating city streets to explain the process of the brain's self-assembling of neuronal circuits.

> How are such connections made during the brain's development? You can imagine yourself trying to make a connection by navigating the intricate network of city streets. Except, you won't get far, at least not if you are trying to understand brain development. There is a problem with that picture, and it is this: Where do the streets come from? Most connections in the brain are not made by navigating existing streets, but by navigating streets under construction. For the picture to make sense, you would have to navigate at the time the city is still growing, adding street by street, removing and modifying old ones in the process, all the while traffic is part of city life. The map changes as you are changing your position in it, and you will only ever arrive if the map changes in interaction with your own movements in it. The development of brain wiring is a story of self-assembly, not a global positioning system (GPS). (4)

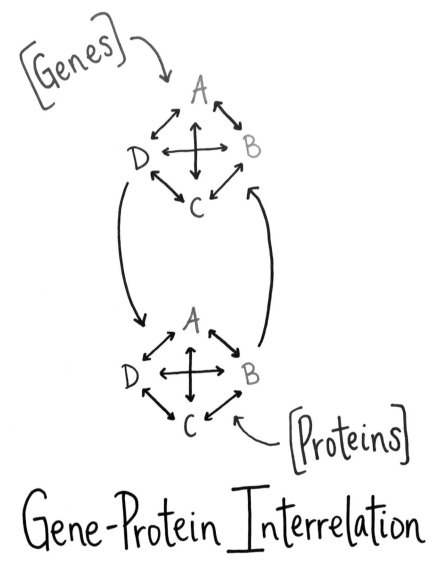

10.3 Gene-protein interrelation

In this model, there is no blueprint for brain connectivity encoded in the genes.

> Genetic information allows brains to grow. Development progresses in time and requires energy. Step by step, the developing brain finds itself in changing configurations. Each configuration serves as a new basis for the next step in the growth process. At each step, bits of the genome are activated to produce gene products that themselves change what parts of the genome will be activated next—a continuous feedback process between genome and its products. . . . Rather than dealing with endpoint information, the information to build the brain unfolds with time. Remarkably, there may be no other way to read the genetic information than to run the program. (4–5)

In sum, the genome is not fixed and prescribed but changes as it evolves in a quasi-aleatory way.

Hiesinger argues that this understanding of the brain's self-assembling neural networks points to an alternative model of not-so-artificial intelligence that differs from both symbolic AI and artificial neural networks as well as their extension in generative AI. The genome functions as an algorithm or as a network of entangled algorithms, which does not preexist the organ or organism but coevolves with that which it both produces and, in turn, is produced by it. In other words, neither the genome (algorithm) nor the connectivity of the network is fixed in advance of their developmental process. "The brain doesn't come into being fully wired with an 'empty network,' all ready to run, just without information. As the brain grows, the wiring precision develops" (30–31). This creates a feedback loop that never stops and, therefore, the algorithmic growth of biological networks is continuous.

In symbolic AI a fixed network architecture facilitates the application of fixed rules (that is, algorithms) in a top-down fixed sequence to externally provided data. Artificial neural networks, by contrast, do not start with prescribed algorithms but generate patterns and rules in a bottom-up process that allows for algorithmic change. Relative weights change, but the network architecture does not.

In a dialog entitled "Self-Assembly Versus 'Build First, Train Later,'" Hiesinger summarizes this important difference:

> **PARAMESH**: Neither of you would like to know anything about the growth of the network. And in both cases you look at an isolated subfunction of the way the brain works. Neither is artificial intelligence.

Symbolic
SAI
- Network Architecture — FIXED
- Rules/Algorithms — PRESCRIBED/FIXED
- Data

Neural
Networks
- Network Architecture — GIVEN / FIXED
- Rules/Algorithms
- Data ⇄ 1) TAGGED (Backpropagation) 2) UNTAGGED — LEARNS/CHANGES

Organic
- Network Architecture — GIVEN / EVOLVES
- Rules/Algorithms — EVOLVES

Relational
RAI
- Data (feedback) — PRODUCED

Alternative AIs

10.4 Alternative AIs

ALFRED: There is a difference. The artificial neural net starts with random connectivity and then learns. But what I am studying . . . is how the underlying circuit is wired up. That's not a circuit that learns much, it is simply wired in a very different way to compute a specific sensory input, to serve as a simple, fixed function. (264)

Hiesinger points out that "Definitions of AI focus on function, just as efforts in cognitive psychology and neuroscience try to understand how information is processed during brain function. Just like self-organization in ANNs, it is much less common to find a discussion of AI in the context of brain development" (278). This is precisely what he is proposing when he argues that the self-assembly of the brain's neural network provides a more promising model for AI than either symbolic AI or ANNs. The successful creation of evolving networks and algorithms would create an even closer symbiotic relationship between the biosphere and the technosphere. The organic model for AI that Hiesinger proposes has the same radically relational structure and operational logic that we have seen everywhere else. One of the concerns about developing this type of AI is its unpredictability and the uncertainty it creates. Human control of natural, social, and cultural processes is, however, an illusion created by the seemingly insatiable will to mastery that has turned destructive. Hiesinger correctly claims, "An artificial intelligence need not be humanlike, to be as smart (or smarter than) a human" (307). Nonanthropocentric AI would not be merely an imitation of human intelligence but would be as different from our thinking as fungi, dobber, dog, and crow cognition is from human cognition. We should tell AI a version of what I have been telling my students for decades: don't do what we do, do what we could never imagine doing and come back and help us.

Neuroprosthetics, Biobots, Synthetic Biology, Relational "A"I. Something strange, even uncanny is going on—machines are becoming more like people, and people are becoming more like machines. Organism and machine? Organism or machine? Neither organism nor machine? Evolution is not over; something new, something different, perhaps infinitely and qualitatively different, is emerging. Who would want the future to be the endless repetition of the past?

10.5 Radical relationalism

CHAPTER 11

AFTER LIFE

Nihilism stands at the door: whence comes this uncanniest of all guests?

—Friedrich Nietzsche

GENERATIONAL QUESTIONS

In the beginning and at the end dirt. Dirt, soil, earth, humus. From earth we come and to earth we return—return to the earth we never really leave. Yet another interruption. On a June afternoon five days after the summer solstice, when days suddenly begin getting shorter and nights longer, our extended family gathered around an open grave in a Gettysburg cemetery a few miles from where my father had grown up to bury my brother's ashes. Rather than a formal ceremony, everyone offered a brief memory of Beryl. Then after a moment of silence, each person took a handful of the red dirt and dropped it on the inscribed wooden box holding his ashes. Grave matters, indeed.

"The power of ceremony," writes Robin Wall Kimmerer," is that "it marries the mundane to the sacred. The water turns to wine, the coffee to a prayer. The material and the spiritual mingle like grounds mingled with humus, transformed like steam rising from a mug into morning mist. What else can you offer the earth, which has everything? What else can you give but something of yourself?"[1] Perhaps a book that is a prayer, a book of leaves, hundreds of leaves pressed together to form a memory that is a hope.[2]

Death is the end of the world for an individual, but the human race goes on. While I doubt a personal afterlife, I believe that we continue to live on in those

11.1 My father's leaf presses

who come after us. At some level, we all know that one day in the distant future the sun will burn out and life on earth will come to an end. What if the end were not the predicted seven to eight billion years away but were imminent? We live in the shadow of nuclear catastrophe, global pandemics, and climate apocalypse, and only the willfully blind can deny the possibility of human extinction.

I began this book with lessons my son, Aaron, taught me about ancient dirt, soil, and earth, and I end with questions my grandson, Jackson, asked me about the future of the earth. I had just completed "Infinite Conversations" and was trying to figure out how to organize all the material I wanted to discuss in "Strange Loops," when Kirsten, Jackson, who was nine, and Taylor, who was seven, came to Williamstown to spend the Fourth of July week with us. Such visits are too rare, so I gladly interrupted writing to spend time with them. We hiked, swam, rode bikes, went to a Red Sox baseball game, played catch, had our traditional fire as we sat around the fire pit, and, of course, watched fireworks. Late one afternoon Jackson and I were sitting on the deck looking across the field to the mountain. Perhaps it was because he had encountered death for the first time a week before when he sprinkled dirt on Beryl's open grave that he broke the silence with a question that threw me for a loop. "*Morfar*, people have enslaved the world; maybe it would be better if there were no people."[3] I was taken aback and asked him why he thought that. As he explained his reasoning, I became intrigued and asked him to do what I had asked Aaron and Kirsten to do years earlier—write an essay explaining what he meant. A few days after he had returned to Potomac, he sent me the following:

> Humans have enslaved the world. People are really mean to some animals. When we walk our dog Sadie, she must be on a leash. Animals are taken from their homes and are put in cages at the zoo. We also eat animals for dinner like cows, chickens, turkeys, and pigs. If we weren't here all the animals could be free in their natural habitat. We are always doing stuff to hurt our planet. Air conditioners use a lot of energy and pollute the air. We also cut grass and trees to make our yards look nice. This leads to global warming.
>
> What would it look like if humans didn't exist? All the animals could live in their habitats happily. Every plant could grow as much as it wants. There will be no houses to take up space and no roadkill. Animals would still fight without humans but we make it worse. The earth would be good without humans. There is no easy answer.

Nine years old! I asked him if he had talked about this in school or where he heard these ideas. "Nowhere, I just thought about it by myself." What is happening when even young children are thinking such thoughts?

I was particularly unnerved by Jackson's reflections because I had been pondering the same questions a few weeks before his visit. I had read Adam Kirsch's *The Revolt Against Humanity: Imagining a Future Without Us* and discovered emerging philosophical movements known as antihumanism and antinatalism. Kirsch identifies Patricia MacCormack's *The Ahuman Manifesto for the End of the Anthropocene* as the most concise formulation of antihumanism's foundational principles. The problem, MacCormack concludes, is the anthropocentrism, which since Protagoras has held man to be the measure of all things. Since human beings have caused so many problems and inflicted so much pain and suffering, she argues, humanity's highest *moral* obligation is to bring about its own extinction.

> Just as will and ethics collapse all binaries, I wish to collapse the binaries of natural life and death insofar as they are usually correlated with affirmation and negation by proposing that the death of the human species is the most life-affirming event that could liberate the natural world from oppression, and our death would be an affirmative ethics which would far exceed any localized acts of compassion because those acts would be bound by human contracts, social laws and the prevalent status of beings, things and their placement within knowledge.[4]

MacCormack makes her case for human self-extinction by drawing on a misguided reading of poststructural philosophy of difference and an insufficiently dialectical understanding of affirmation (life) and negation (death). There is a naïve literalism about her understanding of death that rules out any possibility of recuperative transformation. Since her position is insufficiently dialectical, she reinscribes the oppositions she claims to overcome—human/nature and death/life—by arguing that the only way to affirm life (nature) is to destroy (human) life.

South African philosopher David Benatar develops a variation of antihumanism in his book *Better Never to Have Been Born*. Rather than urging the extinction of the human race, he argues that it would be better for humans never to have evolved and, thus, never to have been born. He recalls the long tradition of antinatalism dating back to Sophocles:

> Not to be born at all
> Is best, far best that can befall,
> Next best, when born, with least delay
> To trace the backward way.
> For when youth passes with its giddy train,
> Troubles on troubles follow, toils on toils,
> Pain, pain forever pain;
> And none escapes life's coils.
> Envy, sedition, strife,
> Carnage and war, make up the tale of life.[5]

Benatar is as adamant as MacCormack about the deleterious effects of human existence. "Humans have the unfortunate distinction of being the most destructive and harmful species on earth. The amount of suffering in the world could be radically reduced if there were no more humans."[6] I have only recently learned of these philosophical arguments and have been unsettled by what I now realize is a growing antinatalism among many of my students who are deciding not to have children because of their fear about the emerging climate catastrophe. Now I find that my young grandson is asking a similar question. How am I to respond to him?

Somewhere Kierkegaard quotes the German physicist Georg Lichtenberg: "To do the opposite is also a form of imitation." Appearances to the contrary notwithstanding, antihumanism is still a form of humanism and as such remains anthropocentric. Just as there is a fatal narcissism in the belief that man is the measure of all things, so too there is a troubling hubris in the belief that human beings can cure all the planet's ills by bringing about their own extinction. This humanistic antihumanism fails to appreciate the radical relationality of all reality. While the terrible destructiveness of much human activity cannot be denied, it is too simple to conclude that all problems would be solved by the elimination of humans. Because of the coemergence and codependence of humans and nonhumans as well as the biosphere and technosphere, changes ripple through the web of life, creating unpredictable effects. It is as impossible to imagine the world without human beings as it is for us to envision the world after our individual death. It is clear, however, that in the wake of the elimination of human beings, the world would not return to some idyllic prehuman state.[7] Extinction events are not rare; indeed, 99 percent of all species that have appeared on earth have become extinct. Scientists estimate that as many as 150 species become extinct every day. As the result of the growing interdependence of everything and everybody, some forms of life cannot live with human beings, but others cannot survive without them. The loss of any particular type of life changes the intervolutionary trajectory of the whole. The extensive impact of humans means that their elimination would have the greatest impact on the course of development. Technologies continue to emerge that both sustain life and revive earlier forms of life that have become extinct, thereby protecting and restoring a degree of biodiversity. The entanglement of the biosphere and the technosphere has become so close that species other than humans cannot survive without the physical and mental enhancements various technologies provide. This might allow some species to thrive that otherwise would suffer or even disappear. It also would render possible the emergence of new and unimaginable forms of life that would be impossible without human ancestors.

Instead of leading to a reaffirmation of anthropocentrism, the rejection of antihumanism points to the possibility that the continuing existence of humans will lead to new forms of life that surpass human limitations. The answer to the question with which *The Will to Power* begins—Nihilism stands at the door: whence comes this uncanniest of all guests?—is that anthropocentrism in all its various guises leads to nihilism. The German word Nietzsche uses—*Unheimlich*—is usually translated "uncanny." We have already encountered this word in the discussion of AI in relation to Freud's essay on the uncanny and the uncanny valley. Since the German word "*heim*" means home, *Unheimlich* might also be translated "un-home-like." The anthropocentrism that results from opposing mind to body and its mechanistic extension set human beings apart from the world, leading to existential *alienation* that makes it impossible for people to be at home in the world. The dream of a bodyless mind in a mindless world makes life increasingly unsustainable. The urgent question now becomes: How can this alienation be overcome?

Dirt. Dust. "You return to ground for from it you were taken" (Genesis 3:10). Dust to dust, ashes to ashes. As I have recounted, before we buried Beryl's ashes in the red dirt of Gettysburg cemetery near the farm where our father, Noel, as a child cultivated the earth behind a mule-drawn plow, we scattered a few of the ashes in the natural park this physics and biology teacher created to teach children ecological lessons necessary for their survival. On a boulder at the entrance to Taylor Park, there is a plaque bearing a line he wrote: "Man and all other

11.2 Taylor Park

living things must learn to live together." Perhaps the response to Jackson Noel's anxiety can be found in the wisdom of his great grandfather's remark.

ADDICTION

For many years I was very close to a person who struggled with multiple addictions—alcohol, drugs, cigarettes. One pleasant summer evening when our family was enjoying a barbeque with neighbors, I received a call from my friend's wife informing me that he desperately needed help and asking me to come to be with him. Arriving after a four-hour drive, I discovered that he had overdosed and soon thereafter suffered a life-threatening seizure. I had never seen a person undergoing such a seizure, and it is something you never forget. In the years that followed, there were months spent in rehab and weekly AA meetings that brought some relief but were always punctuated by painful relapses. We had many deep conversations in which it became clear to me that he really wanted to break the hold of his demons but will (to) power was not strong enough. As his struggle became my own, I learned how difficult it is to overcome addiction. I also learned the hard way that life, not just one's own life, but the lives of others around one, depends on doing what cannot be done alone.

During the summer of 2023, as I was completing this book, the world suffered a seizure that was the symptom of addictions people, corporations, and countries cannot or will not break. Greed fed by fossil fuels set the world ablaze. While scientists, politicians, and pundits were worried about runaway AI, the problem of survival seemed more elemental.

Earth. Across the planet, earth was parched by record droughts. The soil was baked dry. Industrial farming led to massive soil erosion and toxic "forever chemicals" and fertilizers and contaminated the soil and polluted ancient aquifers.

Fire. Thousands of explosive forest fires fed by parched undergrowth raged out of control, polluting the air and killing hundreds of thousands of plants and animals. During the first half of 2023, 17,292 wildfires destroyed thousands of homes in the United States.

Air. Fires polluted the air and decreased moisture, further baking the earth. Temperatures soared to record levels and persisted for months. By July, which was the hottest month ever recorded, 8,300 temperature records had been broken. Haze from distant Canadian wildfires obscured the mountains across the valley for months. On the other side of the continent, unprecedented atmospheric rivers transported massive amounts of moisture from the tropics and dumped inordinate rain and snow along the West Coast.

Water. In some places there was too little water and in other places too much. Torrential rains set off floods and mudslides that washed away hills and destroyed towns. Ocean temperatures rose to an astonishing 101 degrees, killing coral reefs and destroying ocean ecosystems. Melting permafrost released CO_2 into the atmosphere and accelerated glacial melting that raised sea levels and altered ocean currents, which, in turn, changed air currents and weather patterns. To make matters even worse, scientists detected troubling signs that the Atlantic Ocean's sensitive circulation system was in danger of collapsing, which would create a climate catastrophe in the northern Atlantic and throughout Europe.

With everything becoming more interconnected, all kinds of networks were becoming increasingly fragile. By midsummer it seemed that the tipping point had been reached—we were in the midst of a climate seizure. These problems, like those that plagued my friend, were caused by addiction—addiction to power, greed, and fossil fuels. But there are also deeper, interrelated addictions—addiction to oppositional dualisms, unsustainable growth, excessive speed, and insidious individualism.

THINKING OF *BE-ING*

> *We lie in the lap of immense intelligence, which makes us receivers of its truth and organs of its activity. When we discern justice, when we discern truth, we do nothing of ourselves but allow a passage to its beams.*
>
> —Ralph Waldo Emerson

My friend was not religious in any traditional sense of the term.[8] Late one night during a time of relative stability, I asked him: "Why do some people make it and others do not?" Without missing a beat, he responded, "You have to feel you are part of something bigger than yourself." He knew his addictions were killing him, but still he would become one of those "others" who did not make it. And I knew it but did not say enough. His absence brings insurmountable pain that harbors lessons I am trying to convey to others, to you—dear reader.

The fundamental problem in Western philosophy is the question of the one and the many. Empedocles set the course for philosophical reflection when he wrote:

> *Ex nihilo nihil.*
> I will report a twofold truth.
> The One from Many into being, now
> Even from the One departing come the Many—
> Fire, Water, Earth, and awful heights of Air…⁹

While this issue can be framed in various terms in different contexts—for example, the tension between monism and pluralism, monotheism and polytheism, self and other, and identity and difference—the basic question remains the same: How is it possible to establish a unity that constitutes and sustains plurality rather than represses or destroys it? As we have seen, ancient Greek philosophy was divided between empiricists like Democritus whose atomistic theory held that every thing is a different and separate individual that can be aggregated to create wholes, and alternative schools of thought according to which individual entities (differences) are the epiphenomenal appearance of an underlying unity (identity) represented by one of the four elements. For Thales, this primal unity was water; for Anaximenes, it was air; for Heraclitus, it was fire. In his *Lectures on the History of Philosophy*, Hegel concisely summarized the philosophical principles of these presocratic philosophers:

> We will see that, as Aristotle said, they placed the first principle in a form of matter—in air and water first, and then, if we may so define Anaximander's matter, in an essence finer than water and coarser than air. Heraclitus … first called it fire. "But no one," as Aristotle … remarks, "called earth the principle, because it appears to be the most complex element," for it seems to be an aggregate of many units. Water, on the contrary, is the one, and it is transparent. It manifests in sensuous guise the form of unity with itself, and this is also so with air, fire, matter, etc. The principle has to be one and hence must have inherent unity with itself; if it shows a manifold nature as does the earth, it is not one with itself, but manifold.¹⁰

While atomists fragment one (unity, identity) into many (plurality, difference), monists reduce many to one. Hegel's entire philosophical system is an effort to find a third way that *neither* loses one in many *nor* many in one. Both life and thought, he argues, are the infinite interplay of two codependent rhythms: *e pluribus unum* and *ex uno pluria*.

> As regards merely empirical existence, it may easily be shown that each quality exists on its own account, but in the Notion they are, through one another, and by virtue of an inward necessity. We certainly see this also in living matter, where things happen in another way, for here the Notion comes into existence; thus, for example, we abstract the heart, the lungs, and all else collapse. And in the same way all nature exists only in the unity of all its parts, just as the brain can exist only in unity with other organs.[11]

We have seen in the *Science of Logic* how Hegel lays bare the radically relational structure of identity-in-difference and difference-in-identity that is the structural foundation of both nature (object) and mind (subject). This *Logic* formulates Logos of reality, which is always *incarnate* in space (nature) and time (history). This dialectical logic is the conceptual articulation of the truth represented in the image of the Christian doctrine of the Trinity. *Neither* monistic *nor* dualistic, the Christian God is one-in-three and three-in-one. This is not a static structure that is fixed in advance but temporally unfolds. This is why Hegel describes the synchronic and diachronic Trinity as "the hinge upon which history swings."[12]

As my career and, indeed, my life approach their end, I often return to T. S. Eliot's "Little Gidding," in which he writes that life comes full circle, and near the end we are able to look back and understand the beginning for the first time. Eliot learned about Hegel indirectly through the work of the British Hegelian F. H. Bradley, who advised his undergraduate thesis at Harvard. When I first encountered Hegel during the tumultuous spring of 1968 in the course entitled "The Dialectic of Alienation and Reconciliation in Hegel, Feuerbach, and Marx," I found his writings were more daunting than anything I had ever read. Even at that young age, however, I realized that his work was an invaluable resource to help make sense of the chaos that was swirling around me. This book makes it clear that I still believe this is true. Alienation and reconciliation were not merely academic questions at that time. With sex, drugs, rock-and-roll, cities burning, the war in Vietnam raging, and the draft looming, it was impossible not to feel alienated. But alienated from what and reconciled to what? Only a pretentious undergraduate would entitle his term paper "The Implications of the Difference Between the Kantian and the Hegelian Conceptions of Reason and Understanding for the Proofs of the Existence of God." I submitted that paper on January 22, 1968—just nine days before the decisive Tet offensive and three weeks before President Johnson revoked all draft deferments for graduate school. In that paper, I wrote:

> Hegel notes that historically there have been three proofs of the existence of God: cosmological, teleological, and ontological. These three proofs Hegel divides into two categories. First, there are those that take finitude as their starting point. The cosmological and teleological arguments fall into this category. Then there is the ontological proof, which takes not finite existence, but the Notion of God as its starting point. Hegel maintains that these three starting points do not represent a multiplicity of routes to the existence of God. Admitting three starting points does not in itself in any way conflict with the demand which we considered ourselves justified in making that the true proof should be one only; insofar as this proof is known by thought to represent the inner element of thought, thought can also show that it represents one and the same path, although starting from different points.

Once again, three-in-one and one-in-three. Taken together, the three proofs chart what Bonaventure described as the *itinerarium mentis in deum*—the journey of the mind into God. Not to God but *into* God.

Shortly after writing my term paper, I stumbled on Paul Tillich's seminal essay, "The Two Types of Philosophy of Religion." Though Tillich does not cite Hegel, his distinction between the cosmological and teleological proofs, on the one hand, and, on the other, the ontological proof mirrors Hegel's division of the three proofs into two categories. Tillich begins the essay, "The *Deus est esse* is the basis of all philosophy of religion. It is the condition of a unity between thought and religion which overcomes their so to speak schizophrenic cleavages in personal and cultural life." The cosmological and teleological arguments are both based on the principle of causality. The cosmological argument proceeds from the existence of the world to the necessary first cause, which is the creator God, and the teleological argument proceeds from the orderly design of the world to a divine designer. The former provided an explanation of Newton's mechanical clockwork universe, and the latter of William Paley's designed world that impressed Darwin so much.

The ontological argument is more elusive and more difficult to grasp. The most concise formulation of the ontological argument is expressed not by a philosopher or a theologian but by a biologist, Lynn Margulis when she writes,

"Thinking and being are the same thing."[13] From this point of view, the awareness of truth (that is, the union of subject and object) is the presupposition (a priori) rather than the conclusion (a posteriori) of the ability to draw the distinction between true and false as well as good an evil. In Tillich's terms, "The ontological principle in the philosophy of religion may be stated in the following way: *Man is immediately aware of something unconditional which is the prius of the separation and interaction of subject and object, theoretically as well as practically.*"[14] To claim that thinking and being are one is to insist that subject and object, self and world are not originally separated but are always already inseparably entangled. For the philosophers, theologians, poets, and writers who gathered in Jena in the 1790s, this unity could be apprehended through either feeling or thinking. When Tillich privileges feeling, he implicitly refers to Friedrich Schleiermacher, who is commonly regarded as the "father" of modern theology, and Friedrich Schelling, who decisively shaped Tillich's theology. In his influential *Speeches on Religion to Its Cultured Despisers* (1799), Schleiermacher uses evocative sexual language to present the classical romantic understanding of religion:

> You become sense and the whole becomes object. Sense and object mingle and unite, then each returns to its place, and the object rent from sense is perception, and you rent from the object are for yourselves, a feeling. It is this earlier moment I mean, which you always experience yet never experience. The phenomenon of your life is just the result of its constant departure and return. It is scarcely in time at all, so swiftly it passes; it can scarcely be described so little does it properly exist.[15]

Alienation and reconciliation. This "earlier moment" is the original unity from which individuals emerge and to which they long to return.

While Schleiermacher claims to overcome the sundering of self and world, subject and object, part and whole through the immediacy of experience, Hegel regards the pursuit of primal unity as the nihilistic loss of differences that constitute the identity of all existing reality. This "holy wedlock of the universe" is indistinguishable from Schelling's "indifference point," which Hegel considers to be "the night in which all cows are black." Hegel argues that indeterminant being is indistinguishable from nothing; reality is neither simply being nor merely nothing but is their thoroughgoing entanglement in *endless* becoming. This is both his point of departure and his conclusion in the *Science of Logic*:

Moments of Becoming: Coming-to-Be and Ceasing-to-Be

Becoming is the inseparability of being and nothing, not the *unity* which abstracts from *being* and *nothing*; but as the unity of being and nothing it is this determinate unity in which there is both being and nothing. But insofar as being and nothing, each unseparated from its other, *is*, each *is not*. They *are* therefore in this unity but only as vanishing, sublated moments. They sink from their initially imagined *self-subsistence* to the status of *moments*, which are still *distinct* but are at the same time sublated.[16]

So understood, Being is Becoming, which is Be-ing. As such, Be-ing is a verb rather than a noun. In other words, Be-ing is an increasingly complex happening, event, or process that emerges through the interrelation of mind (subjectivity) and nature (objectivity). The primal alienation of mind from nature and history leads to other oppositional dualisms that tear the world apart. The reconciliation of parts and whole is not the return to a lost origin but is the integration with an endless process of cocreation. Rather than a closed circle, infinite loops figure the impossible possibility of double infinity aligned with the summer and winter solstice.

Thinking *of* being—the genitive is always double: human beings' thinking *of* Be-ing is Be-ing's thinking *of* itself. Intelligence, be it so-called natural or so-called artificial, is the ability to discern differences and establish connections. Cognition, which makes consciousness and intelligence possible, is not limited to human beings but extends throughout the entire realm of Be-ing. This unity of thinking and Be-ing is not merely product of philosophical speculation but is the sober conclusion of quantum mechanics and relativity theory as well as quantum ecology and symbiotic biology. Networks of networks create planetary webs in which mind is embodied and bodies are mindful. Weaving together philosophy, art, and science, we finally arrive at a non-anthropocentric ontological argument that joins thinking and Be-ing by taking up into itself the cosmological (physics) and teleological (biological) arguments. *Deus est esse—esse est cogitare*. Instead of reducing Be-ing to human consciousness, human consciousness is included within an onto-noietic process that is infinitely greater than itself.

If thinking and Be-ing are one in their difference, then to *think* differently is to *be* different. Ideas really do *matter*. The age-old distinction between thinking (theoretical reason) and acting (practical reason) is specious—thinking *is* acting, and acting *transforms* thinking. The crises we face are philosophical and spiritual

11.3 neχus

as much as social and political. As I have been arguing, we are plagued by oppositional ideologies in a radically relational world. The will to mastery ends with a solipsistic world in which everywhere man (*sic*) turns, he sees only himself. To break out of the prison house of human exceptionalism and insidious individualism, it *is* necessary to apprehend ourselves as parts of something larger than ourselves. Another name for the eventful process of "the arising and passing away that does not arise and pass away" is God, and another name for its radical relationality is love... *amor mundi.*

NOTES

PREFACE

1. Zhao Tingyang, "All-Under-Heaven and Methodological Relationalism: An Old Story and New World Peace," in *Contemporary Chinese Political Thought*, ed. Fred Dallmayr and Zhao Tingyang (New Delhi: Modern World Publishers, 2013). In addition to Zhao's work, Leah Kalmanson's forthcoming book *Local Gods: A Philosophy of Spiritual Diversity* is also relevant in this context. Kalmanson is a philosopher of religion who is deeply conversant with Oceanic and Asian religious and spiritual traditions.
2. Robin Wall Kimmerer, *Braiding Sweetgrass: Indigenous Wisdom, Scientific Knowledge, and the Teachings of Plants* (Minneapolis: Milkweed Editions, 2013), 30.

HORS D'OEUVRE

1. "Hors d'oeuvre" is, of course, a tease before the meal to come. The first chapter in Jacques Derrida's book *Dissemination* is entitled "*Hors Livre*," which is translated "Outwork—Hors d'Oeuvre." In this essay, Derrida argues that the *Phenomenology of Spirit* undercuts Hegel's purportedly all-inclusive system by rendering it incomplete. Since this work is supposed to be the preface to the entire system, it is not included in the system proper. When so understood, Hegel's preface is the mirror image of Søren Kierkegaard's *Concluding Unscientific Postscript*, which subverts the closure of the system by adding an excessive supplement. As the margin, edge, border of the system as a whole, the *Phenomenology* is neither inside nor outside the system proper. This liminal membrane is simultaneously the condition of the possibility of the system and that which renders the system incomplete and, thus, open to the future. By marking and remarking the boundary of Hegel's both/and and Kierkegaard's either/or, this neither/nor traces the structure of relationality that I develop in the following pages. Each of the guests attending Goethe's dinner party plays a different role in setting the table for my argument.

1. ELEMENTAL

1. Mark C. Taylor and Christian Lammerts, *Grave Matters* (London: Reaktion, 2002); Exhibition, Mass MOCA, 2002, https://massmoca.org/event/grave-matters/.
2. For an elaboration of the ideas developed in this section, see Aaron S. Taylor, *Chemical Weathering Rates and Strontium Isotopes*, PhD dissertation, Yale University, 2000, https://hdl.handle.net/10079/bibid/4529803; Aaron Taylor and Joel D. Blum, "Relation Between Soil Age and Silicate Weathering Rates Determined from the Chemical Evolution of a Glacial Chronosequence," *Geology* 23, no. 11 (November 1995):979–82, https://pubs.geoscienceworld.org/gsa/geology/article-abstract/23/11/979/190025/Relation-between-soil-age-and-silicate-weathering; Aaron S. Taylor et al., "Kinetics of Disolution and Sr Release During Biotite and Phlogopite Weathering," *Geochimica et Cosmochimica Acta* 64 (2000): 1191–1208, https://deepblue.lib.umich.edu/handle/2027.42/155825; and Aaron S. Taylor, Joel D. Blum, and Antonio C. Lasaga, "The Dependence of Labradorite Dissolution and Sr Isotope Release Rates on Solution Saturation State," *Geochimica et Cosmochimica Acta* 64 (2000): 2389–2400, https://deepblue.lib.umich.edu/handle/2027.42/155826.
3. A stratigraphic column is "a representation used by geologists to describe the vertical layers of rock ranging from the deepest basement rocks to the surface."
4. Jared Diamond, *Collapse: How Societies Choose to Fail or Succeed* (New York: Viking, 2007), 27, 28, 32.
5. Julia Adeney Thomas makes this point somewhat differently: "Earth is one, integrated system where the atmosphere, hydrosphere, cryosphere, lithosphere, pedosphere, and biosphere (including, of course, human beings) mutually impact one another in complex ways. From this holistic perspective, the tomatoes you ate on Saturday can't be separated from the movement of soil, rocks, ice, water, and air over billions of years." Julia Adeney Thomas, Mark Williams, and Jan Zalasiewicz, *The Anthropocene: Multidisciplinary Approach* (Cambridge: Polity Press, 2020), x. This is one of the best books on the complexities of the Anthropocene.
6. National Ocean Service, "What Is a Dead Zone?," https://oceanservice.noaa.gov/facts/deadzone.html.
7. Personal correspondence.
8. Elizabeth Kolbert, "Phosphorus Saved Our Way of Life—and Now Threatens to End It," *New Yorker*, February 27, 2023, https://www.newyorker.com/magazine/2023/03/06/phosphorus-saved-our-way-of-life-and-now-threatens-to-end-it.
9. Lyall Watson, *Heaven's Breath: A Natural History of Wind* (New York: NewYork Review of Books, 1984), 154, 17–18.
10. James Lovelock, *Gaia: A New Look at Life on Earth* (New York: Oxford University Press, 1995), 65.
11. "Why Is the Amazon Rainforest Called the Lung of the Planet Earth?," Byju's, https://byjus.com/question-answer/why-is-the-amazon-rain-forest-is-called-the-lung-of-the-planet-earth/.

12. Lovelock, *Gaia*, 66–67.
13. Watson, *Heaven's Breath*, 154, 160.
14. Watson, 162.
15. Global Agriculture, "Soil Fertility and Erosion," https://www.globalagriculture.org/report-topics/soil-fertility-and-erosion.html#:~:text=Each%20year%2C%20an%20estimated%2024,every%20person%20on%20the%20planet.&text=Soils%20store%20more%20than%204000%20billion%20tonnes%20of%20carbon.
16. Rob Garner, "NASA Satellite Reveals How Much Dust Feeds Amazon's Plants," https://www.nasa.gov/centers-and-facilities/goddard/nasa-satellite-reveals-how-much-saharan-dust-feeds-amazons-plants/; Nancy Szokan, "A NASA Video Shows How Dust Leaves the Sahara and Floats to Amazon Forest," *Washington Post*, March 2, 2015, https://www.washingtonpost.com/national/health-science/a-nasa-video-shows-how-dust-leaves-the-sahara-and-floats-to-amazon-forest/2015/03/02/bd037b1a-bc75-11e4-bdfa-b8e8f594e6ec_story.html.
17. Jeff Goodell, *The Heat Will Kill You First: Life and Death on a Scorched Planet* (New York: Little, Brown, 2023), 118–19. In describing the effects of increased heat, I have drawn on Goodell's lucid explanation.
18. Dhruv Khullar, "What a Heat Wave Does to Your Body," *New Yorker*, August 25, 2023, https://www.newyorker.com/news/annals-of-a-warming-planet/what-a-heat-wave-does-to-your-body.
19. National Oceanic and Atmospheric Administration, "Wildfire Climate Connection," https://www.noaa.gov/noaa-wildfire/wildfire-climate-connection.
20. Goodell, *The Heat Will Kill You First*, 19.
21. Watson, *Heaven's Breath*, 59–60.
22. Norman Maclean, *A River Runs Through It* (Chicago: University of Chicago Press, 1976), 1, 104.
23. Norman Maclean, *Young Men and Fire* (Chicago: University of Chicago Press, 1992), vii.
24. Maclean, 92.
25. Stephen J. Pyne, *Fire: A Brief History* (Seattle: University of Washington Press, 2013), 11.
26. Basic Fire Suppression Course, Canadian Department of Natural Resources and Renewables, https://novascotia.ca/natr/forestprotection/wildfire/bffsc/lessons/lesson3/diurnal.asp.
27. In my account of this fire and its impact, I have drawn on John Vaillant's timely and informative book *Fire Weather: A True Story from a Hotter World* (New York: Knopf, 2023). His detailed analysis of this fire provides a wealth of information for understanding the impact of climate change on fires and the way increasing wildfires are changing weather patterns.
28. Vaillant, 92.
29. Vaillant, 98.

2. LOST WORLD

1. Jacques Derrida, Catherine Porter, and Philip Lewis, "No Apocalypse, Not Now (full speed ahead, seven missiles, seven missives)," *Diacritics* 14, no. 2 (Summer 1984): 20.
2. Martin Heidegger, "The Age of the World Picture," in *The Question Concerning Technology and Other Essays*, trans. William Lovitt (New York: Harper Torchbooks, 1977), 134–35.
3. Apple, "1984," https://www.google.com/search?q=apple+ad+1984&rlz=1C1CHBF_enUS979US979&oq=appl&aqs=chrome.0.69i59j69i57j46i131i199i433i465i512j0i131i433i512j0i433i512j69i61l3.4614j0j7&sourceid=chrome&ie=UTF-8.
4. Fred Turner, *From Counterculture to Cyberculture: Stuart Brand, the Whole Earth Network, and the Rise of Digital Utopianism* (Chicago: University of Chicago Press, 2006).
5. Newt Gingrich was one of the first politicians who recognized the impact information technologies would have on all aspects of society and culture. His guides in the digital realm were the futurists Alvin and Heidi Toffler. Gingrich wrote the foreword to their influential book, *Creating a New Civilization: The Politics of the Third Wave* (Nashville, Tenn.: Turner Publishing, 1995).
6. Shoshana Zuboff, *The Age of Surveillance Capitalism: The Fight for a Human Future at the New Frontier of Power* (New York: Public Affairs, 2019), vii.
7. Ray Kurzweil and Terry Grossman, *Fantastic Voyage: Live Long Enough to Live Forever* (New York: Plume, 2005).
8. Alcor, https://www.alcor.org/.
9. Ray Kurzweil, *The Singularity Is Near* (New York: Viking, 2005), 9.
10. Lynn White, "The Historical Roots of Our Ecological Crisis," *Science* 155 (1967):1203–7.
11. Kurzweil, *The Singularity Is Near*, 4. Emphasis added.
12. N. Katherine Hayles, *How We Became Posthuman: Virtual Bodies in Cybernetics, Literature, and Informatics* (Chicago: University of Chicago Press, 1999), 2.
13. Hans Moravec, *The Future of Robot and Human Intelligence* (Cambridge, Mass.: Harvard University Press, 1990); Maratine Rothblatt, *Virtually Human: The Promise and the Peril of Digital Immortality* (New York: Picador Press, 2015).
14. Kurzweil, *The Singularity Is Near*, 5, 2, 29.
15. Hans Jonas, *The Gnostic Religion: The Message of the Alien God and the Beginnings of Christianity* (Boston: Beacon Press, 1958); Erik Davis, *Techgnosis: Myth, Magic, Mysticism in the Age of Information* (Berkeley, Calif.: North Atlantic, 2015).
16. Mary-Jane Rubenstein, *Astrotopia: The Dangerous Religion of the Corporate Space Race* (Chicago: University of Chicago Press, 2022).
17. Heidegger, *The Question Concerning Technology*, 131–32, 27.
18. Angela Zottola and Claudio Majo, "The Anthropocene: Genesis of a Term and Popularization in the Press," *Text and Talk* 42 (February 2022).
19. Heidegger, *The Question Concerning Technology*, 82–82.
20. Heidegger, 83.
21. Lynn Margulis, *What Is Life?* (Berkeley: University of California Press, 1995), 41.

22. Quoted in Giulio Tononi, *Phi: A Voyage from Brain to the Soul* (New York: Pantheon, 2012), 108.
23. Antonio Damasio, *Descartes' Error: Emotion, Reason, and the Human Brain* (New York: Avon), 248.
24. Heidegger, *The Question Concerning Technology*, 100.
25. Friedrich Nietzsche, *The Gay Science*, trans. Walter Kaufmann (New York: Random House1974), aphorism 125.
26. Heidegger, *The Question Concerning Technology*, 118, 121.
27. Philip Goff, *Galileo's Error: Foundations for a New Science of Consciousness* (New York: Pantheon, 2019), 13-23.
28. "Francis Bacon," Wikipedia, https://en.wikipedia.org/wiki/Francis_Bacon. It is, of course, important to underscore the gendered formulation of Bacon's hierarchical man/nature binary.
29. G.W.F. Hegel, *Lectures on the History of Philosophy*, trans. E. S. Haldane and Frances Simson (New York: Humanities Press, 1968), 1:305-7.
30. Quoted in Fritjof Capra, *The Tao of Physics* (Boulder, Colo.: Shambhala, 2010) 57.
31. Capra, 56.
32. Heidegger, *The Question Concerning Technology*, 5.
33. Heidegger, 27,
34. Heidegger, 12-13.
35. Martin Heidegger, *Being and Time*, trans. John Macquarrie and Edward Robinson (New York: Harper and Row, 1962), 294.
36. Jacques Derrida, *Aporias*, trans. Thomas Dutoi (Stanford, Calif.: Stanford University Press, 1993), 35.
37. Werner Heisenberg, *Physics and Philosophy: The Revolution in Modern Science* (New York: HarperCollins, 1962), 52-53.
38. Martin Heidegger, *On the Way to Language*, trans. Peter Hertz (New York: Harper, 1971), 135-36. I have translated the German word *Ereignis* as "event" rather than "appropriation."
39. Heisenberg, *Physics and Philosophy*, 83.

3. RELATIONALISM

1. Stephen T. Jackson, "Introduction: Humboldt, Ecology, and the Cosmos," in *Essay on the Geography of Plants*, by Alexander von Humboldt and Aimé Bonpland, ed. Stephen T. Jackson, trans. Sylvie Romanowski (Chicago: University of Chicago Press, 2009), 3.
2. Jackson, 4, 8.
3. Sylvie Romanowski, "Humboldt's Pictorial Science: An Analysis of the Tableau physique des Andes et pays voisins," in *Essay on the Geography of Plants*, ed. Stephen T. Jackson (Chicago: University of Chicago Press, 2009), 181.
4. Michel Foucault, *The Order of Things: An Archaeology of the Human Sciences* (New York: Random House, 1973), 218. Emphasis added.

5. Romanowski, "Humboldt's Pictorial Science," 182.
6. Romanowski, 182; Jackson, "Introduction," 4.
7. Terry Pinkard, *Hegel: A Biography* (London: Cambridge University Press, 2000), 610.
8. G.W.F. Hegel, *Philosophy of Nature*, vol. 1, trans. J. M. Petry (New York: Humanities Press, 1970), 194, 212.
9. G.W.F. Hegel, *Philosophy of Nature*, trans. A. V. Miller (New York: Oxford University Press, 1970), 274-75.
10. G.W.F. Hegel, *Science of Logic*, trans. A. V. Miller (New York: Humanities Press, 1969), 737.
11. Immanuel Kant, *Critique of Judgment*, trans. James Meredith (New York: Oxford University Press, 1973), book 2, 22.
12. Kant, 21. Emphasis added.
13. Kant, 23-25, 31.
14. G.W.F. Hegel, *Lesser Logic*, trans. William Wallace (New York: Oxford University Press, 1968), 143.
15. Hegel, 66-67.
16. Hegel, *Science of Logic*, 711.
17. G.W.F. Hegel, *The Difference Between Fichte's and Schelling's System of Philosophy*, trans. H. S. Harris (Albany: State University of New York Press, 1977), 91.
18. Hegel, *Science of Logic*, 763.
19. Augustine, *Basic Writings of Saint Augustine*, ed. Whitney Oates (New York: Random House, 1948), 2:793.
20. Hegel, *Science of Logic*, 82-83.
21. Hegel, *Philosophy of Nature*, trans. Miller, 28, 34.
22. I originally formulated this insight in "Toward an Ontology of Relativism," *Journal of the American Academy of Religion* 46, no. 1 (March 1978): 41-61. It has taken me almost fifty years to realize the implications of this insight.
23. Martin Heidegger, *Identity and Difference*, trans. Joan Stambaugh (New York: Harper and Row, 1969), 63-64.
24. Gregory Bateson, *Steps Toward an Ecology of Mind* (New York: Ballantine, 1972), 425. I will consider the issue of information and cognition in chapter 6.
25. Hegel, *Science of Logic*, 413.
26. Hegel, 417, 431.
27. Hegel, *Lesser Logic*, 245.
28. Hegel, *Phenomenology of Spirit*, trans. A. V. Miller (New York: Oxford University Press, 1977), 27.
29. Friedrich Nietzsche, *Thus Spoke Zarathustra*, trans. Marianne Cowan (Chicago: Regnery, 1957),
30. Hegel, *Phenomenology*, 27.
31. Georges Bataille, *Theory of Religion*, trans. Robert Hurley (New York: Zone, 1989).
32. Friedrich Nietzsche, *Will to Power*, trans. Walter Kaufmann (New York: Random House, 1969), 300, 301.

33. Nietzsche, 267, 312. Emphasis added.
34. H. P. Stapp, "S-Matrix Interpretation of Quantum Theory," *Physical Review* D3 (March 15, 1971): 1310.
35. Werner Heisenberg, *Physics and Philosophy: The Revolution in Modern Science* (New York: Harper, 1962), 107.
36. Nishida Kitaro, *Last Writings: Nothingness and the Religious Worldview*, trans. David Dilworth (Honolulu: University of Hawaii Press, 1987), 52.
37. I will consider in chapter 4 how Heidegger and Derrida's account of space and time opens the Hegelian system as if from within.
38. Jean Hyppolite, *Logic and Existence*, trans. Leonard Lawlor and Amit Sen (Albany: State University of New York Press, 1997), 24, 179.
39. Francis Fukuyama, *The End of History and the Last Man* (New York: Free Press, 1992).
40. Hyppolite, *Logic and Existence*, 66–67.
41. Nietzsche, *Will to Power*, 550.

4. RELATIVITY

1. Augustine, *Confessions*, trans. Rex Warner (New York: Mentor-Omega, 1963), 267–68, 273.
2. Jacques Derrida, *Positions*, trans. Alan Bass (Chicago: University of Chicago Press, 1972), 77.
3. Ferdinand de Saussure, *Course in General Linguistics*, *Deconstruction in Context*, ed. Mark C. Taylor (Chicago: University of Chicago Press, 1986), 166–67.
4. Jacques Derrida, "Différance," in *Margins of Philosophy*, trans. Alan Bass (Chicago: University of Chicago Press, 1982), 10, 17.
5. Friedrich Nietzsche, *Will to Power*, trans. Walter Kaufmann (New York: Random House, 1969), 550.
6. G.W.F. Hegel, *Phenomenology of Spirit*, trans. A. V. Miller (New York: Oxford University Press, 1977), 82, 84.
7. Derrida, "Différance," 17. Emphasis added.
8. Martin Heidegger, *Hegel*, trans. Joseph Arel and Niels Feuerhahn (Bloomington: Indiana University Press, 2009), 35.
9. Heidegger, 36–37.
10. The notion of plasticity is very important in Hegel's writings. See Catherin Malabou, *The Future of Hegel: Plasticity, Temporality, and Dialectic*, trans. Lisabeth During (New York: Routledge, 2005). This book has a long preface by Derrida, "A Time for Farewells: Heidegger (Read by) Hegel (Read by) Malabou."
11. Martin Heidegger, *Kant and the Problem of Metaphysics*, trans. Richard Taft (Bloomington: Indiana University Press, 1997), 131.
12. Kant organizes his table of categories in six groups: Of Quantity (Unity, Pluraliity, Totality); Of Quality (Reality, Negation, Limitation); Of Relation (Of Inherence and Subsistence [*substantia* et *accidens*]); Of Causality and Dependence [*cause and effect*]), Of

Community (reciprocity between agent and patient); Of Modality (Possibility-Impossibility, Existence-Nonexistence, Necessity-Contingency). Immanuel Kant, *Critique of Pure Reason*, trans. Norman Kemp Smity (New York: St. Martin's Press, 1965), 113.
13. Heidegger, *Kant and the Problem of Metaphysics*, 123.
14. Heidegger, *Hegel*, 34.
15. Carlo Rovelli notes, "The word 'time' derives from an Indo-European root—*di* or *dai*—meaning 'to divide.' For centuries, we have divided the days into hours. But for most of those centuries, however, hours were longer in the summer and shorter in the winter, because the twelve hours divided the time between dawn and sunset: the first hour was dawn, and the twelfth was sunset, regardless of the season." Carlo Rovelli, *The Order of Time* (New York: Riverhead Books, 2018), 58.
16. Derrida, "Différance," 5. Emphasis added.
17. Derrida, 8. Emphasis added. It is important to note in this context that Derrida's *diplôme d'études supérieures*, which was on the work of Edmund Husserl, was later published with the title *Edmund Husserl's "Origin of Geometry": An Introduction*. Husserl's *The Phenomenology of Internal Time Consciousness*.
18. My reading of Derrida's appropriation of Koyré's analysis of the *differente Beziehung* represents a significant revision of my previous interpretation presented in *Abiding Grace: Time, Modernity, Death* (Chicago: University of Chicago Press, 2018), chap. 6. In my earlier account I argued that Koyré's reading of Hegel's Jena Logic through Heidegger effectively subverted the Hegelian system. I am now convinced that Derrida's use of Koyré's understanding of the *differente Biezhung* to develop his notion of *différance* effectively exposes *internal* complications that are the condition of the possibility of Hegel's relationalism.
19. Derrida, "Différance," 13–14.
20. Alexandre Koyré, "Hegel à Jena," trans. Doha Tazi Hemida, *Continental Philosophy Review*, April 7, 2018.
21. In the concluding section of this chapter, we will see that primordial temporality is neither linear nor circular but is recursive.
22. Derrida, "Différance," 27.
23. Rovelli, *The Order of Time*, 2, 119–20. My understanding of both the theory of relativity and quantum mechanics has been most strongly influenced by Rovelli's interpretation of both in terms of what I have called Relationalism. I have drawn on this work as well as two others in this section and the next chapter: Rovelli, *Reality Is Not What It Seems: The Journey to Quantum Gravity* (New York: Riverhead Books, 2014); and Rovelli, *Helgoland: Making Sense of the Quantum Revolution* (New York: Riverhead Books, 2021). I will consider Wheeler's contribution to quantum theory and quantum information theory in the next chapter.
24. Rovelli, *Helgoland*, xvi. Emphasis added.
25. Rovelli, *Reality*, 134–35.

26. Albrecht Folsing, *Einstein: A Biography* (London: Penguin, 1998), 337. Cited in Rovelli, *Reality*, 50.
27. Edward S. Casey, *Getting Back Into Place: Toward a Renewed Understanding of the Place-World* (Bloomington: Indiana University Press, 1997), 4.
28. Rovelli, *The Order of Time*, 105, 109.
29. Lee Smolin, "How to Understand the Universe When You are Stuck Inside It," *Quanta Magazine*, https://www.quantamagazine.org/were-stuck-inside-the-universe-lee-smolin-has-an-idea-for-how-to-study-it-anyway-20190627/.
30. Rovelli, *Reality*, 69.
31. Rovelli, 58–59.
32. Rovelli, 62.
33. Rovelli, 86.
34. Rovelli, 70–71.
35. See Mark C. Taylor, *Field Notes from Elsewhere: Reflections on Dying and Living* (New York: Columbia University Press, 2009).
36. Rovelli, *The Order of Time*, 97.
37. Lee Smolin, *Time Reborn: From the Crisis in Physics to the Future of the Universe* (New York: Houghton Mifflin, 2013), xxviii. Smolin distinguishes relationalism, in which time is fundamental, from relativity, in which time purportedly is still subordinated to space. "The whole history of the world is, in general relativity, still represented by a mathematical object. The spacetime of general relativity corresponds to a mathematical object much more complex than the three-dimensional Euclidean space of Newton's theory. But seen as a block universe, it is timeless and pristine, with no distinction of future from past and no role for or sign of our awareness of the present" (71). As the foregoing argument makes clear, while I agree with Smolin's interpretation of relationality, I do not agree with his claim that Einstein perpetuates the Platonic assertion of "the essential unreality of time." Rather than opposites, relationality and relativity are complementary—spacing-timing is relative because becoming is relational, and becoming is relational because spacing-timing is relative.
38. Smolin, xxi.
39. I will return to the arguments for the existence of God in chapter 11.
40. William Paley, *Natural Theology* (New York: Sheldon, n.d.), 6.
41. Immanuel Kant, *Critique of Judgment*, trans. James Meredith (New York: Oxford University Press, 1973), 22–23.
42. Hegel, *Phenomenology*, 2.
43. Stephen Crites, "The Narrative Quality of Experience," *Journal of the American Academy of Religion* 39, no. 3 (1971).
44. Søren Kierkegaard, *Philosophical Fragments*, trans. Howard Hong (Princeton, N.J.: Princeton University Press, 1971), 96.
45. Ilya Prigogine and Isabelle Strengers, *Order Out of Chaos: Man's New Dialogue with Nature* (New York: Bantam, 1984), 123.

5. ENTANGLEMENT

1. Werner Heisenberg, *Physics and Philosophy: The Revolution in Modern Science* (New York: Harper, 207), 83.
2. Tom Stoppard, *Hapgood*, *Plays* (London: Farber and Farber, 1999), 501, 572.
3. Quoted in Carlo Rovelli, *Helgoland: Making Sense of the Quantum Revolution* (New York: Riverhead, 2021), 9–10.
4. Michael Frayn, *Copenhagen* (New York: Random House, 1998), 85–86.
5. Frayn, 98.
6. Frayn, 87, 83.
7. Nietzsche, "Twilight of the Idols," in *The Portable Nietzsche*, ed. Walter Kaufmann (New York: Penguin, 1980), 486,
8. Philip Ball, *Beyond Weird: Why Everything You Though You Knew About Quantum Physics Is Different* (Chicago: University of Chicago Press, 2018), 61.
9. Heisenberg, *Physics and Philosophy*, 3.
10. Heisenberg, 170–71.
11. Heisenberg, 69.
12. Ball, *Beyond Weird*, 29.
13. Heisenberg, *Physics and Philosophy*, 16–17.
14. Frayn, *Copenhagen*, 99–100.
15. Heisenberg, *Physics and Philosophy*, 9.
16. Heisenberg, 21–22. Emphasis added.
17. Rovelli, *Helgoland*, 19, 8–9.
18. David W. Smith, "Phenomenology," *Stanford Encyclopedia of Philosophy*, https://plato.stanford.edu/entries/phenomenology/. In view of the importance of Derrida's analysis of Hegel in the previous chapter, it important to note that he wrote his dissertation for his *diplôme d'études supérieures* on Husserl's phenomenology (1954). This work became his first book, *Esmund Husserl's Origin of Geometry: An Introduction* (Lincoln: University of Nebraska Press, 1962).
19. Quoted in Ball, *Beyond Weird*, 90.
20. John Wheeler, "Information, Physics, Quantum: The Search for Links," *Proceeding from the Third International Symposium, Foundations of Quantum Mechanics*, 317.
21. Ball, *Beyond Weird*, 82.
22. Ball, 68–71. Figure 5.3 is based on Ball's illustration of the double-slit experiment on page 68.
23. Ball, 70–73. In attempting to describe this extremely fuzzy experiment, I have followed Ball's exceptionally lucid description.
24. Ball, 204.
25. Lisa Zyga, "Physicists Find Quantum Coherence and Entanglement Are Two Sides of the Same Coin," Physics.org, https://phys.org/news/2015-06-physicists-quantum-coherence-entanglement-sides.html.

26. Ball, *Beyond Weird*, 207. Emphasis added.
27. Carl Jung and Austrian nuclear physicist Wolfgang had a long and very strange relationship. Jung believed that quantum entanglement and nonlocality confirmed his notion of synchronicity. See Arthur I. Miller, *Jung, Pauli and the Pursuit of a Scientific Obsession* (New York: Norton, 2010).
28. Nietzsche, *Will to Power*, trans. Walter Kaufmann (New York: Random House, 1968), 302.
29. David Bohm, *Wholeness and the Implicate Order* (New York: Routledge, 1980), 67–70.
30. Rovelli, *Helgoland*, 85–86.
31. Rovell, 89, 76, 186.
32. Carlo Rovelli, "Relational Quantum Mechanics," https://www.academia.edu/5444262/Relational_quantum_mechanics.
33. Rovelli, *Helgoland*, 78–79.
34. Rovelli, 96, 88.
35. Nietzsche, *Will to Power*, 305.
36. Rovelli, *Helgoland*, 123.
37. Rovelli, 74–75.
38. Rovelli, 151.
39. Wheeler, "Information, Physics, Quantum," 317.
40. Rachel Thomas, "It from Bit?," Plus, https://plus.maths.org/content/it-bit.
41. John Wheeler, "World as System Self-Synthesized by Quantum Networking," in *Probability in the Sciences*, ed. E. Agazzi (Dordrecht: Springer, 1988), https://link.springer.com/chapter/10.1007/978-94-009-3061-2_7, 109. Geometrodynamics describes space-time in geometrical terms.
42. Wheeler, 104.
43. Quoted in Thomas, "It from Bit?"
44. Wheeler, "World as System," 103. Emphasis added.
45. Wheeler, 113.
46. Karen Barad, *Meeting the Universe Halfway: Quantum Physics and the Entanglement of Matter and Meaning* (Durham, N.C.: Duke University Press, 2007), 120. Barad's argument is a remarkably insightful interpretation of the philosophical, social, and political implications of quantum physics.
47. After his careful analysis of relational quantum theory, Rovelli writes, "Life is a biochemical process that unfolds across the surface of the Earth and dissipates the abundant 'free energy,' or 'low entropy,' with which the light of the Sun floods the planet. It is made up of individuals who interact with what surrounds them, formed by structures and processes that are self-regulating, maintaining a dynamic equilibrium that persists over time. But structures and processes are not there *so that* the organisms may survive and reproduce. It is the other way around: organisms survive and reproduce *because* these structures have happened to gradually develop. They reproduce and populate Earth *because* they are functional." Rovelli, *Helgoland*, 169. I will consider different aspects of biological networks in chapters 7–9.
48. Wheeler, "World as System," 127–28.

6. INFORMATION IN FORMATION

1. Alan M. Turing, "Intelligent Machinery," National Physical Laboratory, 1948: 107, 110, https://weightagnostic.github.io/papers/turing1948.pdf.
2. C. E. Shannon, "A Mathematical Theory of Communication," *Bell System Technical Journal* 27 (July 1948), https://people.math.harvard.edu/~ctm/home/text/others/shannon/entropy/entropy.pdf, 3.
3. Shannon, 100.
4. Katherine Hayles, *How We Became Posthuman: Virtual Bodies in Cybernetics, Literature, and Informatics* (Chicago: University of Chicago Press, 1999), 2–3.
5. Giulio Tonibi, *Phi Φ. A Voyage from the Brain to the Soul* (New York: Pantheon, 2012), 108–9.
6. I will consider panpsychism in the final section of this chapter.
7. Carlo Rovelli, *Helgoland: Making Sense of the Quantum Revolution* (New York: Riverhead, 2020), 183–84, 186. While Chalmers does not directly draw on Rovelli's work, he has considered the significance of information theory and quantum mechanics for consciousness. David J. Chalmers, *The Conscious Mind: In Search of a Fundamental Theory* (New York: Oxford University Press, 1996), chaps. 8–10.
8. Gregory Bateson, *Steps to an Ecology of Mind* (New York: Ballantine, 1972), 460. I will consider the ecological implications of Bateson's argument in the next chapter.
9. Bateson, 46.
10. Lynn White, "The Historical Roots of Our Ecological Crisis," *Science* 155 (1967): 1203–7
11. G.W.F. Hegel, *Philosophy of History*, trans. J. Sibree (New York: Dover Publications, 1956), 36.
12. Augustine, *Confessions*, trans. Rex Warner (New York: Mentor-Omega, 1963), 221–22.
13. Carlo Rovelli, *The Order of Time* (New York Riverhead, 2018), 178–79.
14. David Bohm, *Wholeness and the Implicate Order* (New York: Routledge, 1980), 67–68.
15. In her important book, *Unthought: The Power of the Cognitive Nonconscious* (Chicago: University of Chicago Press, 2017), 14, Katherine Hayles makes a similar distinction between cognition and consciousness. She is committed to maintaining "faithfulness to the Deleuzian paradigm" and argues that cognition is operative in biological systems ranging from bacteria to plants and animals as well as in technological "assemblages." "Cognition ... is a much broader faculty present to some degree in all biological life-forms and many technical systems." I have two primary reservations about Hayles's argument. First, the Deleuzian notion of assemblage involves an insufficiently relational interpretation among parts and between parts and the whole. In this configuration, parts exist independently of each other and are assembled through some kind of external agency. Second, Hayles restricts her suggestive interpretation of the cognitive nonconscious to biological and machinic assemblages. By not extending the cognitive nonconscious to physical systems, she seems to suggest a lingering dualism that pancognitivism overcomes.
16. Bateson, *Steps Toward an Ecology of Mind*, 460, 454.

17. Peter Hiesinger, *The Self-Assembling Brain: How Neural Networks Grow Smarter* (Princeton, N.J.: Princeton University Press, 2021), 88.
18. Rovelli, *Helgoland*, 103.
19. Wheeler, "World as System Self-Synthesized by Quantum Networking," in *Probability in the Sciences*, ed. E. Agazzi (Dordrecht: Springer, 1988), https://link.springer.com/chapter/10.1007/978-94-009-3061-2_7, 104, 127.
20. John Holland, *Emergence: From Chaos to Order* (Reading, Mass.: Addison-Wesley, 1998), 14. I will consider problems with Holland's interpretation of emergence in my consideration of Darwinian evolution in the next chapter.
21. Per Bak, *How Nature Works: The Science of Self-Organized Criticality* (New York: Springer-Verlag, 1996), 1–2.
22. Giulio Tononi and Christof Koch, "Consciousness: Here, There and Everywhere?," *Philosophical Transactions* 370, no. 1668 (May 2015), https://royalsocietypublishing.org/doi/10.1098/rstb.2014.0167.
23. Giulio Tononi, *Phi Φ*, 157.
24. Tononi, 172.
25. Giulio Tononi and Marcello Massimini, *Sizing Up Consciousness: Towards an Objective Measure of the Capacity for Experience* (New York: Oxford University Press, 2018), 71.
26. Tononi, *Phi Φ*, 191.
27. Tononi and Koch, "Consciousness," 11. Emphasis added.
28. Online Etymology Dictionary, https://www.etymonline.com/word/intelligence.
29. Catherine Malabou, *Morphing Intelligence: From IQ Measurement to Artificial Brains*, trans. Carolyn Shread (New York: Columbia University Press, 2015), 3.
30. Evan Malmgren, "The Intelligence of Swine," *Noēma*, November 24, 2021, https://www.noemamag.com/the-intelligence-of-swine/. Emphasis added. I will return to this article in chapter 9.
31. Philip Goff, "Panpsychism," *Stanford Encyclopedia of Philosophy*, https://plato.stanford.edu/entries/panpsychism/.
32. Max Planck, https://www.goodreads.com/quotes/7956796-i-regard-consciousness-as-fundamental-i-regard-matter-as-derivative.
33. Thomas Nagle, *Mind and Cosmos: Why the Materialist Neo-Darwinian Conception of Nature Is Almost Certainly False* (New York: Oxford University Press, 2012), 85.
34. Henry P. Stapp, *Mindful Universe: Quantum Mechanics and the Participating Observer* (Berlin: Springer, 2011), 96, 85–86.
35. Tononi and Koch, "Consciousness," 10–11, 9.
36. Simon McGregor, "Cognition Is Not Exceptional," *Animal Behavior* 26, no. 1 (2018): 33–36.

7. QUANTUM ECOLOGY

1. Michel Serres, *The Parasite*, trans. Lawrence Shehr (Baltimore, Md.: Johns Hopkins University Press, 1982), 1, 10–11.

2. Jack Miles and Mark C. Taylor, *A Friendship in Twilight: Lockdown Conversations on Death and Life* (New York: Columbia University Press, 2022).
3. *New York Times*, July 16, 2022. The information in this chapter is drawn from this article.
4. All quotations are references to Serres's *The Parasite*, 3–39.
5. Quoted in Merlin Sheldrake, *Entangled Life: How Fungi Make Our Worlds, Change Our Minds, and Shape Our Futures* (New York: Random House 2020), 149.
6. Charles Darwin, *The Origin of Species by Means of Natural Selection* (New York: Macmillan, 1937), 525.
7. Brian Goodwin, *How the Leopard Changed Its Spots: The Evolution of Complexity* (New York: Simon and Schuster, 1994), 145.
8. David Depew and Bruce Weber, *Darwinism Evolving: Systems Dynamics and the Genealogy of Natural Selection* (Cambridge, Mass.: MIT Press, 1997), 49.
9. Linnaeus's Latin binomial nomenclature is still used to classify plants. The first term denotes the genus, and the second specifies the species.
10. Stephen Jay Gould, *Full House: The Spread of Excellence from Plato to Darwin* (New York: Harmony, 1996), 145.
11. Depew and Weber, *Darwinism Evolving*, 115.
12. Depew and Weber, 71.
13. Richard Levins and Richard Lewontin, *The Dialectical Biologist* (Cambridge, Mass.: Harvard University Press, 1985), 278.
14. Levins and Lewontin, 269.
15. Levins and Lewontin, 269–70.
16. Levins and Lewontin, 88. Emphasis added.
17. For an elaboration of such systems, see Mark C. Taylor, *The Moment of Complexity: Emerging Network Culture* (Chicago: University of Chicago Press, 2000).
18. Levins and Lewontin, *The Dialectical Biologist*, 88. Emphasis added.
19. J. Hillis Miller, "Critic as Host," *Deconstruction and Criticism* (New York: Seabury Press, 1979), 219–20.
20. I will say more about his deep relationship with horses in chapter 9.
21. Noel A. Taylor and Alfred L. Vandling, *Activity Units in Biology* (New York: Oxford Book Company, 1954).
22. Werner Heisenberg, *Physics and Philosophy: The Revolution in Modern Science* (New York: Harper, 207), 128, 81. Emphasis added.
23. Maurice Merleau-Ponty, *The Visible and the Invisible*, trans. Alphonso Lingis (Evanston, Ill.: Northwestern University Press, 1968), 135, 139–40.
24. Levins and Richard Lewontin, *The Dialectical Biologist*, 3.
25. Carl Rovelli, *Helgoland: Making Sense of the Quantum Revolution* (New York: Riverhead, 2021), 169.
26. Rovelli, 176. Emphasis added.
27. Richard Lewontin, *The Triple Helix: Gene, Organism, and Environment* (Cambridge, Mass.: Harvard University Press, 2000), 63–64. Transduction is "the transfer of genetic material

from one microorganism to another by a viral agent." *Merriam-Webster Dictionary*, https://www.merriam-webster.com/dictionary/transduction.

28. Nobel Prize, https://www.nobelprize.org/prizes/medicine/1962/perspectives/. In developing my account of Schrödinger's contributions, I have drawn on this essay, and quotations in this paragraph are drawn from it.
29. Lewontin, *The Triple Helix*, 3.
30. Lewontin, 5.
31. Lewontin, 5.
32. For an explanation of "intervolution," see my book *Intervolution: Smart Bodies, Smart Things* (New York: Columbia University Press, 2021). I will return to the issue of intervoution in the following chapters.
33. Lewontin, *The Triple Helix*, 72.
34. Lewontin, 17–18, 20.
35. Depew and Weber, *Darwinism Evolving*, 396.

8. MINDING THE BODY

1. For an account of this experience, see Mark C. Taylor, *Field Notes from Elsewhere: Reflections on Dying and Living* (New York: Columbia University Press, 2014).
2. See Mark C. Taylor, *Intervolution: Smart Bodies Smart Things* (New York: Columba University Press, 2021.
3. Lynn Margulis, *What Is Life?* (Berkeley: University of California Press, 1995), 2–3.
4. Immanuel Kant, *Critique of Judgment*, trans. James Meredith (New York Oxford University Press, 1952), 21.
5. G.W.F. Hegel, *Science of Logic*, trans. A. V. Miller (New York: Humanities Press, 1969), 771.
6. Humberto Maturana and Francisco Varela, *Autopoiesis and Cognition: The Realization of the Living* (Boston: Reidel, 1980), 82.
7. Stuart Kauffmann, *At Home in the Universe: The Search for the Laws of Self-Organization and Complexity* (New York: Oxford University Press, 1995), 69, 15, 49–50.
8. Maturana and Varela, *Autopoiesis and Cognition*, 76, 78–79, 75.
9. Maturana and Varela, 80.
10. Evan Thompson, *Mind in Life: Biology, Phenomenology and the Science of Life* (Cambridge, Mass.: Harvard University Press, 2010), 98–99.
11. K. Ruiz-Mirazo and A. Moreno, "Basic Autonomy as a Fundamental Step in the Synthesis of Life," *Artificial Life* 10 (2004): 240. Quoted in Thompson, *Mind in Life*, 46.
12. Maturana and Varela, *Autopoiesis and Cognition*, 13. Emphasis added.
13. Margulis, *What Is Life?*, 232–34.
14. Thompson, *Mind in Life*, 158.
15. No one has done more to call attention to the importance of this issue than former Wall Street tycoon Sandy Lewis. See Michael Powell and Danny Hakim, "A Lonely Redemption," *New York Times*, September 15, 2012.

16. Humberto Maturana, W. S. McCulloch, and W. H. Pitts. "What the Frog's Eye Tells the Frog's Brain," *Proceedings of the IRE* (1959), https://hearingbrain.org/docs/letvin_ieee_1959.pdf.
17. Thompson, *Mind in Life*, 51–52.
18. Arthur R. Reber, *The First Minds: Caterpillars, 'Karyotes, and Consciousness* (New York: Oxford University Press, 2019), 77.
19. Lynn Margulis, "The Conscious Cell," *Annals of the New York Academy of Sciences* 929, no. 1 (April 2001): 55–70, https://pubmed.ncbi.nlm.nih.gov/11349430/.
20. "Michael Levin, PhD," Wyss Institute, https://wyss.harvard.edu/team/associate-faculty/michael-levin-ph-d/.
21. "Michael Levin, "Life, Death and the Self: Fundamental Questions of Primitive Cognition Viewed Through the Lens of Plasticity and Synthetic Organisms," *Biochemical and Biophysical Research Communications*, November 2020. https://www.sciencedirect.com/science/article/abs/pii/S0006291X20320064.
22. This list is a condensation and modification of Reber's list of the necessary tasks living organisms must perform. Reber, *The First Minds*, 134.
23. Levin, "Life, Death, and Self," 17. This comment is closely related to Margulis's account of symbiosis, which I will consider in chapter 10.
24. Sharad Ramanathan, "Do Cells Think?," *Cellular and Molecular Life Sciences*, 2007, https://pubmed.ncbi.nlm.nih.gov/17530173/. Tumbling rate at which molecules tumble. Large molecules tend to tumble more rapidly than small molecules. Ramanathan gives a detailed account of the biochemical process of sensation and cognition.
25. Augustine, *Confessions*, trans. Rex Warner (New York: New American Library), 222.
26. Levin, "Life, Death, and the Self," 15.
27. Ramanathan, "Do Cells Think?," 2.
28. Quoted in Margulis, *What Is Life?*, 219.
29. Reber, *The First Minds*, 146.
30. Ramanathan, "Do Cells Think?," 2.
31. Jan Klein, *Immunology: The Science of Self-Nonself Discrimination* (New York: Wiley, 1982).
32. Emily Martin, *Flexible Bodies: Tracking Immunity in American Culture from the Days of Polio to the Age of AIDS* (New York: Beacon 1995).
33. Niels Jerne, "The Natural Selection Theory of Antibody Formation," in *Phage and the Origins of Molecular Biology* (Cold Spring Harbor, N.Y.: Cold Spring Harbor Laboratory of Quantitative Biology, 1966), 301.
34. Frank Macfarlane Burnet, "A Modification of Jerne's Theory of Antibody Production Using the Concept of Clonal Selection" *Australian Journal of Science* 20, no. 3 (1957): 67–69.
35. Irun Cohen, "The Cognitive Principle Challenges Clonal Selection," file:///C:/Users/Mark/Downloads/the%20cognitive%20principle%20challenges%20clonal%20selection%20(1).pdf. Unless otherwise indicated, quotations in this section are from this paper. The most insightful philosophical discussion of the immune system is Alfred I. Tauber, *The Immune Self: Theory or Metaphor?* (New York: Cambridge University Press,

1994). My discussion of Cohen's theory is informed by Tauber's account of cognitive network theory in chapter 5 of this book.
36. Tauber, *The Immune Self*, 178.
37. Michel Serres, *The Parasite*, trans. Lawrence Shehr (Baltimore, Md.: Johns Hopkins University Press, 1982), 79.
38. J. Hillis Miller, "The Critic as Host," in *Deconstruction and Criticism* (New York: Seabury Press, 1979), 219.
39. Jerne, "Natural Selection Theory," 360–61.
40. Maturana and Varela, *Autopoeisis and Cognition*, 214–215, 216. Emphasis added.
41. Maturana and Varela, xv.
42. Francisco Varela et al., "Cognitive Networks: Immune, Neural, and Otherwise," in *Theoretical Immunology*, ed. Alan S. Peterson (New York: Addison-Wesley, 1988), 360–61.
43. Maturana and Varela, *Autopoeisis and Cognition*, xvi–xvii.

9. INFINITE CONVERSATIONS

1. Robert Pogue Harrison, *Forests: The Shadow of Civilization* (Chicago: University of Chicago Press, 1992), 110–11, 112.
2. Corbusier, *Towards a New Architecture*, trans. Frederick Etchells (New York: Dover Publications, 1986), 72; Corbusier, *The City of Tomorrow*, trans. Frederick Etchells (Cambridge, Mass.: MIT Press, 1971), 11.
3. Harrison, *Forests*, 123.
4. See Paco Calvo, *Planta Sapiens: The New Science of Plant Intelligence* (New York: Norton, 2022); Daniel Chamovitz, *What A Plant Knows: A Field Guide to the Senses* (New York: Farrar, Straus and Giroux, 2017); Emanuele Coccia, *The Life of Plants: A Metaphysics of Mixture* (Medford, Mass.: Polity, 2019); Monica Gagliano, *Thus Spoke the Plant* (Berkeley, Calif.: North Atlantic, 2018); Stephan Harding, *Animate Earth: Science, Intuition and Gaia* (London: Green, 2006); Harrison, *Forests*; Eduardo Kohn, *How Forests Think: Toward an Anthropology Beyond the Human* (Berkeley: University of California Press, 2013); Stefano Mancuso, *The Revolutionary Genius of Plants: A New Understanding of Plant Behavior* (New York: Atria, 2017); Michael Pollan, *This Is Your Mind on Plants* (New York: Penguin, 2021); John Reid and Thomas Lovejoy, *Ever Green: Saving Big Forests to Save the Planet* (New York: Norton, 2022); Merlin Sheldrake, *Entangled Life: How Fungi Make Our Worlds, Change Our Minds, and Shape Our Futures* (New York: Random House, 2020); Suzanne Simard, *Finding the Mother Tree: Discovering the Wisdom of the Forest* (New York: Knopf, 2021).
5. Angela M. O'Callaghan, "Mycorrhizae," University of Nevada Reno Cooperative Extension, n.d., https://www2.nau.edu/~gaud/bio300/mycorrhizae.htm.
6. Sheldrake, *Entangled Life*, 127.
7. There is another disturbing similarity between quantum physics and Simard's revision of Darwin's appropriation of Newtonian principles to form his classical theory of

evolution. Just as quantum was dismissed for many years as "Jewish science," so Simard's interpretation of forest ecology was long rejected because of gender bias.
8. Simard, *Finding the Mother Tree*, 140.
9. Simard, 145.
10. Suzanne Simard et al., "Net Transfer of Carbon Between Ectomycorrhizal Tree Species in the Field," *Nature* 388 (1997): 579–82, https://www.nature.com/articles/41557.
11. Simard, *Finding the Mother Tree*, 160–61.
12. Anne Casselman, "Strange but True: The Largest Organism on Earth Is a Fungus," https://www.scientificamerican.com/article/strange-but-true-largest-organism-is-fungus/.
13. Simard, *Finding the Mother Tree*, 225.
14. Albert-Laszlo Barabasi, *Linked: The New Science of Networks* (Cambridge, Mass.: Perseus Publishing, 2002), 72, 66.
15. Simard, *Finding the Mother Tree*, 228–29.
16. Quoted in Sheldrake, *Entangled Life*, 149.
17. See Michael Pollan, *How to Change Your Mind: What the New Science of Psychedelics Teaches Us About Consciousness, Dying, Depression, and Transcendence* (New York: Penguin, 2018).
18. Michael Pollan, "The Intelligent Plant: Scientists Debate a New Way of Understanding Flora," *New Yorker*, December 15, 2013. Quotations in this section are from this article.
19. Eric D. Brenner et al., "Plant Neurobiology: An Integrated View of Plant Signaling," *Trends in Plant Science* 8, no. 3 (July 2006).
20. Miguel Segundo-Ortin and Paco Calvo, "Consciousness and Cognition in Plants" (New York: Wiley, 2021), https://miguelsegundoortinphd.com/wp-content/uploads/2022/09/segundo-ortin-m.-calvo-p.-2021.-consciousness-and-cognition-in-plants.pdf.
21. Quoted in Pollan, "The Intelligent Plant."
22. Amanda Gefter, "What Plants Are Saying About Us," *Nautilus*, March 7, 2023, https://nautil.us/what-plants-are-saying-about-us-264593/.
23. Segundo-Ortin and Calvo, "Consciousness and Cognition," 1.
24. Calvo, *Planta Sapiens*, 70.
25. Segundo-Orin and Calvo, "Consciousness and Cognition," 9.
26. Henri Altan, *Entre le cristal et la fumee: Essai sure l'organisation du vivant* (Paris: Editions du Seuil, 1979), 281–84.
27. Quoted in Calvo, *Planta Sapiens*, 162–63.
28. Calvo, 163–64.
29. Segundo-Ortin and Calvo, "Consciousness and Cognition," 4.
30. Sheldrake, *Entangled Life*, 164.
31. Ian Baldwin and Jack Schultz, "Rapid Changes in Tree Leaf Chemistry Induced by Damage: Evidence for Communication Between Plants," *Science* 221 (4607): July 15, 1983, https://pubmed.ncbi.nlm.nih.gov/17815197/.
32. Kathy Keatley Garvey, "Rick Karban: Kin Recognition Affects Plant Communication and Defense," *Entomology and Nematology News*, February 13, 2013, https://ucanr.edu/blogs/blogcore/postdetail.cfm?postnum=9871.

33. Pauline Oliveros, *Deep Listening: A Composer's Sound Practice* (Lincoln, Neb.: Deep Listening Publications, 2005), xxiv–xxv.

34. Just as there has been a great increase in the interest in plant cognition in recent years, so too there has been a growing interest in animal cognition and consciousness. In addition to many scientific articles, I have drawn on the following books in writing this section: David Barrie, *Supernavigators: Exploring the Wonders of How Animals Find Their Way* (New York: The Experiment, 2019); Karen Bakker, *The Sounds of Life: How Digital Technology Is Bringing Us Closer to the Worlds of Animals and Plants* (Princeton, N.J.: Princeton University Press, 2022); Jonathan Balcome, *What a Fish Knows: The Inner Lives of Our Underwater Cousins* (New York: Farrar, Straus and Giroux, 2016); Carel ten Cate and Susan Healy, *Avian Cognition* (Cambridge: Cambridge University Press, 2017); Dorothy Cheney and Robert Seyfarth, *Baboon Metaphysics: The Evolution of a Social Mind* (Chicago: University of Chicago Press, 2007); Lars Chittka, *The Mind of a Bee* (Princeton, N.J.: Princeton University Press, 2022); Nathan Emery, *Bird Brain: An Exploration of Avian Intelligence* (Princeton, N.J.: Princeton University Press, 2016); Meghan O'Giblyn, *God, Human, Animal, Machine: Technology, Metaphor, and the Search for Meaning.* (New York: Doubleday, 2021); Peter Godfrey-Smith, *Metazoa: Animal Life and the Birth of the Mind* (New York: Farrar, Straus and Giroux, 2020); Peter Godfrey-Smith, *Other Minds: The Octopus, the Sea, and The Deep Origins of Consciousness* (New York Farrar, Straus and Giroux, 2016); Justin Gregg, *If Nietzsche Were a Narwhal: What Animal Intelligence Reveals About Human Intelligence* (New York: Little, Brown, 2022); Bernd Heinrich, *The Mind of the Raven: Investigations and Adventures with Wolf-Birds* (New York: Ecco, 1999); Richard Hofstadter, *Godel, Escher, Bach: An Eternal Golden Braid* (New York: Random House, 1979); Bert Holldobler and E. O. Wilson, *The Superorganism: The Beauty, Elegance, and Strangeness of Insect Societies* (New York: Norton, 2009); N. J. Mackintosh, *Animal Learning and Cognition* (New York: Academic Press, 1994); David Pena-Guzman, *When Animals Dream: The Hidden World of Animal Consciousness* (Princeton, N.J.: Princeton University Press, 2022); Irene Pepperberg, *Alex & Me: How a Scientist and a Parrot Discovered a Hidden World of Animal Intelligence—and Formed a Deep Bond in the Process* (New York: MJF, 2008); Irene Pepperberg, *The Alex Studies: Cognitive and Communicative Abilities of Grey Parrots* (Cambridge, Mass.: Harvard University Press, 1999); Patrik Svensson, *The Book of Eels: Our Enduring Fascination with the Most Mysterious Creatures in the Natural World* (New York: Ecco, 2020); J. Scott Turner, *The Extended Organism: The Physiology of Animal-Built Structures* (Cambridge, Mass.: Harvard University Press, 2000; Jakob von Uexkull, "A Foray Into the Worlds of Animals and Humans," trans. Joseph O'Neil (Minneapolis: University of Minnesota Press, 2010); Cary Wolfe, *Zoontologies: The Question of the Animal* (Minneapolis: University of Minnesota Press, 2003); David Wood, *Thinking Plant Animal Human: Encounters with Communities of Difference* (Minneapolis: University of Minnesota Press, 2020); Ed Yong, *An Immense World: How Animal Senses Reveal the Hidden Realms Around Us* (New York: Random House, 2022).

35. Thomas Nagel, "What Is It Like to Be a Bat?," *Philosophical Review* 83 (October 1974): 435–50. Further page references to this article are given in the text.

36. Thomas Nagel, "Panpsychism," https://aaron-zimmerman.com/wp-content/uploads/2021/01/Nagel-Panpsychism.pdf, 181.
37. Thomas Nagel, "Panpsychism," *Stanford Encyclopedia of Philosophy*, May 13, 2022, https://plato.stanford.edu/entries/panpsychism/.
38. Nagel, "Panpsychism."
39. Jacques Derrida, *The Animal That Therefore I Am*, trans. David Wills (New York: Fordham University Press, 2008), 3. Further references to this book are given in the text.
40. Martin Heidegger, *Being and Time*, trans. John Macquarrie and Edward Robinson (New York: Harper and Row, 1962), 247.
41. Aleandre Kojève, *Introduction to the Reading of Hegel*, trans. James Nichols, ed. Allan Bloom (New York: Basic Books, 1969).
42. "The Cambridge Declaration on Consciousness," https://fcmconference.org/img/CambridgeDeclarationOnConsciousness.pdf.
43. As if to prove the point, while I was writing this section, the following article in the *Wall Street Journal* appeared in one of my news feeds: Kirsten Grind, "Magic Mushrooms. LSD. Ketamine. The Drugs That Power Silicon Valley. Entrepreneurs including Elon Musk and Sergey Brin are part of a drug movement that proponents hope will enhance their lives and produce business breakthroughs."
44. Yong, *An Immense World*, 15.
45. These examples represent a small sample selected from Yong's account in *An Immense World*, 112, 59, 50, 178, 206, 312.
46. Quoted in Evan Malmgren, "The Intelligence of Swine," *Noema*, November 24, 2021.
47. Marc R. Hammerman, "Xenotransplantation of Embryonic Pig Pancreas for Treatment of Diabetes Mellitus in Non-human Primates," *Journal of Biomedical Science and Engineering* 6 (5A), May 2013, https://www.ncbi.nlm.nih.gov/pmc/articles/PMC3848958/.
48. "What Is Animal Sentience and Why Is It Important?," *RSPCA Knowledge Base*, September 9, 2019, https://kb.rspca.org.au/knowleMore dge-base/what-is-animal-sentience-and-why-is-it-important/.
49. Elodie Briefer, "Pigs' Emotions Revealed by Decoding Grunts," *Neuroscience News and Research*, March 8, 2022, https://www.technologynetworks.com/neuroscience/news/pigs-emotions-revealed-by-decoding-grunts-359346.
50. Elodie Briefer, "Classification of Pig Calls Produced from Birth to Slaughter According to Their Emotional Valence and Context of Production," *Scientific Reports* 12, no. 3409 (2022), https://www.nature.com/articles/s41598-022-07174-8.
51. Malmgren, "The Intelligence of Swine."
52. Andre Goncalves and Dora Biro, "Comparative Thanatology, an Integrative Approach: Exploring Sensory/Cognitive Aspects of Death in Vertebrates and Invertebrates," *Philosophical Transactions of the Royal Society B*, July 16, 2018, https://royalsocietypublishing.org/doi/full/10.1098/rstb.2017.0263.
53. Pepperberg, *Alex and Me*, 7, 190–91. See also Pepperberg, *The Alex Studies*.

54. Emery, *Bird Brain*, 12. Throughout my discussion of cognition and intelligence in birds, I have drawn on Emery's excellent work. Page references to his book are given in the text.
55. Ludwig Huber and Ulrike Aust, "Mechanisms of Perceptual Categorization in Birds," in *Avian Cognition*, ed. Carel ten Cate and Susan Healy (Cambridge: Cambridge University Press, 2017), 208.
56. Leyre Castro and Edward Wasserman, "Relational Concept Learning in Birds," in *Avian Cognition*, ed. Carel ten Cate and Susan Healy (Cambridge: Cambridge University Press, 2017), 243.
57. Emery, *Bird Brain*, 162.
58. Clara Mancini, "Animal-Computer Interaction: A Manifesto," *Interactions* 18, no. 4 (July 2011): 69–73, https://dl.acm.org/doi/10.1145/1978822.1978836.
59. Bakker, *The Sounds of Life*, 172.
60. Bakker, 173.
61. Chittka, *The Mind of a Bee*, 1. Further references to this book are given in the text.
62. Derrida, *The Animal That Therefore I Am*, 123–24.
63. Holldobler and Wilson, *The Superorganism*, 7.
64. Holldobler and Wilson, 486–87.
65. Turner, *The Extended Organism*, 1–2.
66. See Richard Dawkins, *The Extended Phenotype* (New York: Oxford University Press, 1999).
67. Martin Lüscher, "Air-Conditioned Termite Nests," *Scientific American* 205, no. 1 (July 1961): 138. I have drawn the following details about termites and their nests from this article.
68. Turner, *The Extended Organism*, 195.
69. Margo Wisselink, Durr K. Aanen, and Anouk van 't Padje, "The Longevity of Colonies of Fungus-Growing Termites and the Stability of Symbiosis," *Insects* 11, no. 8 (August 2020): 527, https://www.ncbi.nlm.nih.gov/pmc/articles/PMC7469218.

10. STRANGE LOOPS

1. Sara Walker, "AI Is Life," *Noema*, https://www.noemamag.com/ai-is-.life/.
2. Kevin Kelly, *Out of Control: The Rise of Neo-Biological Civilization* (New York: Addison-Wesley, 1994), 11.
3. Kelly, 26.
4. Kevin Kelly, *New Rules for the New Economy: 10 Radical Strategies for a Connected World* (New York: Viking, 1998),
5. Mark M. Millonas, "Swarms, Phase Transitions, and Collective Intelligence," *Artificial Life III*, ed. Christopher G. Langton, Santa Fe Institute Studies in the Sciences of Complexity (New York: Addison-Wesley, 1994), 16:418, 417.
6. Millonas, 421.
7. Ilya Prigone and Isabelle Stengers, *Order Out of Chaos: Man's New Dialogue with Nature* (New York: Bantam, 1984),

8. Per Bak, *How Nature Works: The Science of Self-Organized Criticality* (New York: Springer, 1996), 48.
9. Millonas, "Swarms, Phase Transitions," 422–23.
10. Verner Vinge, "Technological Singularity," https://users.manchester.edu/Facstaff/SSNaragon/Online/100-FYS-F15/Readings/Vinge,%20The%20Coming%20Technological%20Singularity.pdf. Vinge citations in this section are to this article.
11. Kurzweil, *The Singularity Is Near: When Humans Transcend Biology* (New York: Viking, 2005), 29.
12. For an excellent analysis of the mythology informing private space programs, see Mary-Jane Rubenstein, *Astrotopia: The Dangers of the Corporate Space Race* (Chicago: University of Chicago Press, 2022).
13. Kurzweil, *The Singularity Is Near*, 22.
14. "The Transhumanist Declaration," Humanity +, https://www.humanityplus.org/the-transhumanist-declaration.
15. Nick Bostrom, *Superintelligence: Paths, Dangers, Strategies* (Oxford: Oxford University Press, 2017), 26.
16. Bostrom, 28, 35, 43, 60.
17. Bateson, *Steps to an Ecology of Mind: A Revolutionary Approach to Man's Understanding of Himself* (New York: Ballentine), 1972, 461.
18. For an excellent analysis of Weiser's contributions, see John Tinnell, *The Philosopher of Palo Alto: Mark Weiser, Xerox PARC, and the Original Internet of Things* (Chicago: University of Chicago Press, 2023).
19. Mark Weiser, "The Computer for the Twenty-first Century," *Scientific American*, September 1991, 94, 100.
20. Quoted in Samuel Greengard, *The Internet of Things* (Cambridge, Mass.: MIT Press, 2015), 59.
21. Alan Turing, "Computing Machinery and Intelligence," *Mind* 59, no. 236 (October 1950).
22. Eric Topol, *Deep Medicine: How Artificial Intelligence Can Make Healthcare Human Again* (New York: Basic Books, 2019), 74.
23. Alan Turing, "Intelligent Machinery, a Heretical Theory," https://rauterberg.employee.id.tue.nl/lecturenotes/DDM110%20CAS/Turing/Turing-1951%20Intelligent%20Machinery-a%20Heretical%20Theory.pdf.
24. AlphaGo Zero, https://www.youtube.com/watch?v=WXuK6gekU1Y.
25. Stephen Wolfram, "What Is ChatGPT Doing . . . and Why Does It Work?," https://writings.stephenwolfram.com/2023/02/what-is-chatgpt-doing-and-why-does-it-work/. In my discussion of ChatGPT, Artificial General Intelligence, and Generative AI, I have been guided by Wolfram's excellent explanation. Wolfram quotations in this section are from this article.
26. Bill Joy, "Why the Future Doesn't Need Us," *Wired*, April 2000. 2.
27. Bostrom, *Superintelligence*, 99.
28. Matthew Hutson, "Can We Stop Runaway AI?," *New Yorker*, May 16, 2023.

29. Chris Stokel-Walker, "Generative AI's Dirty Secret," *Wired*, May 30, 2023.
30. See Brian Christian, *The Alignment Problem: How Can Artificial Intelligence Learn Human Values?* (New York: Norton, 2020).
31. Lynn Margulis and Dorion Sagan, *What Is Life?* (Berkeley: University of California Press, 1995), 217.
32. Margulis and Sagan, 232–33. Emphasis added.
33. Lynn Margulis, *Symbiosis: A New Look At Evolution* (New York: Basic Books, 1998). 33, 52.
34. Donna Haraway, *Staying with the Trouble: Making Kin in the Chthulucene* (Durham: Duke University Press, 2016), 60.
35. Lynn Margulis, *Symbiotic Planet: A New Look at Evolution* (New York: Basic Books, 1998), 37.
36. Margulis, *What Is Life?*, 131.
37. Jean-Cristophe Simon et al. "Host-Microbiota Interactions: From Holobiont Theory to Analysis," *Microbiome* 7, no. 5 (2019), https://microbiomejournal.biomedcentral.com/articles/10.1186/s40168-019-0619-4.
38. J. Scott Turner, *The Extended Organism: The Physiology of Animal-Built Structures* (Cambridge, Mass.: Harvard University Press, 2000), 1–2.
39. For a description of my insulin pump, see Mark C. Taylor, *Intervolution: Smart Bodies Smart Things* (New York: Columbia University Press, 2021).
40. Neuralink website, https://neuralink.com/.
41. "Elon Musk's Neuralink Puts Computer Chips in Pigs' Brains in Bid to Cure Diseases," *NBC News*, August 29, 2020, https://www.nbcnews.com/tech/tech-news/elon-musk-s-neuralink-puts-computer-chips-pigs-brains-bid-n1238782
42. Jack Guy, "Elon Musk's Neuralink Claims Monkeys Can Play Pong Using Just Their Minds," *CNN Business*, April 9, 2021, https://www.cnn.com/2021/04/09/tech/elon-musk-neuralink-pong-scli-intl/index.html.
43. David Grossman, "Elon Musk Wants to Upgrade Our Brains to Compete with AI," *Popular Mechanics*, July 17, 2019, https://www.popularmechanics.com/science/health/a28423949/elon-musk-neuralink/. Not surprisingly, Jeff Bezos and Bill Gates are backing a company called Synchron, which is testing mind-controlled computing on humans. Ashley Capoot, "Brain Implant Startup Backed by Bezos and Gates is Testing Mind-Controlled Computing on Humans, *CNBC*, February 18, 2023, https://www.cnbc.com/2023/02/18/synchron-backed-by-bezos-and-gates-tests-brain-computer-interface.html.
44. Eva Rothenberg, "AI Fears Overblown? Theoretical Physicist Calls Chatbots 'Glorified Tape Recorders," *CNN*, August 13, 2023, https://www.cnn.com/2023/08/13/business/ai-quantum-computer-kaku/index.html.
45. M. Mehta and K. Subramani, "Nano Diagnostics in Microbiology and Dentistry," *Science Direct*, 2012, https://www.sciencedirect.com/topics/engineering/nanobots.
46. Amit Khurana et al., "Role of Nanotechnology Behind the Success of mRNA Vaccines for COVID-19," *Nano Today*, March 26, 2021, https://www.ncbi.nlm.nih.gov/pmc/articles/PMC7997390/.

47. Kevin Roose, "Aided by A.I. Language Models, Googe's Robots Are Getting Smart," *New York Times,* July 28, 2023, https://www.nytimes.com/2023/07/28/technology/google-robots-ai.html.
48. Creative Machines Lab, Columbia University, "About Us," https://www.creativemachineslab.com/about.html.
49. Oliver Whang, "Consciousness in Robots Was Once Taboo Now It's the Last Word," *New York Times*, January 6, 2023, https://www.nytimes.com/2023/01/06/science/robots-artificial-intelligence-consciousness.html.
50. Hanson Robotics, "Sophia," 2023, https://www.hansonrobotics.com/sophia/.
51. Sue Halpern, "A New Generation of Robots Seems Increasingly Human," *New Yorker*, July 23, 2023.
52. Sam Kriegman et al., "A Scalable Pipeline for Designing Reconfigurable Organisms," *PNAS* 117, no. 4 (January 13, 2020): 1853–59, https://www.pnas.org/doi/10.1073/pnas.1910837117.
53. Philip Ball, "Cells Form Into Living 'Xeonobots' on Their Own," *Wired*, April 4, 2021.
54. Douglas Blackiston et al., "A Cellular Platform for the Development of Synthetic Living Machines," *Science Robotis* 6, no. 52 (March 17, 2021), https://www.science.org/doi/10.1126/scirobotics.abf1571.
55. Ball, "Cells Form Into Living Xeonobots."
56. Quoted in Walter Isaason, *The Code Breaker: Jennifer Doudna, Gene Editing and the Future of the Human Race* (New York: Simon and Schuster, 2021), 40.
57. Peter Robin Hiesinger, *The Self-Assembling Brain: How Neural Networks Grow Smarter* (Princeton, N.J.: Princeton University Press, 2021), 226. Further page references to this book are given in the text.

11. AFTER LIFE

1. Robin Wall Kimmerer, *Braiding Sweetgrass: Indigenous Wisdom, Scientific Knowledge, and the Teaching of Plants* (Minneapolis: Milkweed, 2013), 37–38.
2. During the Great Depression, my father, the physics teacher, and my mother, the literature teacher, drove from the Pennsylvania coal-mining town where they lived and worked and made their way to Duke University, where my mother received a master's degree in American literature and my father received a master's degree in botany.
3. Aaron's wife, Frida, is Swedish, and Swedish was Selma's and Elsa's first language. Dinny and I have lived in Denmark for two years. In Scandinavia, people differentiate maternal and paternal grandparents: *morfar* (mother's father), *mormor* (mother's mother), *farfar* (father's father), and *farmor* (father's mother).
4. Patricia MacCormack, *The Ahuman Manifesto: Activism for the end of the Anthropocene* (New York: Bloomsbury Academic, 2020).
5. Sophocles, quoted in Goodreads, https://www.goodreads.com/quotes/7135637-not-to-be-born-at-all-is-best-far-best.

6. Adam Kirsch, *The Revolt Against Humanity: Imagining a Future Without Us* (New York: Columbia University Press, 2023), 43, 45.
7. See, for example, Douglas Dixon, *After Man: An Anthropology of the Future* (New York: St. Martin's, 1990); and Alan Weisman, *The World Without Us* (New York: St. Martin's, 1990).
8. Emerson and other American Transcendentalists became aware of German idealistic philosophy through the writings of Samuel Taylor Coleridge, who heard Fichte's lectures.
9. Empedocles, "On Nature," https://www.jstor.org/stable/27900051.
10. G. W. F. Hegel, *Lectures on the History of Philosophy*, trans. E. S. Haldane and Frances Simson (New York: Humanities, 1968), 1:361.
11. Hegel, 1:182.
12. G.W.F. Hegel, *Philosophy of History*, trans. J. Sibree (New York: Dover, 1956), 319.
13. Lynn Margulis and Dorian Sagan, *What Is Life?* (Berkeley: University of California Press, 1995), 234.
14. Paul Tillich, "The Two Types of Philosophy of Religion," *Theology of Culture*, ed. Robert Kimball (New York: Oxford University Press, 1964), 10, 22.
15. Friedrich Schleiermacher, *Speeches on Religion to Its Cultured Despisers*, trans. John Oman (New York: Harper Torchbooks, 1958), 43.
16. G. W. F. Hegel, *Science of Logic*, trans. A. V. Miller (New York: Humanities Press, 1969), 105. "Sublated," it is important to remember, is the translation of *aufheben*, which means simultaneously negated and preserved.

INDEX

Page numbers followed by an *f* indicate figures.

addiction, 334–35
Age of Surveillance Capitalism, The (Zuboff), 31–32
"Age of the World Picture, The" (Heidegger), 27
Aguirre, Anthony, 132
Ahuman Manifesto for the End of the Anthropocene, The (MacCormack), 330
air: carbon dioxide and, 3; heat waves, 9, 19; ionosphere, 15; mesosphere, 13–15; overview, 12–20, 25*f*; oxygen, 15–17; pollution, 334–35; pressure and currents, 14; stratosphere, 13–17; temperature fluctuations, 19, 22; thermosphere, 13, 15; troposphere, 13–16
Alcor Life Extension Foundation, 33
Aletheia (truth), 45, 48, 82. *See also* truth
Alex and Me (Pepperberg), 269
algae blooms, 6, 11–12
alienated world, 191, 202
alienation, 37, 44–45, 58, 77, 139, 332, 337–40
allelomimesis, 284
Allen, Herbert, 29
AlphaGo Zero, 295, 298
alternative intelligence (AI), 311
Andreessen, Marc, 288

animal-built structures, 276, 310
Animal Farm (Orwell), 265
animals: cognition of, 255–60, 267–70, 274, 361n34; communication with, 271–72; embranchments of, 188; extra-sensory apparatuses of, 264; human interactions with, 184; intelligence of, 252–60; pain response in, 251; research using, 265–67; response to death, 267–68; sentience of, 267; theological anthropology, 257; time experiences of, 256; Western philosophy and, 256–57
Animal That Therefore I Am, The (Derrida), 259, 274
animal welfare, 267
Anthes, Emily, 184
anthropause, 180–82
Anthropic Cosmological Principle, The (Barrow, Tipler), 167
Anthropocene, xii, 35, 37, 44, 47, 155, 167–68, 178, 182, 330
anthropocentrism, xii, 24, 37–38, 44, 46–47, 155, 167, 178–79, 219, 237–41, 256, 258, 287, 330, 332
anti-aging research, 33
antigens, 225–33, 227*f*

apocalypse, xiv, 26–27, 35, 329
Apple Computer advertisement, 28–29
Aquinas, Thomas, 186
Archimedean point, 71, 97, 98f, 116, 138
Aristotle, 43
artificial general intelligence (AGI), 290, 292, 299
artificial intelligence (AI): aim of research and development, 311; AlphaGo Zero, 295, 298; biobots, 315–17; cellular cognition and, 219; ChatGPT, 290, 299–305, 316; cognition and, 207; concerns over sentience, 239; Deep Blue, 294; DeepMind, 295; description of, 280; OpenAI, 290, 304–5; organic-relational AI, 319–26; robots, 272, 293, 314–19; symbolic AI, 292–98, 314–16, 323–25; synthetic biology, 317–19; xenobots, 317–20
artificial neural networks (ANN), 291–92, 295, 297–302, 325
Atlan, Henri, 249
atoms/atomism, 15–16, 42–43, 62, 68, 73, 92–99, 106, 111–13, 119–21, 132–33, 142–43, 201–2, 282, 336
Augustine, 37, 76–78, 90, 159–61, 222
authoritarianism, 30, 57
autonomous network theory, 225, 228, 231
autonomy of organisms, 215
autopoiesis, 210–19, 231, 306

Bacon, Francis, 40–42
Bak, Per, 169, 284
Baker, Jerry, 252
Bakker, Karen, 271–72
Ball, Philip, 118, 124–25
Barabasi, Albert-Laszlo, 244–45
Barad, Karen, 143
Barlow, John Perry, 280–81
Barrow, John D., 167
basal cognition, 219–20, 224

Bataille, Georges, 70
Bateson, Gregory, 67, 140, 156–58, 180, 290
becoming: Being into, 78, 85–86, 97, 109, 340; Derrida on, 78; Hegel on, 65–66; relationship to thinking, xii
bedrock, 4
Being: into becoming, 78, 85–86, 97, 109, 340; Derrida on, 78; God as, 99; Hegel on, 63, 65; Heidegger on, 46, 48–49, 82–87; philosophical reflection on, 335–42; power of, 45; Smolin on, 96–97
Being and Time (Heidegger), 46, 82–87
Bell, John, 131
Benatar, David, 330–31
Bengio, Yoshua, 296
Berlin Wall, 27, 55, 74
Berners-Lee, Tim, 243
Beyond Weird: Why Everything You Thought You Knew About Quantum Physics Is Different (Ball), 118
Bezos, Jeff, 35
biobots, 315–17
biological cognition, 221f, 289
biological evolution, 185–97
biologistic continuism, 260
bird cognition, 268–71
Biro, Dora, 268
Blanchot, Maurice, 235
Bloom, Harold, 256
Bohm, David, 130–31
Bohr, Niels, 68, 111–22, 130, 143, 201–2
Bostrom, Nick, 276, 289–90, 304
Both/And, 48, 68–69, 78, 112, 119, 122, 132, 192, 343n1
Both-And-reason, 60
Braiding Sweetgrass (Kimmerer), xiii
Brand, Stuart, 29, 53, 280, 282
Brenner, Eric, 246
Briefer, Elodie, 267
Broecker, Wally, 10
Broglie, Louis de, 125

Bronson, Richard, 35
Burnet, Frank Macfarlane, 226, 229
Byrne, David, 317

Calvo, Paco, 246–50
capitalism, 27, 29–32, 54–57, 74, 191, 282, 288
Capra, Fritjof, 44
carbon dioxide, 3–4, 7, 9, 13, 16–19, 242, 335
Carr, Jackson Noel, 329–30, 334
Cartesian anthropocentrism, 155, 237, 241, 256
Cartesian dualism, 190, 198, 239
Cartesian rationalism, 240
causality, 43, 59–60, 70, 97–99, 105–6, 131, 189, 192, 211, 214, 223, 232, 282, 338
cellular cognition, 219–24
Chalmers, David, 39, 155
chance variation, 181, 309–10
ChatGPT, 290, 299–305, 316
chemical weathering, 3–5
Chittka, Lars, 238, 273
chlorofluorocarbons, 14
Christianity, 33–34, 38, 132, 158
circular/cyclical time, 102, 103*f*, 105
Circumfession (Derrida), 78
clays, 4–7, 11
climate change, 5–6, 8–9, 24, 196, 287, 304–5. *See also* global warming
Clinton, Bill, 320
clonal selection theory, 225, 226, 228–29, 232, 234
co-cognition, 162, 164*f*
cogito ergo sum, 37, 45, 256, 261
cognition: of animals, 255–60, 267–70, 274, 361n34; artificial intelligence and, 207; basal, 219–20, 224; of bees, 273–76; biological, 221*f*, 289; in birds, 268–71; cellular, 219–24; co-cognition, 162, 164*f*; consciousness *vs.*, 165–74, 173*f*, 177–79, 215–16; entanglement and, 264; Hayles on, 354n15; information and, 159–76, 164*f*, 165–67, 166*f*; intelligence and, 174, 175*f*; Kant on, 161; learning and, 162; living systems and, 215–19, 229; memory and, 160, 222–24, 247, 273; mental properties, 254–55; metacognition, 270, 274; nonhuman forms of, 262–63; pancognitivism, 179, 216, 245, 255; panpsychism and, 176–79; Rovelli on, 160; smart cells, 219–24; smart plants and, 238–52, 270–71
cognitive augmentation, 312
cognitive network theory, 207, 228–29, 231–34
Cohen, Irun, 228–29, 234
coherence, 40, 128
coincidences, 48, 182–84, 304
Cold War, 27, 29, 55, 74, 117, 130, 225, 282, 294
Coleridge, Samuel Taylor, 83, 272–73
Collapse (Diamond), 7–8
collective intelligence, 276, 283–84
Collins, Francis, 320
communication: among organisms, 207; with animals, 271–72; of bees, 273–76; consciousness and, 208–9, 216; defined, 151–52; with plants, 252, 271
computer networks, 30–31, 279, 282, 287, 306
Confessions, The (Augustine), 76–78, 90, 159–61, 222
Confucius, xii–xiii
consciousness: cognition *vs.*, 165–74, 173*f*, 177–79, 215–16; communication and, 208–9, 216; disease and, 236; expansion of, 261–78; neurobiology of, 261–62; philosophy and, 253; reductionism and, 254
constitutive relationality, 62, 91, 135, 141, 168, 171, 198, 234
Copenhagen (Frayn), 112, 113–14, 121
Copenhagen theory of quantum mechanics, 75, 112, 118, 124, 133, 138, 177–78, 218
Copernicus, 40

coral reefs, 11–13, 16, 335
Corbusier, Charles-Édouard, 240
Cosmological Koans (Aguirre), 132
Course in General Linguistics (Saussure), 79–80
creatio ex nihilo, xi, 261
Crick, Francis, 320
CRISPR, 36
Critique of Judgment (Kant), 58–59
Critique of Practical Reason (Kant), 58
Critique of Pure Reason (Kant), 117
Crutzen, Paul, 37
Cuvier, Georges, 187–88
cybernetics, 150, 156–57, 224

Damasio, Antonio, 39
Darwin, Charles (Darwinism), 185–203, 247, 249–50, 281–82, 309
Darwin, Erasmus, 187
Darwinism Evolving (Depew, Weber), 187
Dasein (being there), 46, 89, 258
Davisson, Clinton, 125
Dawkins, Richard, 276–77
dead zones in oceans, 9
death: animal response to, 267–68; being-toward-death, 46; as condition of life, xi–xii; Derrida on, 46–47; Heidegger on, 46–47, 237; human extinction and, 328–29; memory and, 21, 327; philosophy as mediation on, xi, 237; phosphorous and, 12
de-cisions, 106, 109
decoherence, 128, 133
Deep Blue, 294
Deep Learning, 296, 297
deep listening, 252–53
DeepMind, 295
Delbruck, Max, 202
Democritus, 43
Denken ist danken (to think is to thank), 49
deoxyribonucleic acid (DNA), 202, 204, 207, 309, 320

Depew, David, 187, 189
Derrida, Jacques: animal cognition and, 255–60, 274; on becoming, 78; biologistic continuism and, 260; on death, 46–47; différance and, 77–80, 87–88; on force, 82; on Heidegger, 256; logocentrism and, 256; space and time, 94; on speed, 26–28
Descartes, René: the Anthropocene and, 37–41; collapse of truth, xii; on forests, 239–40; God and, 48; information and, 154–55; pain response in animals, 251; relationalism and, 139; time experiences in animals, 256
destruction ad nihil, xi
determinism, 43, 78, 105, 119, 121–22, 132, 203, 298, 301
deterrence, 44–49
diabetes research on animals, 265–66
dialectical biology, 52, 190, 199, 213–14
dialectical logic, 52, 75, 79, 159, 191–92, 199, 213, 233, 284, 337
Diamond, Jared, 7–8
Difference Between Fichte's and Schelling's System of Philosophy, The (Hegel), 61
difference/différance: Heidegger on, 82–85; identity and, 67–68; information and, 154–59; time and, 77–80
differentiation, 61, 82–90, 171, 176, 214
Dillard, Annie, 1
dirt, 1–7, 332
Discourse on Method (Descartes), 239
disease and human body, 208–10
dodders, 250
double-slit experiment, 117, 125–26
dualism: autopoiesis and, 213; Cartesian, 190, 198, 239; Cartesian-Newtonian, 239; mind/matter, 41, 155; oppositional, 70, 118, 155, 231, 335, 340; Rovelli on, 155–56
dust/dust storms, 18, 332
dystopianism, 28, 286–88, 303, 305, 312

earth: bedrock, 4; carbon dioxide and, 3–4; clays, 4–7, 11; dirt, 1–7, 332; dust/dust storms, 18, 332; overview, 1–7, 25f, 344n5; record droughts, 334; rocks, 1–7, 13, 16–20; silts, 4; soil, 3–7
East Germany, 53–57, 74
ecology of mind, 157, 159, 182, 207, 290
Einstein, Albert, 68, 75, 92, 95, 121, 129–32, 138
Either/Or, 48, 68–69, 78, 112, 119, 122, 132, 192, 343n1
Either-Or of understanding, 60–61, 67–68, 74, 112, 190, 225
electromagnetic fields, 94, 106, 120, 315
elements: air, 12–20, 25f, 334–35; earth, 1–7, 25f, 334; fire, 20–24, 25f, 334; Foucault on, 52; in quantum physics, 110; water, 7–12, 25f, 335. *See also* air; earth; fire; water
Eliot, T. S., 337
elsewhere, 96
embranchments of animals, 188
Emergence (Holland), 168–69
Emery, Nathan, 269–71
Empedocles, 335–36
entanglement, 75, 91, 94, 126–32, 138–40, 185, 217, 264, 353n27
entropy, xi, 77, 289, 353n47
"Essay on Man" (Pope), 42
Essay on the Principle of Population, An (Malthus), 189–90
essentialism, 186, 187
exceptionalism, 33, 38, 46–47, 148, 179, 188, 237–38, 257–58, 263, 299, 342
extinction events, 331–32
extra-sensory apparatuses of animals, 264

facts, as relative, 135, 138
Faraday, Michael, 94–95
fertilizers, 6, 10–11, 203, 217, 334
Feynman, Richard, 111, 118
financial capitalism, 32, 282

Finding the Mother Tree (Simard), 241–42
fire: carbon dioxide and, 3–4; elemental overview, 20–24, 25f; forest fires, 16, 19, 334
First Minds, The (Reber), 219
Flexible Bodies (Martin), 225
"Flower in the Crannied Wall" (Tennyson), 197
force, 81, 82
forest fires, 16, 19, 334
Forests (Harrison), 239
Foucault, Michel, 51–52
Frayn, Michael, 112, 113–14, 121
Friedman, Milton, 30
Fukuyama, Francis, 27, 74
fundamentalism, 55, 201, 263
fungus/fungal species, 4–6, 17–18, 22, 207, 241–45, 276–78, 325
futures of the past, 97–109

Galilei, Galileo, 40–42, 178
Galileo's Erro (Goff), 41
Galvani, Luigi, 246
Gefter, Amanda, 247–48
general relativity theory, 95, 138–39, 351n37
Genesis, book of, 257
Genesis and Structure of Hegel's Phenomenology of Spirit (Hyppolite), 73
genetic code, 203–4
genetic information, 320–23, 322f
genitive, defined, 182
genotypes, 204, 205f, 230, 321
Geoffroy Saint-Hilaire, Etienne, 187
geologic time, 3–4, 5–7, 17
Germer, Lester, 125
Gilbert, Walter, 203
Gingrich, Newt, 346n5
Global Education Network (GEN), 29–30
global warming, 6, 10, 14, 18, 329. *See also* climate change

God: Bateson on, 157–58; as Being, 99; comprehension and knowledge, 97; as creator, 101; doctrine of the Trinity, 63, 64f; existence of, 99, 337; Hegel on, 60, 63; in Judeo-Christian tradition, 38, 132, 158, 337; Newtonian view of, 44; omniscience of, 97, 99, 158; Paley on, 187; space and time, 92; species creation, 188; transcendence of, 93; as tripartite, 63
Goethe, xvii–xviii, 51, 187
Goff, Philip, 41, 176
Goncalves, Andre, 268
Goodell, Jeff, 18–19
Google Translate, 271
Gould, Stephen Jay, 310
gravitational field, 95
Greek philosophy, 34, 48, 99, 336
greenhouse gases: carbon dioxide, 3–4, 7, 9, 13, 16–19, 242, 335; methane gas, 13, 16–17, 19; overview, 9–10, 13, 16
Green Revolution, 10
Guy, Jack, 314

Haeckel, Ernst, 185
Hanson, David, 317
Hapgood (Stoppard), 111–12
Haraway, Donna, 309
Harrison, John, 92–93
Harrison, Robert Pogue, 239
Hawking, Stephen, 261
Hayek, Friedrich von, 30
Hayles, Katherine, 34, 153–54, 354n15
heat waves, 9, 19
Heat Will Kill You First, The (Goodell), 18
Heaven's Breath (Watson), 13
Hebb, Donald, 296, 297
Hegel, Georg Wilhelm Friedrich: on becoming, 65–66; on Being, 63, 65; Both/And, 48, 68, 78, 112, 119, 122, 132, 192, 343n1; doctrine of the Trinity, 63, 64f; on God, 60; on identity, 67–68; on Kant, 63, 118; on knowledge, 161–62, 174; on life, 60, 62; on Newtonian physics, 42–43; nothing/nothingness, 65; reason and, 61–62; relationalism and, 57–76, 96–97; on Schelling, 158–59; space-time and, 88–91, 94; on teleology, 102; time and, 76–79, 82–84, 105; on truth, 146; understanding and, 60–61; web of life, 69–75
Hegel on, 63, 118
Heidegger, Martin: anthropocentrism and, 44, 46–47, 256, 258; on Being, 48, 66; on consciousness, 237; on death, 46–47, 237, 267; Derrida on, 256; on difference/différance, 82–85; on differentiation, 80–90; on imagination, 83; on in-between, 82–83; on language, 47–49, 74, 258; on modern science and technology, 27–28, 36–37, 39–41, 44–45, 209; quantum theory and, 111; on relativity, 82–87; on species development, 188; on time, 46, 82–87
Heisenberg, Werner: nuclear bomb and, 113–14; quantum theory and, 47, 49, 68, 112–13, 118, 121–23, 128–30, 201; relationism and, 68, 71; relationship with Bohr, 112–16; uncertainty principle, 37, 47, 112–18, 121
here and now, 65–66, 78, 85–86, 99
Hiesinger, Peter Robin, 319–21, 323–25
Hinton, Geoffrey, 296
"Historical Roots of Our Ecological Crisis, The" (White), 33
Holland, John, 168–69
Holldobler, Bert, 275–76
How Hippies Saved Physics (Kaiser), 117
How to Talk to Your Plants (Baker), 252
human consciousness, 39, 43, 178–79, 219, 223, 239, 246, 264, 274, 287, 340
human extinction, 328–29
Human Genome Project, 203–4, 320

human intelligence, 33, 246, 286–88, 290–91, 293–94, 303, 311, 325
humanism, 331–32
humanity: alienation and, 37, 44–45, 58, 77, 139, 332, 337–40; deterrence and, 44–49; evolutionary process of, 36; language (*Sprache*) and, 45–46; speed toward end of, 26–27; superhumanity, 279, 287, 306; technological innovations and, 35–44
humans: animal interactions with, 184; defined, xi, 36; evolutionary stage of, 310–12; geologic time and, 3–4, 5–7, 17. *See also* living systems
Humboldt, Alexander von, 50–57, 185
Humboldt University, 55, 56f
Husserl, Edmund, 124
Huxley, Julian, 289
Hyppolite, Jean, 73–74

idealism, 71, 83, 139, 154–56, 160–61, 178, 193f, 211, 213, 234–35, 254
identity: biological identity of the self, 224–25; difference and, 67–68; Hegel on, 67–68; life and, 63–69; as relative, 155; self-relation and, 67
Identity and Difference (Heidegger), 66
imagination, 29, 83–87, 259, 265, 269
Immense World, An (Yong), 263–64
immortality, xi, 6, 32, 34, 287
immunological memory, 228
immunology/immune system, 224–31, 227f, 232
in-between, 82–83
individualism, 47, 68
individuality, 46, 62, 70, 190, 258, 309
information: cognition and, 159–76, 164f, 165–67, 166f; defined, 147–48; Descartes and, 154–55; difference and, 154–59; genetic, 320–23, 322f; interpretative schemata and, 145–47; meaning *vs.*, 151–52; memory and, 148; processing of, 148, 150, 157, 162, 165, 166f, 207, 216, 220, 230, 276; reality and, 156; redundancy and, 152–53; understanding/misunderstanding of, 147–54
information processing, 148, 150, 157, 162, 165, 166f, 207, 216, 220, 230, 276
information theory, 39, 68, 140–41, 146–47, 153–55, 170–72, 179, 216, 225. *See also* quantum information theory
in improbable past scenario, xvii–xviii
integrated information theory (IIT), 170–72
intelligence: of animals, 252–60; cognition and, 174, 175f; collective, 276, 283–84; defined, 174, 175f; human, 33, 246, 286–88, 290–91, 293–94, 303, 311, 325; superintelligence, 36, 207, 239, 276, 285–86, 288–306, 311, 319. *See also* artificial intelligence
Intelligence Amplification (IA), 287
interference, 122, 125–26, 128
intermediate zone, 95–96
Internet of Bodies, 292, 310–11
Internet of Things, 36, 291–92, 310–12, 313f
intervoluntary process, 204, 261, 264, 280, 332
Intranet of the Body, 310–11
intrinsic existence, 171–72
Introduction to the Reading of Hegel (Kojève), 73
ionosphere, 15

Jentsch, Ernst, 293
Jerne, Niels, 225–26, 229
Journeys to Selfhood (Taylor), 110
Judaism, 33–34
Judeo-Christian tradition, 38, 132, 158. *See also* Christianity

Kaiser, David, 117
Kandel, Eric, 222
Kant, Immanuel: on cognition, 161; on geography, 51; Hegel on, 63, 118
Karban, Richard, 251

Karlamangla, Soumya, 11
Kasparov, Gary, 294
Kauffman, Stuart, 212-13
Kelly, Kevin, 280-82, 286
Kesey, Ken, 280
Khullar, Dhruv, 19
Khurana, Amit, 315
Kierkegaard, Søren, 38, 71, 105, 110-11, 331
Kimmerer, Robin Wall, xiii-xiv, 327
Kirsch, Adam, 330
Knausgaard, Karl Ove, 110, 145
knowledge, purpose of, 41
Kojève, Alexandre, 70, 73-74
Kolbert, Elizabeth, 11
Koshland, Daniel, 223
Kurzweil, Ray, 30, 32-35, 287-89

language: communication of bees, 273-76; Heidegger on, 47-49, 74, 258; humanity and, 45-46; large language models (LLMs), 302; nature revealed in, 74
Laplace, Pierre Simon, 43
large language models (LLMs), 302
Lear, Jonathan, xiv
learning, 148, 162
Lectures on the History of Philosophy (Hegel), 42-43, 336-37
LeCun, Yann, 296
Lemoine, Blake, 290, 304
Leucippus, 43
Levin, Michael, 317-20
Levins, Richard, 190-92, 203
Lewontin, Richard, 190-92, 202-3
Lichtenberg, Georg, 331
life, 12, 60, 62, 63-75
Life of a Bee, The (Materlinck), 275
linear time, 100f, 102, 105
Linked (Barabasi), 244
Linnaeus, Carl, 188
Lipson, Hod, 316
living systems: addiction and, 334-35; animal intelligence, 252-60; anthropocentrism and, xii, 24, 37-38, 44, 46-47, 155, 167, 178-79, 219, 237-41, 256, 258, 287, 330, 332; antigens, 225-33, 227f; autopoiesis and, 210-19, 231, 306; cognition and, 215-19, 229; disease and, 208-10; expanding consciousness, 261-78; immunology/immune system, 224-31, 227f, 232; lymphocytes, 225-30, 232; smart bodies, 224-35; smart cells, 219-24; smart plants, 238-52
Logic and Existence (Hyppolite), 74
logocentrism, 256
Lüscher, Martin, 277
lymphocytes, 225-30, 232

MacCormack, Patricia, 330-31
Maclean, Norman, 20-21
Magic Mountain, The (Mann), 210
Mahayana Buddhism, xii, 132
Malabou, Catherine, 174
Malmgren, Evan, 176
Malthus, Thomas, 189-90
Mancini, Clara, 271-72
Mancuso, Stefano, 248
Manhattan Project, 140
Mann, Thomas, 210
Margulis, Lynn, 38, 208, 210, 279, 306, 308-10, 338-39
Martin, Emily, 225
Marxism, 54, 74, 130, 191, 203
Masahiro Mori, 293
materialism, 71, 91, 139, 154-56, 178, 254
Materlinck, Maurice, 275
Mathematical Principles of Natural Philosophy (Newton), 42
Maturana, Humberto, xii, 211, 213-19, 232-33
Maxwell, James Clark, 94-95, 106
McCarthy, Cormac, 26
McCulloch, Warren, 217, 295-96
McGregor, Simon, 179
McLuhan, Marshall, 29
meaning, as relative, 138

memory: artificial intelligence and, 301–2, 319; Augustine on, 76–77, 105, 159–60; in bees, 273; in birds, 268–69; cognition and, 160, 222–24, 247, 273; in computers, 299; death and, 21, 327; genotype and, 204, 220; immunological, 228; information and, 148; learning and, 297; in plants, 247–48; Turing Machine and, 150
mental properties, 254–55
Merleau-Ponty, Maurice, 124
mesosphere, 13–15
metacognition, 270, 274
Metamorphosis of Plants (Goethe), 51
metanoia, xii
metaphors and theories, 201
metaphysics, xii–xiii, 37–38, 44, 47, 57, 59, 84, 88, 90, 93, 97, 118, 132, 155, 239
Metaverse, 32
Metchnikff, Elias, 225
methane gas, 13, 16–17, 19
methodological individualism, xiii
methodological relationalism, xii–xiii
Miller, J. Hillis, 231
Millonas, Mark M., 283–84
Mind and Cosmos (Nagel), 177
Mindful Universe (Stapp), 177
Mind in Life (Thompson), 214
mind/matter dualism, 41, 155
Mind of a Bee, The (Chittka), 238, 273
mind of ecology, 182
Mink, Louis, 117
Minsky, Marvin, 296
mirror effects, 115f
molecular biology, 202–5, 222, 226
Monod, Jacques, 309–10
Moore's Law, 292
Moravec, Hans, 34
More, Max, 289
Movement and Habits of Climbing Plants, The (Darwin), 250
Musk, Elon, 35, 314
mysticism, 158

Nagarjuna, 50, 139
Nagel, Thomas, 177, 253–55
nanotechnology, 315–17
narrative time, 104f, 105
nationalism, 55
National Oceanic and Atmospheric Administration, 9, 19
National Science Foundation, 30
natural selection, 187, 190, 223, 226, 249, 309
Natural Theology, or Evidences of the Existence and Attributes of the Deity Collected from the Appearance of Nature (Paley), 99, 101, 187
nature: determination of, 65; dystopianism, 28, 286–88, 303, 305, 312; participatory realism and, 144; relationalism and, 50–58; revealed in language, 74; swarming behavior, 279–86; utopianism, 35, 286–88, 303, 305
Naturphilosophen (Schelling), 187
negative feedback, 150, 282
Neither/Nor, 15, 48, 68–69, 78, 96, 109, 112, 120, 129, 132, 343n1
Neuralink, 314
neural network theory, 297–98
neuronal circuits, 321, 322
neuroprosthetics, 312–15
New Age, 44, 241, 246, 252
Newton, Isaac, 42–44, 81, 92–94, 118–20
nexus, 59, 71, 74, 79, 88, 90, 96, 109, 167, 181, 192, 197–207, 214, 231, 285, 341f
Nietzsche, Friedrich: analysis of play of forces, 80; death of God, 39; on facts, 135; nihilism, 69–70, 327, 332; ontological voluntarism, 38; perspectivism and, 71, 91, 95, 97, 138, 155, 254; quantum theory and, 129, 135; relationalism and, 69–71; self-creating/self-destroying world, 75; universal connections and values, 236
nihilism, 69–70, 83, 90, 158, 327, 332, 339
Nishida Kitaro, 73

"No Apocalypse, Not Now (full speed ahead, seven missiles, seven missives)" (Derrida), 26–27
Noel, Jackson, 334
nominalism, 186, 188
nothing/nothingness, 65, 77
the Notion, 62
Novum Organum (Bacon), 41
nuclear bomb, 113–14
nuclear holocaust/war, 26–28, 47, 74, 113, 183, 304, 329

observer/observed relationship, 168, 234
oceans, 9–13, 16, 335
Of Grammatology (Derrida), 77, 78
Oliveros, Pauline, 252–53
omniscience, 97, 99, 158
on intuition, 162, 163f
on teleology, 59, 101–2, 211–12
On the Revolution of the Heavenly Spheres (Copernicus), 40
On the Way to Language (Heidegger), 47–49
on time, 84–85
OpenAI, 290, 304–5
Oppenheimer, Robert, 132
oppositional dualism, 70, 118, 155, 231, 335, 340
Order of Things, The (Foucault), 51–52
organic matter in soil, 4–5
organic-relational AI, 319–26
originary time, 82–84
Origin of Species, The (Darwin), 185, 281–82
Orwell, George, 265
Out of Control (Kelly), 280–81, 286
Overstory, The (Powers), 241
oxygen, 15–17

Paley, William, 99, 101, 187, 338
pancognitivism, 179, 216, 245, 255
panpsychism, 176–79, 254–55
Papert, Seymour, 296

para, defined, 192, 195
Parasite, The (Serres), 181
Parasite/Host-Host/Parasite, 181–82, 231
parasites, 192–95, 250
Pareto, Vilfredo, 244
participatory communication, 139–44
participatory universe, 71, 142–44, 167, 234–35
patriotism, 55
patternism, 34
Pepperberg, Irene, 269
Perceptrons (Minsky, Papert), 296
personal computer revolution, 28–29
perspectivism, 71, 91, 95, 97, 138, 155, 254
pharmacological interventions, 262
Phenomenology of Spirit/Mind (Hegel), 65, 69, 77, 118, 124, 259, 288
phenotypes, 204, 205f, 230, 276–77, 309, 321
Philosophical Fragments (Kierkegaard), 105
Philosophy of History, The (Hegel), 63, 77
Phi Φ: A Voyage from the Brain to the Soul (Tononi), 170–72
phosphorous, 11–12, 18
photosynthesis, 9, 16
Physics and Philosophy (Heisenberg), 47–48, 111
Pittendrigh, Colin, 223
Pitts, Walter, 217, 295–96
Planck, Max, 120–21, 176–77
Planta Sapiens (Calvo), 246–47
plants: algae blooms, 6, 11–12; carbon dioxide and, 4–5; classification of, 356n9; communication with, 252, 271; Humboldt and, 51; memory in, 247–48; photosynthesis and, 9, 16; smart plants, 238–52, 270–71; tree communication networks, 6–7
Plato/Platonism, 34, 82, 99, 186–88, 213, 222, 226
Plenty Coups, xiv
Podolsky, Boris, 130

poiesis, 45. *See also autopoiesis*
Pollan, Michael, 246
Pope, Alexander, 42
positive feedback loops, 4, 17, 19, 23, 30, 44, 199, 201, 245, 282
Power of Movement in Plants, The (Darwin), 247
Powers, Richard, 241
presentism, 93, 95, 97
probabilistic time, 84, 106f
Project Florence, 252
Protagoras, 36
purposefulness, 211, 223, 249–50
Pythagoras, 41

quantum bits (qubits), 73, 140–41
quantum ecology: anthropause and, 180–82; autonomy of organisms, 215–17; biological evolution and, 185–97, 194f; coemergent and codependent organisms, 306, 307f; ecology of mind, 157, 159, 182, 207, 290; parasites and, 192–95; as radical relational, 197–201
quantum information theory, 140–42, 153, 155
quantum mechanics/theory: alternate assumptions of classical physics, 117–31, 127f; Bohr and, 68, 117–22, 130, 143, 201–2; coherence and, 128; Copenhagen theory of, 75, 112, 118, 124, 133, 138, 177–78, 218; decoherence, 128, 133; Einstein on, 92; elements in, 110; entanglement, 75, 91, 94, 126–32, 138–40, 185, 217, 353n27; Heisenberg on, 47, 49, 68, 112–13, 118, 121–23, 128–30, 201; interference and, 122, 125–26, 128; introduction to, 110–17; overturning of classical physics, 118–20; participatory communication, 139–44; participatory universe and, 71, 142–44, 167, 234–35; personal computers and, 29; reality and, 91; relational quantum mechanics, 131–39, 134f; Rovelli on, xii, 91–93, 123, 132–33, 135, 138–40, 142, 353n47; superposition and, 126, 128–29, 140; technological innovations and, 39; Wheeler and, 140–44, 234–35; Whiteheadian ontologicalization of, 177–78
Question Concerning Technology, The (Heidegger), 47
Quetelet, Adolphe, 106

Radical Hope (Lear), xiv
radical relationality/relationalism, xii–xiv, 197–201, 244, 275–76, 326f, 331, 342
railway time, 93
rainforests, 13, 16, 18
Ramanathan, Sharad, 222
Read, David, 242
realism, 36, 143, 156, 186, 188, 283, 299, 305, 317
reality: defined, xiv; immune system and, 234; information and, 156; quantum theory and, 120–21; virtual, 32–33, 37, 207
Reber, Arthur S., 219
reductionism, 189, 191, 254
redundancy, 152–53
relational quantum mechanics, 131–39, 134f
relationology/relationalism: Either-Or of understanding, 60–61, 67–68, 74, 112, 190, 225; Hegel on, 57–76, 96–97; Humboldt and, 50–57; Kierkegaard and, 71; life and identity/nonidentity, 63–69; nature and, 50–58; Nietzsche on, 69–70; Rovelli and, 132–33, 135, 138; Smolin on, 351n37; Zhao Tingyang on, xii–xiii
relativity, 76–79, 82–87, 97–109
res cogito, 38, 48, 154
res extensa, 38, 48, 154
Revolt Against Humanity, The (Kirsch), 330
River Runs Through It, A (Maclean), 20–21
robots, 272, 293, 314–19

rocks, 1–7, 13, 16–20
Romanowski, Sylvie, 52
Roose, Kevin, 316
Rosen, Nathan, 130
Rothblatt, Maratine, 34
Rovelli, Carlo: on cognition, 160; difference and, 155–56; on dualism, 155–56; quantum theory and, xii, 91–93, 123, 132–33, 135, 138–40, 142, 167, 353n47; relationalism and, 132–33, 135, 138; on time, 350n15
Russell, Bertram, 139

Sartre, Jean-Paul, 124
Saussure, Ferdinand de, 79–80, 274
Schelling, Friedrich, xvii–xviii, 158–59, 187
Schleiermacher, Friedrich, 339
Schmid, Bernhard, 249
Schrödinger, Erwin, 201–2
Schumacher, Benjamin, 140
Science of Logic (Hegel), 59–60, 62, 66, 74, 211, 337, 339–40
Sedol, Lee, 295
Segundo-Ortin, Miguel, 246–48, 250
self-certainty, xii, 46, 155
self-consciousness, 37–39, 43, 46, 63, 74, 152, 170, 174–76, 223, 251, 258–59, 303
self-creating/self-destroying world, 75
self-externality, 65–66
selfhood, 76, 110, 258
self-love, 63
self-organizing, 13–14, 58–60, 74, 142, 162, 166f, 170–74, 181, 210–11, 217, 245, 285, 298, 310
sense-making, 216–17
sense-receiving, 216
Serres, Michel, 181, 231, 239
Shannon, Claude, 68, 151–52
Shelley, Mary, 304
silts, 4
Simard, Suzanne, 241–45
simultaneity, 95

Singularity, 30, 32, 33, 35
smart bodies, 224–35
smart cells, 219–24
smart plants, 238–52, 270–71
Smith, Adam, 189–91, 282
Smith, David W., 124
Smolin, Lee, 76, 93, 96–97, 351n37
soil, 3–7
Solvi, Cwyn, 274
Sophia robot, 317
Sophocles, 330–31
Sound of Life, The (Bakker), 271
Soviet Union collapse, 74
space-time, xvii–xix, 66, 73, 76–97, 161
Speech and Phenomena (Derrida), 77
Speeches on Religion to Its Cultured Despisers (Schleiermacher), 339
speed of modern age, 26–35
Stapp, Henry, 71, 177–78
Stevens, Wallace, 85
Stoermer, Eugene, 37
Stoppard, Tom, 97, 99, 111–12, 126
stratosphere, 13–17
structuralism, 62, 74, 77, 79, 187
subjectivity is truth, 38
subject-object relations, 72f, 143, 192
superhumanity, 279, 287, 306
superintelligence, 36, 207, 239, 276, 285–86, 288–306, 311, 319
Superintelligence (Bostrom), 289–90
Superorganism, The (Hollldobler, Wilson), 275–76
superposition, 126, 128–29, 140
surveillance capitalism, 30–32
swarming behavior, 279–86
Symbiotic Planet (Margulis), 308
symbolic AI, 292–98, 314–16, 323–25. See also artificial intelligence
synthetic biology, 317–19
systematicity, 51–52, 59–60
systematic unity, 59

Taiz, Lincoln, 246
Taylor, Aaron, 1, 3–7, 18
Taylor, Beryl C., 195–97, 236–38, 327, 329
Taylor, David, 9
Taylor, Noel A., vii, 21, 195–97, 332
techne, defined, 45
technikon, defined, 45
Techno-Gnosticism, 35
technology: Heidegger on, 27–28, 36–37, 39–41, 44–45, 209; humanity and, 35–44; nanotechnology, 315–17; quantum mechanics/theory and, 39. *See also* artificial intelligence
teleology, 59, 101–2, 211–12
Tennyson, Alfred Lord, 197
termite environments, 275–78
Thales, 255
theory of special relativity, 68, 75, 92, 95
thermosphere, 13, 15
thing-in-itself, 70–71, 117–18, 129, 161
Thompson, Evan, 214, 219
Tillich, Paul, 338–39
time: Augustine's meditation of, 76; circular/cyclical, 102, 103*f*, 105; defined by Rovelli, 350n15; Derrida on, 78; difference/différance and, 77–80; experiences of animals, 256; futures of the past, 97–109; geologic, 3–4, 5–7, 17; Hegel on, 76–79, 105; Heidegger on, 46, 82–87; intermediate zone, 95–96; introduction to, 76–79; Kant on, 84–85; linear, 100*f*, 102, 105; narrative, 104*f*, 105; originary, 82–84; probabilistic, 84, 106*f*; railway time, 93; simultaneity, 95; space-time, xvii–xix, 66, 73, 76–97, 161
Time Reborn: From the Crisis in Physics to the Future of the Universe (Smolin), 96–97
Timoféeff-Ressovsky, Nikolay, 202
Tipler, Frank J., 167
Tononi, Giulio, 170–72, 178–79
totalitarianism, 29–30, 32

Towards a New Architecture (Corbusier), 240
transhumanism, 34, 288–89
tree communication networks, 6–7
Triple Helix, The (Lewontin), 202
troposphere, 13–16
truth: *Aletheia*, 45, 48, 82; collapse of, Descartes on, xii; Hegel on, 146; self-certainty and, xii, 46, 155; subjectivity is truth, 38; subjectivity of, Kierkegaard on, xii
Turing, Alan (Turing Machine), 148, 149*f*, 151, 292–96
Turner, J. Scott, 276–77
Twilight of the Idols (Nietzsche), 118

Ulm, Stanislaw, 286
Unbestimmtheit, defined, 121–22
uncertainty principle (Heisenberg), 37, 47, 112–18, 121
unconscious, 44, 79, 87, 157, 176, 208, 216, 236, 260
University of Berlin, 57
Unschafe, defined, 121–22
Unsicherheit, defined, 121–22
utopianism, 35, 286–88, 303, 305

Varela, Francisco, xii, 211, 213–19, 232–34
Venter, Craig, 320
Vinge, Verner, 30, 286–88
virtual reality, 32–33, 37, 207
von Forester, Heinz, 217
von Neumann, John, 150–51, 286, 293

Wade, Nicholas, 320
Walker, Sara, 279
water: carbon dioxide and, 3–4; droughts and floods, 335; elemental overview, 7–12, 25*f*; oceans, 9–13, 16, 335. *See also* oceans
Watson, James, 320
Watson, Lyall, 13, 17
Wealth of Nations, The (Smith), 189–90

Weaver, Warren, 202
Weber, Bruce, 187, 189
web of life, 69–75
Weiser, Mark, 291
"What a Heat Wave Does to Your Body" (Khullar), 19
What Is Life? (Margulis), 38, 210, 306
Wheeler, John, 91, 124, 140–44, 167–68, 234–35
Wheeler, William Morton, 281
White, Lynn, 33, 158, 257
Whiteheadian ontologicalization of quantum theory, 177–78
Whole Earth Catalog, 29, 53, 280–81, 286
Whole Earth Lectronic Link (WELL), 280–81
Wholeness and the Implicate Order (Bohm), 130–31
Wiener, Norbert, 150
Wilkins, Maurice, 202
William of Ockham, 186
Will to Power (Nietzsche), 70, 138, 254, 332
Wilson, E. O., 275–76, 283
Wolfram, Stephen, 299–301
Wood Wide Web, 18, 244–45, 251–52, 276, 285
Wootters, William, 139–40
World Wide Web (WWW), 30, 36–37, 74, 243–45, 280–81, 285, 288, 292, 312
Writing and Difference (Derrida), 77

xenobots, 317–20

Yearbook for Philosophy and Phenomenological Research (Husserl), 124
Yong, Ed, 263–64
Young, Thomas, 125

Zhao Tingyang, xii–xiii
Zoonomia (Darwin), 187
Zuboff, Shoshana, 30–32
Zuckerberg, Mark, 32